Introduction to Nuclear and Particle Physics

Introduction to
Nuclear and Particle Physics

Third Edition

V.K. MITTAL
Professor of Physics
Punjabi University, Patiala

R.C. VERMA
Former Professor and Head
Department of Physics
Punjabi University, Patiala

S.C. GUPTA
Former Professor and Head
Department of Physics
Punjabi University, Patiala

PHI Learning Private Limited
Delhi-110092
2017

[redacted]

INTRODUCTION TO NUCLEAR AND PARTICLE PHYSICS, Third Edition
V.K. Mittal, R.C. Verma and S.C. Gupta

ISBN-978-81-203-4738-0

The export rights of this book are vested solely with the publisher.

Fifth Printing (Third Edition) ... ... January, 2017

Published by Asoke K. Ghosh, PHI Learning Private Limited, Rimjhim House, 111, Patparganj Industrial Estate, Delhi-110092 and Printed by Raj Press, New Delhi-110012.

Contents

Note: Some sections of this book marked with asterisk (*) may be skipped in the first reading.

Preface

We feel encouraged by the wide spread acceptability this book has received in its first and second editions. This prompted us to bring out the next edition. In preparing this edition, we have been fortunate in receiving constructive suggestions and comments from several teachers and other readers, which we have now incorporated. In particular, some sections of Chapters 3 to 6 have been revised. We highly appreciate the time and efforts put by the readers in reading the book critically which has helped us in making it more useful. We take this opportunity to express our sincere thanks to all of them, particularly Dr. S.K. Arora, Professor of Physics, D.N. College, Hissar. We also express our thanks to the publisher for bringing out this edition in time.

V.K. Mittal
R.C. Verma
S.C. Gupta

Preface to the First Edition

This comprehensive textbook is the outcome of wide teaching and research experience of the authors. It is designed for B.Sc. (Physics) students of Indian universities. It can also be used as an introductory review material at Masters level. This well-organized and concise text discusses in detail the principles and applications of nuclear and particle physics explaining the latest developments. The text assumes knowledge of the fundamental concepts of quantum mechanics, electricity, magnetism and modern physics.

The book has been divided into nine chapters. Chapter 1 deals with the basic concepts and definitions. Chapter 2 discusses nuclear models including liquid drop model, shell model, and fermi gas model.

Chapter 3 covers the phenomenon of radioactivity, laws of disintegration and applications of radioactivity. Chapter 4 presents nuclear reactions. It covers various types of reactions including reactions occurring in sun and stars.

Chapter 5 emphasizes the interaction of radiations with matter and covers photoelectric effects, Compton effect and pair production. In Chapter 6, particle accelerators are discussed.

In Chapter 7, different types of radiation detectors are explained. Chapter 8 introduces the elementary particles and their interactions in detail. This chapter is capped with the latest advances. Since it is a very active and frontier area of research, it will help the reader to get a bird's eye view of the latest developments. This chapter also provides new terms and concepts. The final chapter deals with cosmic rays mainly due to historical reasons.

The discussion on the topics is simple and lucid avoiding complicated mathematical derivations. Actual data, graphs and figures are displayed to elaborate clarification of the topics. A variety of solved and unsolved numerical problems are included topic-wise to enhance the student's understanding of the subject. The book also contains short and long answer type questions, which are taken from various examinations of Indian universities. Some sections of this book in the contents marked with asterisk (*) may be skipped in the first reading.

Finally, we owe a deep sense of gratitude to our families for showing patience, which enabled us to complete this work. Also during our teaching tenures, several concepts were clarified by interactions with the students and colleagues. We owe a word of gratitude to all our students and colleagues who inspired us to complete the book.

V.K. Mittal
R.C. Verma
S.C. Gupta

Chapter 1

The Nucleus

1.1 INTRODUCTION

In this chapter, we present the concept of the nucleus, nuclear constituents and nuclear properties, such as mass, size, angular momentum, etc. To understand the concept of nucleus, it is worthwhile to review the development of various atomic models.

1.2 HISTORICAL DEVELOPMENTS

The beginning of the nuclear physics may be traced back to the studies on atomic structure started with discovery of radioactivity in 1896 by Henry Becquerel. Further, three different modes of radioactivity were observed emitting α-particles, β-particles and γ-rays. It is well known that α-particles are helium nuclei, β-particles are either electrons or positrons and γ-rays are high-energy electromagnetic radiations. Scattering of α-particles with matter revealed the existence of nucleus inside the atom. In the following paragraphs, various models are presented for establishing the structure of atom and nucleus.

Dalton's Atomic Hypothesis

The human beings have always been inquisitive to know the basic nature of the matter present in the surroundings. Breaking a piece of matter into smaller and smaller parts, search was made for its constituents, which are capable of independent existence. The atomic picture of matter is generally credited to Greeks. Around 400 B.C., the Greek philosopher Democritus postulated that all matter is made up of minute particles, which could not be broken up. Around the same times, perhaps earlier Kanada (an Indian philosopher) proposed the concept of *parmanu*, which in the similar sense, means ultimate constituents of matter. Continuing this search, Dalton in 1803 proposed an atomic theory of matter, which has the following postulates:

1. All elements consist of discrete particles, called atoms. These atoms cannot be subdivided by any known chemical process.
2. All atoms of the same element are identical in all respects, especially in weight. Atoms of different elements differ in weight.

3. A compound is formed by the union of atoms of different elements in simple numerical proportions, for example, 1 atom of sodium combines with 1 atom of chlorine to produce one molecule of sodium chloride. Similarly, 2 atoms of hydrogen and 1 atom of oxygen combine to give a molecule of water.

Till the end of the nineteenth century, scientists did not have any clue about the structure of atom. All that they knew about the atom was that it is neutral. Works of Faraday, Maxwell and J.J. Thomson provided the first insight into the atomic structure with the discovery of electron in 1897. It was also realized that all atoms contain electrons, which carry negative charge. Since atom is neutral, so positive charge must be present in the atom. The question arises, what kind of positive charge is present and how is positive charge arranged in the atom?

Thomson Model of Atom

Discovery of electron in the famous cathode ray experiment by J.J. Thomson provided a starting point for theories of atomic structure. Since electrons are found to carry negative charge, and atom on the whole is electrically neutral, question arose about the magnitude and distribution of positive charge inside the atom. One suggestion about the arrangement of positive charge in the atom was made by J.J. Thomson, who suggested that atoms are simply lumps of matter with positive charge and electrons are embedded in these lumps. This model is popularly known as *Plum-Pudding model*, where plums are negative electrons which are placed in the pudding of smeared positive charge. This model is also called *watermelon model*, where seeds are negative electrons. The pictorial representation is given in Figure 1.1. This model was consistent with the observed overall neutrality of the matter and explained the flow of current in matter. As it turned out later, that the real picture of the atom is quite different from Thomson's model.

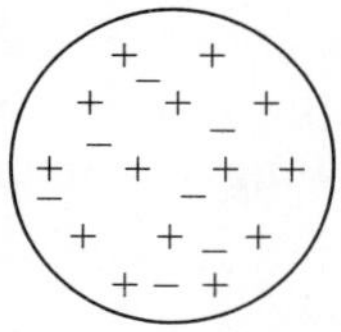

Figure 1.1 Plum-pudding model of the nucleus as given by J.J. Thomson. Plums are represented by – sign and pudding is represented by + sign.

Rutherford Model of Atom

The failure of the Thomson model was finally established in its inability to explain scattering of α-particles by atoms. The scattering of α-particles can be described in terms of the electrostatic Coulomb force between the α-particles and the charges which make up atoms. In the case of Thomson model, the average deflection of α-particles by a single atom should be very small in contrast to the experimental results. In 1911, Ernst Rutherford and his co-workers studied the way the α-particles got scattered when they passed through thin gold foil. They made the following observations:

1. Most of the α-particles went straight without getting deflected by the presence of gold foil.

2. Some α-particles, whose numbers were comparatively smaller, suffered small angle deviations.
3. Still smaller number of α-particles, about 1 in 8000, was back-scattered, i.e. through angles lying between 90° and 180°.

Rutherford failed to explain these observations on the basis of Thomson model. From these observations, it was concluded that atom has large empty space as most of α-particles pass through the gold foil undeflected. So, he proposed a new picture to explain the structure of atom. He assumed that the positive charge of the atom was concentrated in a minute centre, called nucleus, instead of being distributed uniformly in whole of the atom. Thus, Rutherford transformed the static model of Thomson into a dynamic model, also known as the *Planetary Model*. He also concludes that atom contains a positively charged sphere whose size was estimated to be 10^{-14} metres and he also concluded that atomic mass is concentrated in this sphere. This sphere was later on known as nucleus of the atom. There are just enough electrons around the nucleus to make atom electrically neutral. Rutherford developed a detailed theory of the scattering of α-particles by matter, which was tested in 1913 by Geiger and Marden. They investigated dependence of the α-scattering on various quantities like angle of scattering, thickness of scattering material, velocity of α-particles, nuclear charge, etc. The remarkable agreement between the predictions of Rutherford's theory and the experimental results established the concept of the nuclear atom. In fact, such experiments using faster α-particles showed deviations from the Coulomb law of scattering, which provided the first evidence of the existence of the nuclear force.

1.3 CONSTITUENTS OF THE NUCLEUS

So far, the atomic nucleus remains a vague concept. In the Rutherford model, it has been described as very small, in fact point-like, and is also supposed to contain practically all of the mass of the atom. So, in later experiments more quantitative information was sought about the size and structure of the nucleus. In the following sections, constitution of the nucleus is discussed.

As stated above, the nucleus of the atom has a very small size (its volume is 10^{-15} times the volume of the atom i.e. volume of the nucleus is of the order of 10^{-45} m^3). This compact size of the nucleus contains whole of the positive charge and practically the whole mass of the atom. Following are the two hypotheses to explain the compact structure of the nucleus:

1. Proton–electron hypothesis.
2. Proton–neutron hypothesis.

1.3.1 Proton–Electron Hypothesis

In order to explain observed properties of the nucleus, firstly it was proposed that the nucleus constitutes protons and electrons. This model is known as proton–electron model. The concept of the build-up of the nucleus in terms of the elementary constituents was based on the fact that certain atoms emit α- and β-rays, which are corpuscular in nature. As proposed by Prout, atomic weights A of the elements are close to integers. The fractional parts are contributed by the isotopes of the elements. The mass of the proton is approximately equal to the mass of the hydrogen atom.

In fact, the hydrogen nucleus was given the name proton, which shows its importance as a fundamental constituent of nuclei of all atoms.

To account for mass of the nucleus whose atomic weight is close to integer A, called the mass number, it is necessary to assume that nucleus contains A protons. But if this was the case the charge of the nucleus will be equal to A, nearly the same as atomic weight and not equal to atomic number Z. As is well known, value of Z is half or less than half of the atomic weight. To get over this difficulty, it was assumed that in addition to protons the nuclei contain $A - Z$ electrons. The presence of electrons would not contribute to the mass of nucleus but would make the charge Z as required. Thus, it was possible to consider atom made up of a nucleus containing A protons and $A - Z$ electrons surrounded by Z extra-nuclear electrons.

Success of the Proton–Electron Hypothesis

This hypothesis seems to be consistent with the emission of α- and β-particles in radioactive elements. The presence of electrons directly ensures the emission of β-particles, and emission of α-particles is assumed by the combination of 4 protons and 2 electrons in the nucleus. These α-particles may exist as such or may be formed at the instant of emission.

Failure of Proton–Electron Hypothesis

Following are the arguments, which led to the failure of proton–electron hypothesis.

1. **Spin and statistics:** The statistical nature of nuclei can be built up from rotational spectra of diatomic molecules. If the nucleus (A, Z) contains A protons and ($A - Z$) electrons, the spin of odd–odd nucleus or odd–even nucleus would not agree with experimental results. Take the case of odd–odd nucleus ^{14}N. An even number of protons (i.e. 14) produces an integral spin, while an odd number of electrons (i.e. 7) gives half-integral spin. So the total spin of ^{14}N in the proton–electron hypothesis would be half-integral spin and thus ^{14}N would be a fermion system. But the experimental results show that ^{14}N is a boson, i.e. it carries integral spin, which could not be explained in the proton–electron hypothesis.
2. **Binding energy:** As discussed later in Chapter 8, electron being a lepton cannot take part in strong nuclear interactions. Strong interactions bind the nucleons together in the nucleus. If electrons were present in nucleus, the bound state is caused by Coulomb interaction and the binding energy is of the order of

$$E \approx -\frac{Ze^2}{r}$$

where r is the radius of the nucleus and is given by $1.2 \times 10^{-15} A^{1/3}$ m or $1.2 \times A^{1/3}$ fm (1 fm = 10^{-15} m). Negative sign means that the system is bound. Thus,

$$E = -Z\frac{e^2}{\hbar c}\frac{\hbar c}{r}$$

$$\frac{e^2}{\hbar c} = \alpha = \frac{1}{137} \quad \text{(Fine structure constant)}$$

and $\hbar = 1.055 \times 10^{-34}$ J s

and $c = 3 \times 10^8$ m/s

Thus,

$$E = -Z \times \frac{1}{137} \times \frac{1.055 \times 10^{-34} \times 3 \times 10^{8}}{1.2 \times 10^{-15} \times A^{1/3}} \text{ J}$$

$$\approx -\frac{1.925 \times 10^{-13} \times Z}{A^{1/3}} \text{ J}$$

If $A \approx 125$ and $Z \approx 62$, then

$$E = -\frac{1.925 \times 10^{-13} \times 62}{125^{1/3}} = 2.387 \times 10^{-12} \text{ J}$$

$$\approx -15 \text{ MeV}$$

The de Broglie wavelength of the electrons having energy = 15 MeV would be

$$\lambda = \frac{\hbar}{p} = \frac{c\hbar}{E}$$

Here $cp \approx E$ is the energy of the electron in the nucleus

$$\lambda = \frac{3 \times 10^{8} \times 1.055 \times 10^{-34}}{15 \times 1.6 \times 10^{-13}}$$

$$= 13 \times 10^{-15} \text{ m}$$

$$= 13 \text{ fm}$$

This value is much greater than the size of the nucleus. Hence, electrons cannot exist inside the nucleus as bound electrons.

3. **Nuclear magnetic moment:** Magnetic moment of an electron is one Bohr magneton ($\mu_B = e\hbar/2m_e$), while that of a proton is one nuclear magneton $\mu_N = e\hbar/2m_P$. Thus, $\mu_B \approx 1850 \times \mu_N$, where $m_N/m_e \approx 1850$. If the nucleus consisted of protons and electrons, the nuclear magnetic moment should be of the order of μ_B while experimentally nuclear magnetic moment is of the order of μ_N. Thus, electrons do not exist inside the nucleus.
4. **Beta decay:** As discussed later, β-decay is a three-body process and not a two-body process. The electrons emitted in proton–electron hypothesis would have a line spectrum, i.e. all electrons carry same energy, while in actual β-decay, β-energy varies continuously. This means that in β-decay, β-particles are accompanied by a third particle. This contradicts proton–electron hypothesis.

1.3.2 Proton–Neutron Hypothesis

After the discovery of neutrons by Chadwick in 1932 through research on transmutation of nuclei by α-particles, Heisenberg had earlier proposed that nuclei might be composed of protons and neutrons, collectively called nucleons. The neutron carries mass slightly greater than that of the proton, but is electrically neutral. Due to no charge, neutron was hard to detect and several unsuccessful efforts were made before it was finally observed in 1932. Thus, $^{16}_{8}O$ nucleus contains

8 protons and 16 – 8 = 8 neutrons. $^{16}_{8}O$ nucleus is surrounded by 8 electrons to balance the nuclear charge. In general, $^{A}_{Z}X$ nucleus will contain Z protons and (A – Z) neutrons. To balance the nuclear charge it will be surrounded by Z electrons. This model obviously avoids the failures of proton–electron hypothesis.

Following facts support the proton–neutron hypothesis:

1. **Spin:** Both protons and neutrons have spin quantum number 1/2. According to quantum mechanics, if the number of nucleons in a nucleus is even, the resultant spin will be an integral multiple of $\hbar$. And when the number of nucleons in the nucleus is odd, the spin will be half integral multiple of $\hbar$. This observation is in agreement with the experimental observations without any exception.
2. **Magnetic moment:** According to proton–neutron hypothesis, there are no electrons inside the nucleus. Hence, we do not expect the magnetic moment of the nucleus to be of the order of Bohr magneton. On the other hand, the nuclear magnetic moment is of the order of nuclear magneton. This is in agreement with the experimental values.
3. **Finite size:** Since the mass of neutron is approximately equal to the mass of proton, it is possible for the neutron to reside inside the nucleus, according to the uncertainty principle.
4. **Wave mechanical considerations:** According to the uncertainty principle

$$\Delta x \Delta p \geq h \tag{1.1}$$

where Δx and Δp are the uncertainties in the position and momentum of proton/neutron respectively. Radius of a typical nucleus of mass ~ 200 is of the order of 0.6×10^{-14} m, the uncertainty in position will be $\Delta x = 2R = 1.2 \times 10^{-14}$ m. Therefore, uncertainty in momentum Δp is given by Eq. (1.1) as

$$\Delta p \approx \frac{h}{\Delta x}$$

$$= \frac{6.6 \times 10^{-34}\ \text{J s}}{1.2 \times 10^{-14}\ \text{m}} = 5.5 \times 10^{-20}\ \text{J s/m}$$

Let us estimate the energy E of a nucleon in the nucleus using the relativistic relation

$$E^2 = p^2c^2 + m^2c^4$$

$$= (5.5 \times 10^{-20})^2 (3 \times 10^8)^2 + (1.67 \times 10^{-27})^2 (3 \times 10^8)^4$$

$$= 2.29 \times 10^{-20}$$

which gives

$$E = 1.51 \times 10^{-10}\ \text{J} \approx 945\ \text{MeV}$$

This value is slightly greater than the rest mass energy of proton, which is about 938 MeV. Hence, the kinetic energy of neutron or proton in the nucleus is of the order of few MeV (i.e. $945 - 938 \approx 7$ MeV) and it should be possible for a free proton or neutron to be confined in the nucleus.

5. **Isotopic masses:** It is possible to explain the existence of isotopes of different elements. Different isotopes of an element have same number of protons but different number of neutrons in the nucleus.

6. **Beta decay:** The process of β^--decay can be explained by the fact that a free neutron is transformed into a proton as follows:

$$n \rightarrow p + \beta^- + \overline{\nu}$$

The process of β^+-decay can be explained by conversion of a proton into a neutron as given by

$$p \rightarrow n + \beta^+ + \nu$$

In the above decays ν and $\overline{\nu}$ are neutrino and antineutrino respectively.

1.3.3 Terms Associated with the Nucleus

Atomic Number (Z)

It is the number of protons present in the nucleus. For example, nitrogen has 7 protons, so Z for nitrogen is 7, Z for uranium is 92 and for hydrogen, Z is 1.

Mass Number (A)

It is the total number of protons and neutrons present in the nucleus. For example, carbon has 6 protons and 6 neutrons in its nucleus, so its mass number is 12; uranium has 92 protons and 143 neutrons, therefore, mass number of uranium is 235, ordinary hydrogen has only 1 proton in its nucleus, so its mass number is 1. It is obvious that A can never be less than Z.

Neutron Number (N)

It is the total number of neutrons present in the nucleus and is equal to $A - Z$.

A nucleus X with atomic number Z, mass number A and neutron number N is represented as ${}^{A}_{Z}X_N$, for example uranium with $Z = 92$, $A = 235$ and $N = 143$ is written as ${}^{235}_{92}\text{U}_{143}$.

Isotopes

Nuclei of an element having the same atomic number but different mass numbers are called isotopes of the element. There are two types of isotopes: (i) stable and (ii) radioactive. Stable isotopes are those which do not show radioactivity, for example, ^{14}N and ^{15}N. Some radioactive isotopes are ^{40}K, ^{60}Co, etc.

Isotones

Nuclei having same N and different Z are known as isotones. The stable isotones with $N = 1$ are ${}^{2}_{1}\text{H}_1$ and ${}^{3}_{2}\text{He}_1$.

Nucleons

The term nucleons refers to protons or neutrons present inside the nucleus. Thus, the nucleus with mass number A having N neutrons and Z protons has A nucleons.

1.4 QUANTITATIVE FACTS ABOUT NUCLEUS

Here, we introduce some facts such as size, mass, density of the nucleus and charge on the nucleus. Also the energy units commonly used in nuclear physics are discussed.

1.4.1 Size

First estimates of nuclear size were made by Rutherford. According to him, α-particles with a given kinetic energy can come closer to the nucleus till they feel a force of repulsion. The kinetic energy is converted into Coulomb potential energy. Thus, kinetic energy gets reduced and potential energy continues to increase. At a certain distance d, the kinetic energy becomes zero and potential energy becomes maximum. After this the α-particle turns back and kinetic energy starts increasing and potential energy starts decreasing. The point where kinetic energy is zero is known as distance of closest approach and it is denoted by letter d and it provides a rough estimate of the nuclear size. It is well known that in a head-on collision, the minimum distance between the projectile and the target nucleus is equal to the sum of the radii of the projectile and the target nucleus. As whole of kinetic energy gets converted into potential energy at a distance d, so the balancing equation of the kinetic energy and the potential energy is

$$\frac{1}{2}mv^2 = \frac{ZZ'e^2}{4\pi\varepsilon_0 d}$$

$$d = \frac{2ZZ'e^2}{4\pi\varepsilon_0\, mv^2} \tag{1.2}$$

where m is the mass, v is the velocity and Ze is the charge of α-particle. $Z'e$ is the charge of the target nucleus and d is the distance of the closest approach. The physical meaning of d is shown in Figure 1.2.

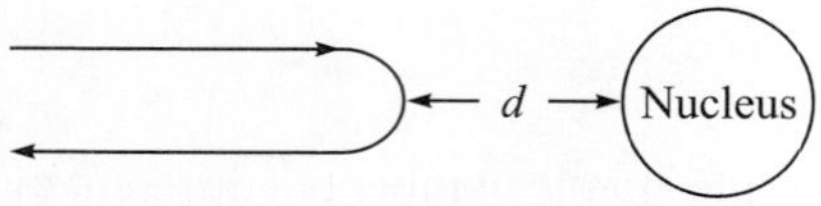

Figure 1.2 Particles being back-scattered from the nucleus, d is the distance of closed approach.

Velocity of α-particles from radon is about 1.6×10^7 m/s. The mass of α-particle is 4 times the hydrogen mass, i.e. $4 \times 1.67 \times 10^{-27}$ kg. Z is 2 and Z' is 79 for gold. e is 1.6×10^{-19} C. The resultant value of d [from Eq. (1.2)] is 4.26×10^{-14} m. By increasing the velocity of the α-particles one could go up to the minimum distance d equal to 3.2×10^{-14} m for gold. For scattering of α-particles from copper nuclei the value of d is approximately 1.2×10^{-14} m. The value of d is larger for the case of gold compared to copper. This is evident from Eq. (1.2) as d

increases with Z', the atomic number of the target. In case of gold because of higher Z', d is larger compared to that of copper. It has been observed that size of the nuclei depends upon A. Presently, the accepted nuclear radius formula is

$$\boxed{r = r_0 A^{1/3}} \tag{1.3}$$

where

$$\boxed{r_0 = 1.2 \times 10^{-15} \text{ m} = 1.2 \text{ fm}} \tag{1.4}$$

1 fm = 10^{-15} m

Formula (1.3) is an *empirical formula*. Its derivation is not based on any theory but purely on experimental facts.

1.4.2 Mass

Nuclear masses are comparatively much smaller as compared to that of physically occurring objects. Therefore, it is generally expressed in terms of a new unit called Atomic Mass Unit (amu). To estimate the nuclear masses we use the Avogadro's hypothesis. According to this hypothesis, a mole of a substance is that amount which has mass in grams equal to its atomic weight. For example, 1 mole of ^{12}C has 12 g. So, 1 mole of ^{12}C contains $N_A = 6.02214199 \times 10^{23}$ atoms of ^{12}C.

Thus, the weight of 1 atom of ^{12}C is

$$\frac{12}{N_A} = \frac{12}{6.02214199 \times 10^{23}}$$

$$= 1.992646 \times 10^{-23} \text{ g}$$

$$= 1.992646 \times 10^{-26} \text{ kg}$$

For the sake of simplicity, the weight of ^{12}C atom/12 is defined as 1 Atomic Mass Unit or in short 1 amu.

Thus,

$$1 \text{ amu} = \frac{1.992646 \times 10^{-23}}{12}$$

$$= 1.660538 \times 10^{-24} \text{ g}$$

$$= 1.660538 \times 10^{-27} \text{ kg}$$

Therefore,

$$\boxed{1 \text{ amu} = 1.660538 \times 10^{-27} \text{ kg}} \tag{1.5}$$

Thus, weight of 1 atom of ^{12}C is 12 amu. This is taken as standard for all the atomic weight measurements. In this unit, proton and neutron masses are given by

$$\boxed{\text{Mass of proton} = 1.0072764668 \text{ amu} = 1.67262 \times 10^{-27} \text{ kg}} \tag{1.6}$$

$$\boxed{\text{Mass of neutron} = 1.0086649157 \text{ amu} = 1.6749264 \times 10^{-27} \text{ kg}} \tag{1.7}$$

For comparison mass of an electron in this unit

$$\text{Mass of electron} = 0.000549 \text{ amu} = 9.109 \times 10^{-31} \text{ kg}$$

1.4.3 Density

Density of a substance is defined as mass per unit volume.

We calculate nuclear densities for two specific cases, i.e. ^{12}C and ^{197}Au.

$$\text{Mass of } ^{12}\text{C} = 1.992646 \times 10^{-26} \text{ kg}$$

$$\text{Radius } R \text{ of } ^{12}\text{C} = r_0 A^{1/3}$$

$$= 1.2 \times 10^{-15} \times 12^{1/3} \text{ m}$$

$$= 2.7473 \times 10^{-15} \text{ m}$$

$$\text{Volume of } ^{12}\text{C} = \frac{4}{3}\pi R^3$$

$$= 8.6859 \times 10^{-44} \text{ m}^3$$

Therefore,

$$\text{density of } ^{12}\text{C} = \frac{1.992646 \times 10^{-26} \text{ kg}}{8.6859 \times 10^{-44} \text{ m}^3}$$

$$= 2.294 \times 10^{17} \text{ kg/m}^3$$

$$\text{Mass of } ^{197}\text{Au} = 3.2707 \times 10^{-25} \text{ kg}$$

$$\text{Radius of } ^{197}\text{Au} = r_0 A^{1/3}$$

$$= 1.2 \times 10^{-15} \times 197^{1/3} \text{ m}$$

$$= 6.9824 \times 10^{-15} \text{ m}$$

$$\text{Volume of } ^{197}\text{Au} = \frac{4}{3}\pi r^3$$

$$= 1.4259 \times 10^{-42} \text{ m}^3$$

Therefore,

$$\text{density of } ^{197}\text{Au} = \frac{3.2707 \times 10^{-25} \text{ kg}}{1.4259 \times 10^{-42} \text{ m}^3}$$

$$= 2.294 \times 10^{17} \text{ kg/m}^3.$$

Thus we observe that the nuclear densities are the same for carbon and gold. Or the nuclear density is constant and is of the order of 10^{17} kg/m^3. The nuclear density is extremely large. It is comparable to the density of neutron star.

In comparison to these nuclear densities, the density of carbon atom is 2.26×10^3 kg/m^3 and that of gold atom is 1.92×10^4 kg/m^3. Thus, nuclear densities are larger than atomic density by an order of 10^{14}.

Such a high density can be visualized by the following example. If we take 150 million cubic metres of water and compress it to 10^{-6} m^3 (or 1cc), we get value of density, which is of the order of nuclear density.

The nuclear density is independent of the mass number A. It can be explained as under:

$$\text{Mass of nucleus with } A \text{ nucleons} = A \times m \text{ kg}$$

where m is the mass of one nucleon.

$$\text{Volume of a nucleus with } A \text{ nucleons} = \frac{4}{3}\pi r_0^3 A$$

Therefore,

$$\text{density} = \frac{m}{\frac{4}{3}\pi r_0^3}$$

Substituting for m and r_0, we get

$$\boxed{\text{Nuclear density} = 2.294 \times 10^{17} \text{ kg/m}^3} \quad (1.8)$$

1.4.4 Energy

In nuclear physics, the unit of energy is taken to be electron volt (eV). The larger units of energy are keV and MeV. However, the SI unit of energy is joule. But SI unit of energy is not used in nuclear physics.

Electron volt (eV) is defined as the energy acquired by an electron, when it is accelerated through a potential difference of one volt.

Relation between eV and Joule

$$\begin{aligned} 1 \text{ electron volt } (1 \text{ eV}) &= \text{charge on 1 electron} \times 1 \text{ volt} \\ &= 1.602 \times 10^{-19} \text{ C} \times 1 \text{ V} = 1.602 \times 10^{-19} \text{ J} \end{aligned}$$

i.e.

$$\boxed{1 \text{ eV} = 1.602 \times 10^{-19} \text{ J}} \quad (1.9)$$

Therefore,

$$\boxed{1 \text{ keV} = 1.602 \times 10^{-16} \text{ J}} \quad (1.10)$$

$$\boxed{1 \text{ MeV} = 1.602 \times 10^{-13} \text{ J}} \quad (1.11)$$

In Table 1.1, commonly encountered energies are shown.

TABLE 1.1 Commonly encountered energies

Room temperature thermal energy of a molecule	0.025 eV
Visible light photons	1.5–3.5 eV
Ionization energy of atomic hydrogen	13.6 eV
Approximate energy of an electron striking a colour television screen	20,000 eV (= 20 keV)
High energy diagnostic medical X-ray photons	2,00,000 eV (= 0.2 MeV)
γ-decay	0–3 MeV
β-decay	0–3 MeV
α-decay	2–10 MeV
Cosmic-ray energies	1 MeV–1000 TeV

These energy units are related to amu through mass–energy relation $E = mc^2$ as under. Using this relation, the value of 1 amu in terms of energy is

$$\begin{aligned} 1 \text{ amu} &= 1.660538 \times 10^{-27} \times (2.9979 \times 10^8)^2 \\ &= 1.49239 \times 10^{-10} \text{ J} \end{aligned}$$

$$\boxed{1 \text{ amu} = 1.49239 \times 10^{-10} \text{ J}} \tag{1.12}$$

$$\boxed{1 \text{ amu} = 931.47 \text{ MeV}} \tag{1.13}$$

1.4.5 Charge

Rutherford as a result of α-particle scattering experiments concluded that all the positive charge on an atom is confined to a tiny central region called nucleus. Later on from α-particle and X-ray scattering from the atoms, it was found that the number of unit charges on the nucleus of any atom is approximately half of its atomic weight. Rutherford also concluded that proton was identical with a hydrogen ion (electron removed) that carried a single unit positive charge. Since hydrogen atom is neutral, so the charge on proton must be equal to that of electron, but of opposite sign. Similarly, α-particle is actually a helium nucleus, that is a helium atom minus its two electrons. Therefore, helium nucleus carries positive charge equal to $2e$, where e is the charge present on one electron. Thus, the charge on a nucleus carrying Z protons is Ze units.

1.5 BINDING ENERGY

Here we define various terms such as mass defect, packing fraction, nuclear binding energy and binding energy per nucleon. Also packing fraction has been plotted versus A and is known as *packing fraction curve*. Fission and fusion have been discussed on the basis of this curve.

1.5.1 Mass Defect

Mass of the atom is concentrated in the nucleus, which is at the centre of the atom. As discussed earlier, nucleus is constituted by neutrons and protons. It has been observed that the mass of the nucleus is always less than the sum of the masses of all nucleons present in the nucleus. The difference in the sum of masses of all the nucleons present in the nucleus and the nuclear mass is known as the *mass defect* (Δm).

If $M'(Z, N)$ is the mass of the bare nucleus consisting of Z protons and N neutrons, the mass defect Δm is given as

$$\Delta m = ZM_p + NM_n - M'(Z, N)$$

or

$$\boxed{\Delta m = ZM_p + (A - Z)M_n - M'(Z, N)} \tag{1.14}$$

It is convenient to introduce the mass of Z atomic electrons into the right-hand side of this equation, so that it becomes

$$\boxed{\Delta m = ZM_{\text{H}} + (A - Z)M_n - M(Z, N)} \tag{1.15}$$

In this equation M_H and $M(Z, N)$ are the masses of neutral hydrogen and the nucleus under investigation.

Let us calculate the mass defect of deuterium ${}^2_1H$. It has one proton and one neutron in its nucleus. Thus, we would expect that the mass of deuterium to be equal to the mass of one neutron plus mass of ordinary hydrogen atom.

$$\text{Mass of } {}^1_1\text{H atom} = 1.007825 \text{ amu}$$
$$\text{Mass of neutron} = 1.008665 \text{ amu}$$

Therefore, expected mass of deuterium = 2.016490 amu

The measured mass of ${}^2_1H$ is found to be 2.014102 amu. The mass difference in these masses is 0.002388 amu. Therefore, in deuterium mass defect is 0.002388 amu. This missing mass may be regarded as the mass, which would be converted into energy, if a particular atom is to be formed from the requisite number of electrons, protons and neutrons. This is also equal to the amount of energy required to break up the atom into its constituents. Therefore, mass defect is a measure of binding energy of an atom. More the mass defect, more tightly the nucleons are bound in the nucleus.

1.5.2 Packing Fraction

Mass defect does not convey much information about nuclear stability, and it is misleading to say that higher the mass defect, more tightly bound nucleons exist in the nucleus. For example, mass defect for ${}^4_2He$ is 0.002604 amu (= 2.4249 MeV), while that for ${}^{235}_{92}U$ it is 0.04396 amu (= 40.930 MeV). But ${}^4_2He$ is much more stable than ${}^{235}_{92}U$. The term packing fraction was introduced by Aston in 1926 which gives better information about the nuclear stability. Packing fraction is defined as

$$\boxed{\text{Packing fraction } f = \frac{\text{Atomic mass} - \text{Mass number}}{\text{Mass number}}} \tag{1.16}$$

or

$$\boxed{f = \frac{\Delta m}{A} = \frac{\text{Mass defect}}{A}} \tag{1.17}$$

The smaller the value of packing fraction the more stable is the nucleus and vice versa. A plot of packing fraction f versus mass number A is shown in Figure 1.3, from which we can draw the following conclusions:

1. For very light nuclei, like ${}^2_1H$, ${}^3_1H$, etc. packing fraction is very large, hence nucleons in these nuclei are loosely bound. In fact, out of all known stable nuclei, nucleons in ${}^2_1H$ are most loosely bound.
2. As mass number A increases (up to $A = 16$), the packing fraction goes on decreasing till it becomes zero for $A = 16$, i.e., for ${}^{16}_8O$ nucleus.
3. As A further increases, for nuclei with $16 < A < 180$, packing fraction becomes negative. The nucleons in these nuclei are strongly bound in the nucleus. There is a flat minimum for $60 < A < 80$. Negative packing fraction also means that in order to break these nuclei into the constituents, we must supply external energy.
4. Beyond $A > 180$, packing fraction again is positive. Thus, most of the nuclei with $A > 235$ are unstable.

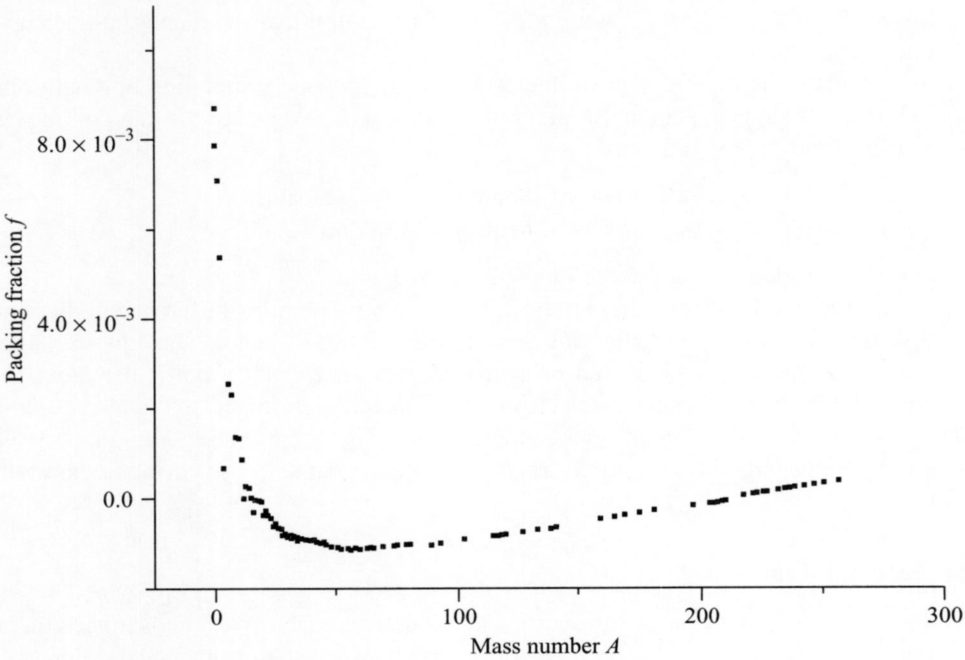

Figure 1.3 Packing fraction f versus mass number A.

1.5.3 Fusion and Fission

The phenomena of fusion and fission can also be explained on the basis of the packing fraction curve. It can be stated that each nucleus tries to reduce its packing fraction. If two light nuclei are fused together to form heavier nucleus, like

$$^{2}_{1}H + ^{3}_{1}H \rightarrow ^{4}_{2}He + n + \text{energy}$$

The packing fraction of heavier nucleus is less compared to that of the lighter nuclei. So in lighter nuclei, fusion is possible.

Similarly, if a heavy nucleus breaks into two lighter nuclei, like

$$^{235}_{92}U + ^{1}_{0}n \rightarrow ^{90}_{36}Kr + ^{144}_{56}Ba + ^{1}_{0}n$$

$$^{252}_{98}Cf \rightarrow ^{106}_{41}Nb + ^{142}_{57}La + 4^{1}_{0}n$$

The packing fraction of lighter nuclei is small compared to the heavier nuclei. So in heavy nuclei, fission is possible.

1.5.4 Binding Energy per Nucleon

To form an atom from requisite number of electrons, protons and neutrons, some amount of energy is required. The energy comes from the mass defect. The energy that keeps the nucleons together in a bound state is known as the binding energy of the nucleus. It can be calculated as:

The mass of constituent particles of an atom ${}_Z^AX$ is sum of masses of Z protons, $A - Z$ neutrons and Z electrons. Masses of one proton and one electron can be written as mass of one hydrogen atom (M_H). So masses of Z protons and Z electrons can be written as ZM_H. Therefore, the binding energy of ${}_Z^AX$ with mass $M(A, Z)$ is given as

$$BE = [ZM_H + (A - Z)M_n - M(A, Z)]\,c^2 \tag{1.18}$$

In driving this equation, we have neglected binding energies of the electrons in an atom, as these energies are comparatively very small. This binding energy is in amu. For converting it into MeV, we use the fact that 1 amu = 931.47 MeV. So, binding energy BE in MeV is

$$BE = [ZM_H + (A - Z)M_n - M(A, Z)] \times 931.47 \text{ MeV} \tag{1.19}$$

Now binding energy per nucleon is given as

$$\frac{BE}{A} = \frac{[ZM_H + (A - Z)M_n - M(A, Z)] \times 931.47}{A} \text{ MeV/nucleon} \tag{1.20}$$

Binding energy per nucleon versus mass number A has been plotted in Figure 1.4, from which the following conclusions can be drawn:

1. Barring few exceptions, like ^{4}He, ^{12}C, ^{16}O, etc., the values of binding energy per nucleon lie on a smooth curve.

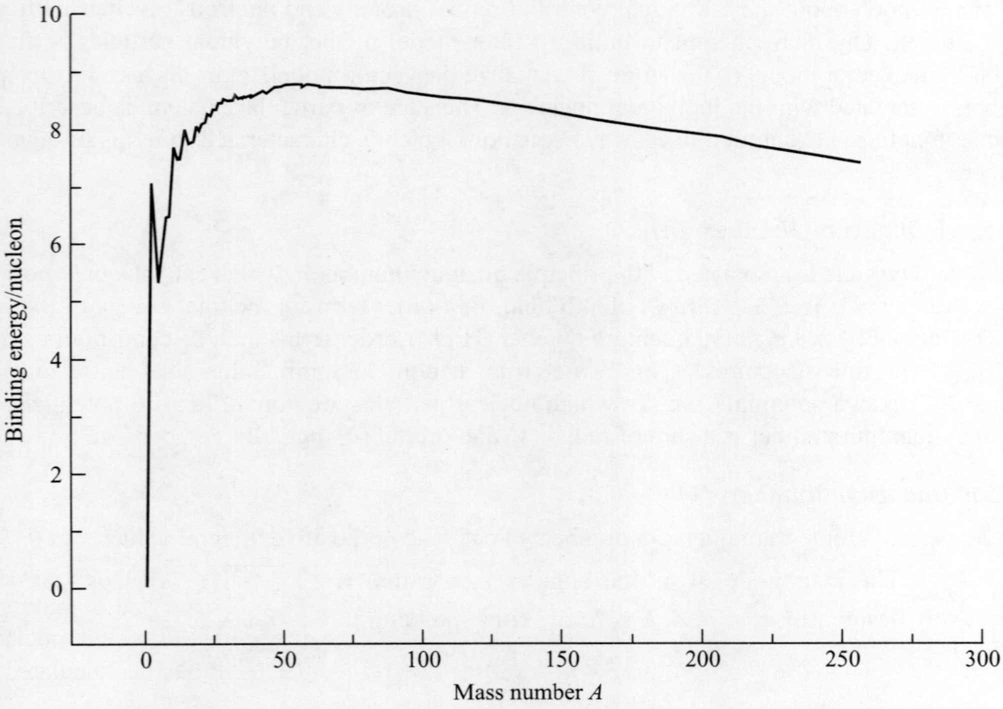

Figure 1.4 Binding energy per nucleon (in units of MeV) versus mass number A.

2. When mass number is small, i.e. $A < 12$ the binding energy per nucleon is less and it rises rapidly with increasing A.
3. Around $A = 50$, there is a flat maximum, where binding energy per nucleon is approximately 8.8 MeV. It slowly drops down to 8.4 MeV at $A = 140$. The average value of binding energy per nucleon between $A = 50$ and $A = 140$ is close to 8.5 MeV.
4. Above $A = 140$, binding energy per nucleon starts decreasing and at $A = 238$, its value is 7.6 MeV. It further reduces as A increases.
5. There are sharp peaks for ${}^{4}_{2}He$, ${}^{8}_{4}Be$, ${}^{12}_{6}C$, ${}^{16}_{8}O$, etc. nuclei. This indicates that these nuclei are more stable than the neighbouring nuclei.
6. If we take two lighter nuclei (say ${}^{2}_{1}H$, binding energy ~ 1.1 MeV/n) and fuse them together to form ${}^{4}_{2}He$, (binding energy ~ 7 MeV/n), there is a gain in binding energy. This indicates that fusion of lighter nuclei is energetically feasible.
7. If we take a heavy nucleus say $A = 240$ (binding energy ~ 7.6 MeV/n) and break it into two lighter nuclei of $A \sim 120$ (binding energy ~ 8.2 MeV/n), again there is a net gain in binding energy. Therefore, fission of heavy nuclei is again energetically feasible. It is left as an exercise to show that fusion of heavy nuclei is not feasible.

1.6 NUCLEAR ANGULAR MOMENTUM

Total angular momentum of the nucleus taken about its own axis, is easily measurable. But we cannot say much about the individual contributions of protons and neutrons, as their motion is very complex. One may attempt to build a vector model of the individual particles, which is similar to the vector model of the atom. To visualize the vector model, let us discuss the quantum numbers associated with the individual nucleons. The state of particular nucleon is described by its wave function, i.e. solution of its wave equation, which is characterized in terms of quantum numbers:

Principal Quantum Number (n)

Each bound particle is associated with principle quantum number n, which can take only positive integral values, $n = 1, 2, 3, \ldots$. In Coulomb field, first-order term for the total energy of the state is characterized by n, i.e. radial quantum number (Higher order terms include contributions from fine and hyperfine structures). This is not true in non-Coulomb fields such as rectangular potentials, Yukawa potentials, etc., in which nuclear particles are bound. In such potentials, the principal quantum number is a sum of radial (ν) and orbital (ℓ) quantum numbers, i.e. $n = \nu + \ell$.

Orbital Quantum Number (ℓ)

Orbital angular momentum quantum number (ℓ) can take on positive integral values, i.e. 0, 1, 2, ... $(n - 1)$. The magnitude of orbital angular momentum is $\sqrt{\ell(\ell+1)}\hbar$. Various states are respectively designated as $s, p, d, f, g, h, \ldots$, corresponding to $\ell = 0$, $\ell = 1$, $\ell = 2$, $\ell = 3$, $\ell = 4$, $\ell = 5, \ldots$

Orbital Magnetic Quantum Number (m_ℓ)

Orbital magnetic quantum number is the component of ℓ in the specified direction such as that of an applied magnetic field. It can take $(2\ell + 1)$ possible values starting from $-\ell$ to $+\ell$ differing by unity.

Spin Quantum Number (s)

Spin quantum number (s) has value 1/2 for proton, neutron and electron, which follows from Fermi–Dirac statistics and thereby obey the Pauli exclusion principle. The magnitude of spin angular momentum is $\sqrt{s(s+1)}\,\hbar$.

Magnetic Spin Quantum Number (m_s)

Magnetic spin quantum number is the component of s in the specified direction say that of the applied magnetic field. It can take $2s + 1$ values from $-s$ to $+s$. For particles with $s = 1/2$, $m_s = 1/2$ and $-1/2$.

Total Angular Momentum Quantum Number ($\vec{j}$)

It is the vector sum of the orbital ($\vec{\ell}$) and spin ($\vec{s}$) angular momenta, i.e. $\vec{j} = \vec{\ell} + \vec{s}$. The magnitude of the total angular momentum is $\sqrt{j(j+1)}\,\hbar$. For particles with $s = 1/2$, only two values of j are permitted as $j = \ell + 1/2$ and $j = \ell - 1/2$. If $\ell = 0$, only one value of j is allowed as $j = 1/2$.

Radial Quantum Number (v)

As stated earlier, in a non-Coulomb field, the principle quantum number n does not represent the energy of the state. In the radial wave equation solution, a radial quantum number v arises. v represents the number of radial nodes in the wave function and can have values $v = 1, 2, 3, \ldots$

Total Angular Momentum

It is the vector sum of the individual angular momenta of the constituent nucleons and is represented by J or I. It is also called nuclear spin and its magnitude is given by $\sqrt{j(j+1)}\,\hbar$. The value of J can be calculated in two different ways depending upon the type of coupling between angular momenta of the nucleons. The two types of coupling are:

(i) L–S coupling.
(ii) j–j coupling.

(i) *L–S* coupling: In this there is negligibly weak coupling between the orbital and spin angular momenta of the individual nucleons. Orbital angular momenta of all the nucleons couple together to give a resultant total angular momentum L. Similarly, the spin angular momenta of all the nucleons couple together to give resultant total spin angular momentum S.

$$\vec{L} = \sum_i \vec{\ell}_i$$

and
$$\vec{S} = \sum_i \vec{s}_i$$

Then the resultant L and S couple strongly to give the total nuclear spin J, i.e. $J = L + S$.

(ii) ***j–j* coupling:** In this scheme orbital (ℓ_i) and spin (s_i) angular momenta of individual nucleons couple together to give resultant angular momentum (j_i), or

$$\vec{j}_i = \vec{\ell}_i + \vec{s}_i$$

The different values of j_i couple together to give nuclear spin J.

$$\vec{J} = \sum_i \vec{j}_i$$

Value of J for nuclei having even number of protons and even number of neutrons is always zero. For nuclei having odd Z and even N or even Z and odd N have half integer nuclear spin (J). All odd Z and odd N nuclei have integral spin (J).

1.7 NUCLEAR MOMENTS

In this section, we discuss magnetic dipole moment of the nucleus, Bohr magneton, nuclear magneton and spin g-factor. An expression for electric quadrupole moment, which arises due to non-spherical charge distribution in the nucleus, has been derived. Prolate and oblate shapes of the nuclei have been discussed.

1.7.1 Magnetic Dipole Moment

Magnetic dipole moment of the nucleus arises due to the motion of charged particles. Orbital and spin angular momenta of protons produce magnetic field within the nucleus. This field can be described in terms of resultant magnetic dipole moment located at the centre of the nucleus.

Particle having a charge q and mass m circulates with speed v in a circular orbit of radius r. If it has a time period t, then the current i associated with the charge q is

$$i = \frac{q}{t} \tag{1.21}$$

Now t can be calculated as under

$$t = \frac{2\pi r}{v}$$

Substituting this value of t

$$i = \frac{qv}{2\pi r}$$

Now magnetic dipole moment μ for current i around an area of the loop A is given as

$$\mu = i \times A$$
$$A = \pi r^2$$

Substituting the values of i and A, we have

$$\mu = \frac{qv}{2\pi r}\pi r^2$$

$$= \frac{qvr}{2} \tag{1.22}$$

Considering the case of a single electron atom and taking q to be electronic charge e and multiplying and dividing the above equation by electron mass m, we have

$$\mu = \frac{emvr}{2m} \tag{1.23}$$

Since mvr is the orbital angular momentum, so μ is written as

$$\mu = \frac{e\ell}{2m} \tag{1.24}$$

where $\ell = mvr$. Since ℓ and μ are vector quantities, so in vector notation, the above equation is

$$\vec{\mu} = \frac{e\vec{\ell}}{2m} \tag{1.25}$$

However, in general, Eq. (1.25) is modified for total angular momentum

$$\vec{\mu} = \frac{e\vec{J}}{2m}g \tag{1.26}$$

where g is a factor called gyromagnetic ratio or g-factor.

As stated earlier, J has the units of $\hbar$, so $\frac{J}{\hbar}$ is dimensionless. Using this fact, Eq. (1.26) can be written as

$$\mu = g\mu_0\frac{J}{\hbar} \tag{1.27}$$

where

$$\mu_0 = \frac{e\hbar}{2m}$$

The constant μ_0 is called Bohr magneton and is taken as the unit to measure magnetic moments for atoms

$$\mu_B = \frac{e\hbar}{2m_e} \tag{1.28}$$

Substituting various constants, we get

$$\mu_B = \frac{1.602\times10^{-19}\ \text{C}\times1.0546\times10^{-34}\ \text{J s}}{2\times9.109\times10^{-31}\ \text{kg}}$$

$$= 9.27\times10^{-24}\ \text{J/tesla}$$

For nuclei, we define magnetic moments in terms of the nuclear magneton μ_N, which is defined as

$$\mu_N = \frac{e\hbar}{2m_p} \tag{1.29}$$

where m_p is the mass of proton. Substituting various constants, we get

$$\mu_N = \frac{1.602 \times 10^{-19}\ \text{C} \times 1.0546 \times 10^{-34}\ \text{J s}}{2 \times 1.67 \times 10^{-27}\ \text{kg}}$$

$$= 5.058 \times 10^{-27}\ \text{J/tesla}$$

which is smaller by a factor $\frac{m_e}{m_p}$ than the Bohr magneton.

Equation (1.25) tells us that $\vec{\mu} \propto \vec{\ell}$. However, the nucleus has neutrons and protons. They possess spin angular momentum in additional to orbital angular momentum. This spin gives an addition contribution to total magnetic moment of the nucleus. The spin angular momentum contribution is written as

$$\vec{\mu}_S = g_S \frac{e\hbar}{2m}\vec{s}$$

$$= g_S \times \vec{s} \times \mu_N \tag{1.30}$$

where g_s is known as spin g-factor. Its value for proton is 5.586 and for neutron it is –3.826. Orbital angular momentum contribution to magnetic moment appears only from the protons, thus in Eq. (1.26) $g = 1$ for protons and $g = 0$ for neutrons.

Total nuclear magnetic dipole moment for the nucleus μ is then given by

$$\mu = \mu_N \left[\sum_{k=1}^{A} g_k \vec{s}_k + \sum_{k=1}^{Z} g_k \vec{\ell}_k \right]$$

$$= \mu_N [g_s \vec{S} + g_e \vec{L}]$$

$$= g\mu_N \vec{J} \tag{1.31}$$

where g is the gyromagnetic ratio of the nucleus, and $\vec{J} = \vec{L} + \vec{S}$. Magnetic dipole moment of odd Z and even N or even Z and odd N nuclei is due to unpaired single nucleon.

1.7.2 Electric Quadrupole Moment

This property arises due to non-spherical charge distribution in the nucleus. It is worthnoting that for a nucleus of spherical shape, neither dipole moment exists nor quadrupole moment exists because centre of mass of the nucleus coincides with centre of charge of the nucleus.

In non-spherical nuclei, assume that centre of mass is at the origin of the nucleus and centre of charge is not at the origin. The potential at any point P situated outside the nucleus at a distance r from the origin is given by

$$V(r) = \frac{1}{4\pi\varepsilon_o} \frac{e}{|\vec{r} - \vec{r}'|} \tag{1.32}$$

where charge e is located at a distance r' from the origin.

Now,

$$|r - r'| = [r^2 + (r')^2 - 2rr' \cos\theta]^{1/2}$$

where θ is the angle between $\bar{r}$ and $\bar{r}'$, as shown in Figure 1.5. Substituting $|r - r'|$ in Eq. (1.32), we get

$$V = \frac{1}{4\pi\varepsilon_0} \frac{e}{[r^2 + (r')^2 - 2rr'\cos\theta]^{1/2}}$$

$$= \frac{e}{4\pi\varepsilon_0 r}\left(1 + \frac{(r')^2}{r^2} - \frac{2r'}{r}\cos\theta\right)^{-1/2} \tag{1.33}$$

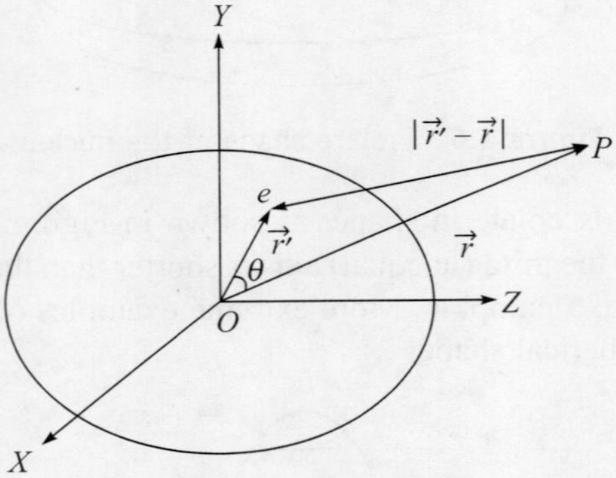

Figure 1.5 Non-spherical charge distribution. The nuclear charge e is located at a distance r' from the origin and the potential at point P outside the nucleus at a distance r' from the origin.

Since r' is much smaller than r, we can carry out a binomial expansion in Eq. (1.33) and obtain

$$\left(1 + \frac{(r')^2}{r^2} - \frac{2r'}{r}\cos\theta\right)^{-1/2} = 1 + \frac{r'}{r}\cos\theta + \left(\frac{r'}{r}\right)^2\left(\frac{3}{2}\cos^2\theta - \frac{1}{2}\right) + \left(\frac{r'}{r}\right)^3\left(\frac{5}{2}\cos^3\theta - \frac{3}{2}\cos\theta\right) + \ldots$$

Substituting this expansion in Eq. (1.33), we have

$$V = \frac{e}{4\pi\varepsilon_0 r}\left[1 + \frac{r'}{r}\cos\theta + \left(\frac{r'}{r}\right)^2\left(\frac{3}{2}\cos^2\theta - \frac{1}{2}\right) + \left(\frac{r'}{r}\right)^3\left(\frac{5}{2}\cos^3\theta - \frac{3}{2}\cos\theta\right) + \ldots\right]$$

$$= \frac{e}{4\pi\varepsilon_0 r} + \frac{er'}{4\pi\varepsilon_0 r^2}\cos\theta + \frac{e(r')^2}{4\pi\varepsilon_0 r^3}\left(\frac{3}{2}\cos^2\theta - \frac{1}{2}\right) + \ldots \tag{1.34}$$

In this equation, the first term on the right-hand side represents the contribution to potential from single charge, the second term represents the contribution due to a dipole, whose electric dipole moment is

$$er'\cos\theta$$

and the third term represents the contribution due to electric quadrupole whose quadrupole moment Q is given as

$$e(r')^2\left(\frac{3}{2}\cos^2\theta-\frac{1}{2}\right)$$

If $Q > 0$, then nucleus is prolate in shape, as shown in Figure 1.6. In prolate shape, when two out of three principal axes are equal and the third (unequal) axis is longer than the other two axes, the nucleus will have a shape like that of a fully inflated football. More extreme examples of prolate shapes are cigar or a hot dog.

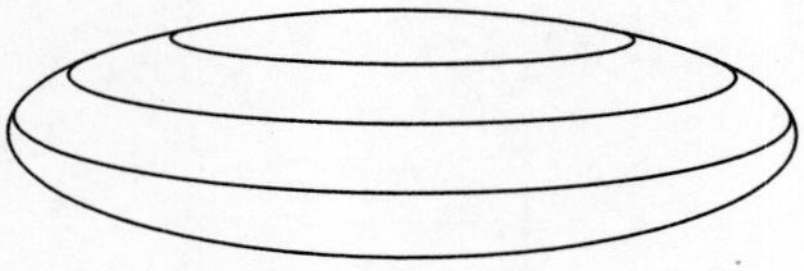

Figure 1.6 Prolate shape of the nucleus.

If $Q < 0$, then nucleus is oblate in shape, as shown in Figure 1.7. When two out of three principal axes are equal and the third (unequal) axis is shorter than the other two axes, the nucleus will have a shape like that of a pumpkin. More extreme examples of oblate shapes are burger or discus. $Q = 0$ represents spherical shape.

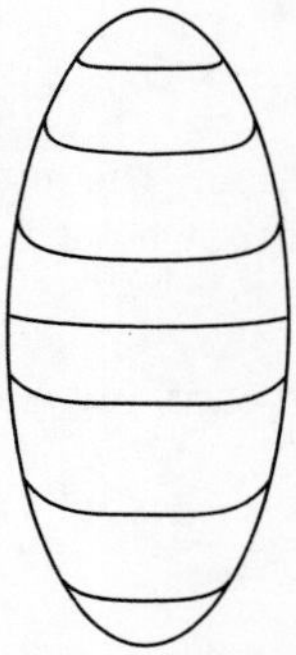

Figure 1.7 Oblate shape of the nucleus.

1.8 WAVE MECHANICAL PROPERTIES

Efforts were made to construct a theoretical model of the nucleus on the basis of classical physics, but they all failed. It was finally realized that classical physics (Newtonian mechanics, Maxwell theory of electromagnetism) can not explain atomic and subatomic phenomena. For instance, classical electromagnetic theory, where electron, being a charged particle should radiate energy while moving around the nucleus, which would eventually collapse in the atom. Similarly, classical physics failed to explain other phenomena, like black body radiations, photoelectric effect, atomic spectra, Compton effect, etc. Initially, semiclassical–semiquantum ideas like Bohr atomic model, were developed, which could only pave the way to develop full quantum theory in two forms called wave mechanical (1927) and matrix mechanics (1925).

The fundamental idea of wave mechanics is based on the wave nature of matter. The dual wave and particle nature of matter is expressed by means of a wave function, which satisfies the following equation, first developed by Schrödinger

$$\nabla^2 \psi + \frac{2m}{\hbar^2}(E - V)\psi = 0 \tag{1.35}$$

where E and m represent energy and mass of the particle moving in a potential $V(\vec{r})$. This equation successfully explained various features of atomic structure, and also introduced new concepts like parity and statistics in quantum physics. All micro-systems, also called quantum systems are expected to satisfy Schrödinger equation or some kind of wave mechanical equation. Nucleus containing nucleons is one such example, though potential experienced by nucleons within the nucleus, could be more complicated than the Coulomb potential, which worked well for the atomic spectra.

In general, any quantum system is always associated with a wave function, which depends on space–time coordinates (x, y, z, t).

Generally, ψ is a complex quantity but $\psi\psi^*$ is always real. The wave function has two important properties:

1. Parity
2. Statistics

1.8.1 Parity

Under reflection, also called space-inversion, through origin x will change to $-x$, y to $-y$ and z to $-z$. If the new wave function $\psi(-x, -y, -z)$ is identical to $\psi(x, y, z)$, then we say that the wave function has even parity. But, if the new wave function is related to old wave function with a change of sign, i.e. $\psi(-x, -y, -z) = -\psi(x, y, z)$, then the wave function has odd parity. Alternatively, this can be represented as

$$P\psi(x, y, z) = \psi(-x, -y, -z)$$

Applying parity operator once more

$$P^2\psi(x, y, z) = P\psi(-x, -y, -z) = \psi(x, y, z) \tag{1.36}$$

Thus,

$$P^2 = 1$$

i.e.

$$P = \pm 1 \tag{1.37}$$

where P, the parity operator has values $+1$ or -1. The parity is a quantum concept and it does not have any classical analogue.

In hydrogen-like system, $P = (-1)^\ell$, where ℓ is the orbital quantum number. Depending upon ℓ, P has the value $+1$ or -1.

A system having even number of odd parity particles has even parity and a system having odd number of odd parity particles has odd parity. Because parity of the system is the product of individual parities.

Except for weak interactions, discussed in Chapter 8, parity is always conserved in a nuclear transformation, e.g. if a even parity excited nuclear state decays by γ-emission, then the product of parities of γ-ray and the daughter nucleus must be even. Similarly, in α-decay if the parent nucleus has even parity, then the product of parities of the daughter nucleus and the α-particle must be even.

1.8.2 Statistics

The concept of statistics is related to the behaviour of large number of particles. For instance, Maxwell–Boltzmann statistics works well in the classical physics, while describing the distribution of energy among the molecules of a gas. However, in quantum physics, two new forms of statistics have been developed which are the Bose–Einstein statistics and the Fermi–Dirac statistics. These statistics reflect in the behaviour of wave function describing systems of identical particles under their interchange.

Bose–Einstein Statistics

If the particles have integral spins, i.e. 0, 1, 2, ... , they are known as Bosons, e.g. photons, pions, etc. The wave function of the system consisting of Bosons is symmetric when we interchange any two identical Bosons, i.e. $\psi(x_1, x_2) = \psi(x_2, x_1)$. Similar results can be written for more than two particles.

Fermi–Dirac Statistics

The particles having half integral spins are known as Fermions. Examples are protons, neutrons, etc. All Fermions obey Pauli's exclusion principle. The wave function of a Fermions system is antisymmetric under interchange of two identical particles, i.e. $\psi(x_1, x_2) = -\psi(x_2, x_1)$.

1.9 NATURE OF NUCLEAR FORCES

The forces known in the beginning of the twentieth century—gravitational and electromagnetic Coulomb forces could not account for the stability of the nucleus. Protons carry positive charge and neutrons are neutral particles. The electrostatic force will be repulsive between two protons. So, this force will cause the nucleus to break and thus it cannot account for the binding of the nucleus. The gravitational force of attraction between two nucleons is extremely weak, it also cannot explain the binding energy of the nucleus. This force is significant only on the macroscopic scale, such as for astronomical objects.

The small size of the nucleus (~ a few fermi) and its great stability shows that the forces operating among the nucleons are short-range forces, i.e. they are effective over very short distances only (~ few fermi). This short-range character distinguishes the nuclear forces from gravitational and electromagnetic forces, which act over large distances. If the forces operating among the nucleons were long range, there would be interactions among distant nucleons in the nucleus, as well as between those in close proximity; the total binding energy would increase roughly as A^2. However, the actual binding energy is approximately proportional to A. This characteristics of the forces is called saturation property. Each nucleon attracts those in its immediate vicinity, but is relatively indifferent to the others. This new force, called nuclear force was introduced to account for the stability of the nucleus. This force was assumed to be much stronger than the repulsive Coulomb force between two protons. This force is attractive between

2 protons, 2 neutrons or 1 proton and 1 neutron. Later on, these nuclear forces were found to be highly complex. Some important characteristics of nuclear forces are:

Nuclear Forces are Charge-Independent

The force between *n–n* is same as that between *p–p* and between *n–p*. This is termed as charge-independence of nuclear forces. This is clear from the formation of mirror nuclei. For example, ^3_1H and ^3_2He are mirror nuclei. The former contains 1 proton and 2 neutrons, while the later contains 2 protons and 1 neutron. The binding energy of ^3_1H is 8.5 MeV and that of ^3_2He is 7.7 MeV. The three types of forces (*n–n*, *p–p*, *n–p*) are equal in magnitude, the difference of 0.8 MeV in ^3_1H and ^3_2He is due to repulsive Coulomb energy between proton–proton in ^3_2He nuclide.

Nuclear Forces are Very Strong in Nature

They are much stronger than electromagnetic and gravitational forces. They are stronger by a factor of 137 from the electromagnetic force and are stronger by a factor of 10^{40} from gravitational forces.

Nuclear Forces have Short Range

Nuclear forces are short-range forces and the range is of the order of nuclear dimensions. If the range of the nuclear forces is larger than the dimensions of the nucleus, they would have definitely affected molecular interactions, but no such effect has ever been observed in molecular structure.

Nuclear Forces are Repulsive at Very Short Distances (~0.5 fm)

If this was not so, the nucleons inside the nucleus would collapse to the size of nuclear force range.

Nuclear Forces are Saturated

The range of nuclear forces is limited. Saturation of nuclear forces implies that their interaction is only with the nearest neighbours. The effect of saturation property of nuclear forces is that the binding energy is proportional to *A*.

Nuclear Forces are Non-Central

Electrostatic and gravitational forces are central forces, i.e. these forces act along the line joining the two particles while nuclear forces depend not only on the distance between the nucleons, but also on the angle that their spins make with the line joining the two nucleons.

Nuclear Forces are Spin and Angular Momentum-Dependent

Two nucleons having parallel spin have stronger force between them than those having anti-parallel spin. This force depends on the angles between the spin and orbital angular momentum vectors.

Nuclear Forces are Velocity-Dependent

Nuclear forces depend upon the velocity of the nucleons.

Nuclear Forces are Exchange Forces

It means that the force is produced by exchange of force carrier. It is described in detail in the next section.

1.10 YUKAWA THEORY OF NUCLEAR FORCES

Yukawa in 1935 proposed a theory to explain the nature of nuclear forces. Yukawa proposed that when two nucleons are interacting, there is a field associated with this system and the cause of the force between the two nucleons is an exchange of a quantum between them. A quantum is emitted by one nucleon and is absorbed by another.

This assumption is similar to the case of Coulomb force between two charged particles. According to quantum field theory, the Coulomb interaction between the two is explained by saying that one charged particle emits a virtual photon that is absorbed by the other. The zero mass of photon explains the long-range nature of the Coulomb forces. According to Yukawa, because nuclear forces being of short-range kind, the quantum exchanged between the nucleons must have a finite mass. The mass of the quantum can be deduced from the range of nuclear forces and was calculated by Yukawa as $\sim 300\ m_e$, where m_e is the mass of electron. The quanta involved in such interactions have been given name as π-mesons or pions. It was proposed that there exist three pions and they were denoted as positively charged pion π^+, neutral pion π^0 and negatively charged pion π^-. All these three pions were discovered and their properties have been measured.

To explain neutron–proton interactions, a neutron decays by emission proton and π^-, i.e. $n \rightarrow p + \pi^-$ and this π^- is absorbed by another proton or $\pi^- + p \rightarrow n$. Similarly, proton–neutron interaction is explained by the emission of π^+, i.e. $p \rightarrow n + \pi^+$ and $n + \pi^+ \rightarrow p$.

For *p–p* interaction, $p \rightarrow p' + \pi^0$, where p' is the new proton by emitting π^0 and this π^0 is absorbed by another proton as $\pi^0 + p \rightarrow p'$ where p' is the new proton formed by absorbing π^0.

Similar equations can be written for *n–n* interaction.

In the case of single nucleon, the pions are constantly emitted and absorbed by the same nucleon, i.e. a nucleon is surrounded by a cloud of pions.

There is another question we may ask. When a pion is emitted, will there be an apparent non-conservation of energy by an amount $\Delta E = m_\pi c^2$? The answer lies in the uncertainty principle, which allows non-conservation of energy by an amount given by ΔE

$$\Delta E\ \Delta t \sim \hbar$$

or

$$\Delta t \sim \frac{\hbar}{\Delta E} \approx \frac{\hbar}{m_\pi c^2} \leq 4 \times 10^{-24}\ \text{s} \tag{1.38}$$

If pion is travelling with approximately equal to that of light or c, then in time Δt, it will cover a distance

$$\Delta t \times c = \frac{\hbar}{m_\pi c} \approx 1.2 \times 10^{-13}\ \text{cm} \tag{1.39}$$

which is precisely the range of nuclear forces.

1.11 MASS SPECTROMETRY

Mass spectrometry also known as *mass spectroscopy* is an analytical technique used to measure the mass-to-charge ratio of ions. It is most generally used to find the composition of a physical sample by generating a mass spectrum representing the masses of sample components. The mass spectrum is obtained by a *mass spectrometer*.

Although the general term mass spectroscope is often used, a distinction can be made between *spectrometer* in which a metre measures the ion current, and a *spectrograph* in which the record appears on a photographic plate. The term mass spectrometer, however, is often used for all such instruments. A typical mass spectrometer consists of three parts, as shown in Figure 1.8, an ion source, a mass analyzer, and a detector and data analysis system. The stages within the mass spectrometer are:

1. Producing ions from the sample.
2. Separating ions of differing masses.
3. Detecting the number of ions of each mass produced.
4. Collecting the data and generating the mass spectrum.

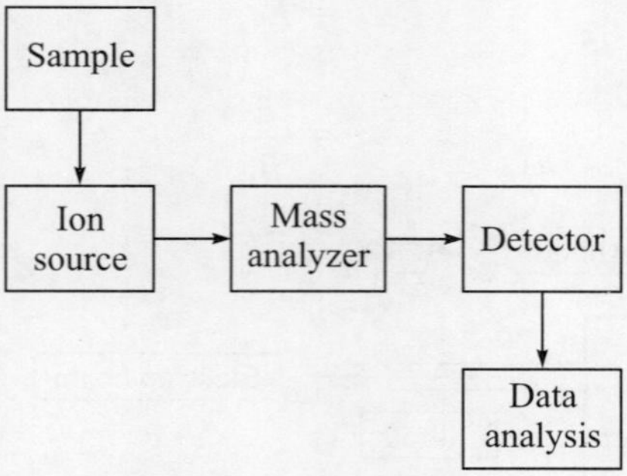

Figure 1.8 Block diagram of the mass spectrometer.

The foundations of mass spectrometry lie in the work of Thomson and Aston at the Cavendish Laboratories, Cambridge University. From 1897, the work was carried out by Thomson and his co-workers. Thomson's original work on the existence and properties of canal rays (positive ions) was taken up by Aston who demonstrated the existence of several isotopes of non-radioactive elements. Aston used electrostatic and magnetic fields to separate isotopes and focus them onto a photographic plate. Over the next few years a number of workers took up the development of mass spectrometry, including Dempster, Herzog, Bainbridge and Nier. By the end of the 1930s, mass spectrometry had become an established technique for two purposes:

(i) Determination of number of isotopes in a given element.
(ii) Determination of masses of various isotopes.

In the present discussion, we focus on two types of mass spectrographs:

(a) Bainbridge spectrograph.
(b) Bainbridge and Jordan mass spectrograph.

1.11.1 Bainbridge Spectrograph

In 1933, Bainbridge modified the existing mass spectrographs (like Aston's mass spectrograph, Dempster's mass spectrometer, etc.) and developed a new spectrograph known as Bainbridge spectrograph. It is based on the principle of velocity selecting/filtering. It contains an ion source, in which an anode D is fitted as shown in Figure 1.9. Cathode is perforated. The gas whose isotopes are to be investigated is admitted into the ion source through a gas tube marked as gas in the figure. A high potential difference (5 to 20 kV) is applied between the anode and the cathode (not shown in figure). The high voltage between the anode and the cathode ionizes the gas present in the ion source. Ionization is further improved by bombarding the gaseous isotopes with a beam of electrons. Positive ions are accelerated towards the cathode. Because of the perforation in the cathode the accelerated ions come out and are allowed to pass through a slit S_1. Then they are made to travel through two plates A and B, between which an electric field (E) is applied. Here a magnetic field (strength B_1) is also applied at right angle to the electrostatic field. These two are adjusted in such a way that electrostatic and electromagnetic forces act in opposite direction to each other. Finally, positive ions pass through the second slit S_2. The force on the ions due to electric field is Ee and due to magnetic field B_1ev, where e is the charge present on the ions and v is the velocity of the ions. As these two forces are equal and opposite, therefore,

$$B_1ev = Ee \tag{1.40}$$

or

$$v = \frac{E}{B_1} \tag{1.41}$$

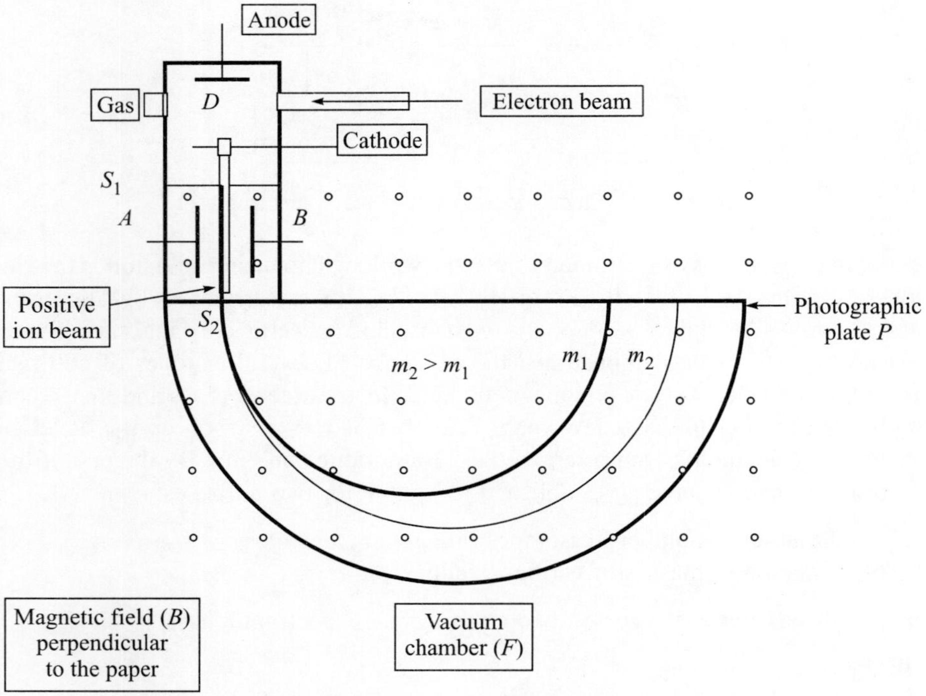

Figure 1.9 Schematic diagram of Bainbridge spectrograph.

Only those positive ions, whose velocity is given by Eq. (1.41), continue to move forward and pass through the slit S_2. Therefore, in this arrangement of cross magnetic and electric fields, positive ions having exactly the same velocity v are selected. For this reason, the combination of slits S_1 and S_2 and deflecting plates A and B is called velocity selector/filter. The ions moving with velocity v enter a vacuum chamber F, which is also called analyzing or deflecting chamber, where a constant magnetic field B is maintained at right angle to the direction of motion of ions. Under the influence of this magnetic field, the ions move in a semicircular path of radius r, given by

$$\frac{Mv^2}{r} = Bev$$

or

$$\frac{e}{M} = \frac{v}{rB}$$

Substituting v from Eq (1.41), we get

$$\frac{e}{M} = \frac{E}{BB_1 r} \tag{1.42}$$

Equation (1.42) shows that ions having different $\frac{e}{M}$ values, describe semicircular paths of different radii. Ultimately these ions strike photographic plate P. Equation (1.42) can also be written as

$$r = \frac{E}{BB_1}\frac{M}{e} = KM$$

where $K = E/BB_1e$ is constant, therefore, the radius of curvature of the path is directly proportional to the mass of the positive ions, provided the charge is same for all the ions. Suppose a given element contains two isotopes with masses m_1 and m_2, with $m_2 > m_1$. Atoms of the isotope with mass m_1 will describe a semicircle of smaller radius, while the atoms of the isotope with mass m_2 will move in a semicircle of larger radius. Thus, for a given element, particles having the same mass will follow a particular path and strike the photographic plate P at the same point.

In the first experiment, Bainbridge analyzed germanium vapours. He used a strong magnetic field (~15 kG) across the analyzing chamber and the semicircular path of the ions was about 40 cm. With this arrangement, he was able to identify five isotopes of germanium with mass numbers 70, 72, 73, 74 and 76.

Some of the advantages of Bainbridge mass spectrograph are as under:

- Since $r = KM$, so the mass scale is linear.
- The ion beam strikes the photographic plate normally, the error due to irregularities in the plate is negligible.
- It has a high precision.

1.11.2 Bainbridge and Jordan Mass Spectrograph

In 1936, Bainbridge and Jordan developed a mass spectrograph capable of mass determination with an accuracy of 1 in 30,000 parts. The earlier spectrographs were designed either to focus ions having the same velocity or moving in same direction to one point. However, in this mass spectrograph, all the ions having same e/M value are focused at the same point irrespective of the direction or velocity.

A schematic diagram of this mass spectrograph is shown in Figure 1.10. The ions from the discharge tube having anode A and a perforated cathode C are collimated by slits S_1 and S_2. They are then deflected through $\pi/\sqrt{2}$ radians (~127°), using the radial magnetic field to produce an energy spectrum in front of the third slit S_3. Beams of ions of all e/M values of this energy spectrum diverging from the focal plane of the electrical deflector are further deflected through a mean angle of $\pi/3$ radians by magnetic field, which forms a mass spectrum of focused ion beams on the photographic plate.

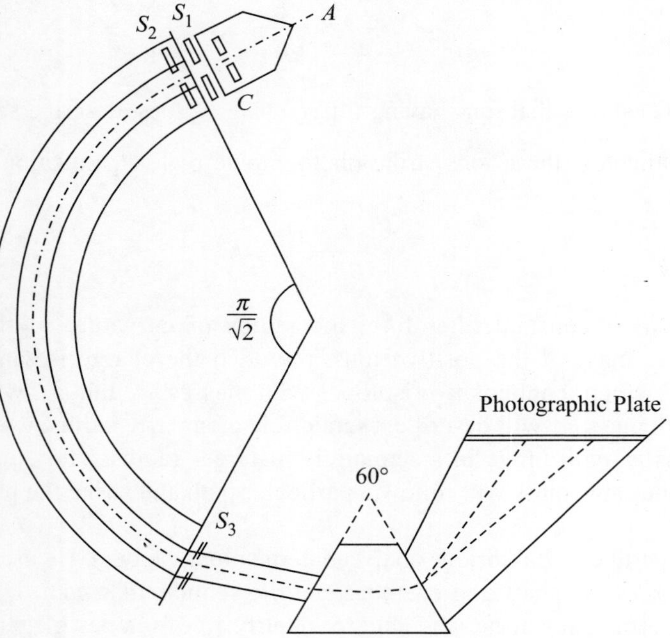

Figure 1.10 Schematic diagram of Bainbridge and Jordan mass spectrograph.

The focusing of ions of different velocities is obtained as a result of the geometrical arrangement of the electric and magnetic deflecting fields which are so placed that the dispersion of the ions produced by one field is cancelled by the dispersion produced by the other field for a given velocity difference. Under these conditions ions of the same e/M values having all velocities within a certain range are focused on the recording plate.

1.12 DETERMINATION OF CHARGE BY MOSELEY LAW

While working in Rutherford's laboratory at University of Manchester, in 1913, H. Moseley demonstrated that charge present on a nucleus could be determined by analyzing its X-ray spectrum. At that time, it was discovered that crystal could act as diffraction grating and hence crystals could be used to compare the wavelengths of X-rays. Moseley studied the characteristic X-rays of a number of elements. The 10-consecutive elements chosen by Moseley were $_{20}Ca$, $_{22}Ti$, $_{23}V$, $_{24}Cr$, $_{25}Mn$, $_{26}Fe$, $_{27}Co$, $_{28}Ni$, $_{29}Cu$ and $_{30}Zn$. He obtained the photographs of X-rays lines. From the position of the lines on photographs, Moseley determined the frequencies of the corresponding X-rays. From this data, he concluded that there was one fundamental quantity Q which increases in regular steps as one passes from one element to the next, using the chemical order of the elements in the periodic table. He found that square root of the frequency of most intense X-ray (known as K_α line) is directly proportional to a quantity which Moseley called Q. Later on the quantity Q was called Z atomic number of the element. Moseley formulated a simple relation between Z and square root of the X-ray frequency as

$$\sqrt{f} = k(Z - k_2) \tag{1.43}$$

where f is the frequency of the most intense X-ray line, k and k_2 are constants. This formula is known as Moseley's law.

Equation (1.43) can also be written as

$$f = k_1(Z - k_2)^2 \tag{1.44}$$

where k_1 is another constant.

The constants k_1 and k_2 depend on the type of X-ray line. For K_α X-ray line (K_α is X-ray emitted when electron makes a transition from L shell to K shell), Moseley found $k_1 = 2.47 \times 10^{15}$ Hz, and $k_2 \approx 1$. Later on it was found that k_1 is actually product of Rydberg constant (3.29×10^{15}) and 3/4. The constants k_1 and k_2 are same for all K_α X-ray lines irrespective of which element is emitting the X-ray. Substituting these constants, Eq. (1.44) becomes

$$f_{K\alpha} = 2.47 \times 10^{15}(Z - 1)^2 \text{ Hz.} \tag{1.45}$$

Similarly, for L_α X-ray line (L_α X-ray is emitted when an electron makes a transition from M shell to L shell), constants k_1 and k_2 were found to be $k_1 = 0.457 \times 10^{15}$ Hz, and $k_2 \approx 7.4$. Here constant k_1 is the product of Rydberg constant and 5/36. Therefore, for L_α X-ray line Moseley law becomes

$$f_{L\alpha} = 0.457 \times 10^{15}(Z - 7.4)^2 \text{ Hz} \tag{1.46}$$

In terms of energy (as with modern X-ray spectrometers it is easy to measure energy than to calculate frequency of X-rays), Eqs. (1.45) and (1.46) become

$$E_{K\alpha}(\text{keV}) = 1.042 \times 10^{-2}(Z - 1)^2$$

Similarly, for L_α line (which is generally less intense than K_α) the above relation is

$$E_{L\alpha}(\text{keV}) = 1.494 \times 10^{-3}(Z - 7.4)^2$$

Thus, by experimentally measuring the energy (E) or frequency (f), Z—the atomic number of the element can be calculated. This method was first applied to newly discovered element hafnium (Hf), which is chemically identical to zirconium (Zr). C. Coster and G. von Hevesy in 1923 found the atomic number of hafnium as 72 from its X-ray spectrum.

NUMERICAL PROBLEMS

Section 1.3

Solved Problems

1. Calculate the momentum of a neutron having de Broglie wavelength of 5 fm.

Solution:

$$\text{Momentum } p = \frac{h}{\lambda}$$

$$\hbar = 1.055 \times 10^{-21} \text{ MeV s}$$

$$\lambda = 5 \text{ fm} = 5 \times 10^{-15} \text{ m}$$

Therefore,

$$p = \frac{1.055 \times 10^{-21}}{5 \times 10^{-15}}$$

$$= 0.211 \times 10^{-6} \text{ MeV s/m}$$

2. Group the following nuclides as isotopes, isotones and isobars:

$$^{12}_{6}\text{C},\ ^{13}_{6}\text{C},\ ^{14}_{6}\text{C},\ ^{14}_{7}\text{N},\ ^{14}_{8}\text{O},\ ^{15}_{7}\text{N},\ ^{15}_{8}\text{O},\ ^{16}_{6}\text{C},\ ^{16}_{7}\text{N},\ ^{16}_{8}\text{O},\ ^{17}_{7}\text{N},\ ^{17}_{8}\text{O}$$

Solution:

Isotopes:

$^{12}_{6}\text{C}$, $^{13}_{6}\text{C}$, $^{14}_{6}\text{C}$, $^{16}_{6}\text{C}$

$^{14}_{7}\text{N}$, $^{15}_{7}\text{N}$, $^{16}_{7}\text{N}$, $^{17}_{7}\text{N}$

$^{14}_{8}\text{O}$, $^{15}_{8}\text{O}$, $^{16}_{8}\text{O}$, $^{17}_{8}\text{O}$

Isotones:

$^{12}_{6}\text{C}$, $^{14}_{8}\text{O}$,

$^{13}_{6}\text{C}$, $^{14}_{7}\text{N}$, $^{15}_{8}\text{O}$

$^{14}_{6}\text{C}$, $^{15}_{7}\text{N}$, $^{16}_{8}\text{O}$

$^{16}_{7}\text{N}$, $^{17}_{8}\text{O}$

Isobars:

$^{14}_{6}\text{C}$, $^{14}_{7}\text{N}$, $^{14}_{8}\text{O}$

$^{15}_{7}\text{N}$, $^{15}_{8}\text{O}$

$^{16}_{6}\text{C}$, $^{16}_{7}\text{N}$, $^{16}_{8}\text{O}$

$^{17}_{7}\text{N}$, $^{17}_{8}\text{O}$

Unsolved Problems

1. Calculate de Broglie wavelength for a proton having energy 150 MeV.

[**Ans.** 0.596 fm]

2. Calculate the ratio of de Broglie wavelengths of a proton and electron having energy equal to 150 MeV. [**Ans.** 0.462]
3. Calculate the ratio of momenta of a proton and neutron having the same de Broglie wavelength. [**Ans.** 0.999]
4. Find the number of neutrons present in $^{238}_{92}U$ nucleus. [**Ans.** 146]
5. Find the number of electrons present on $^{238}_{92}U$ atom. [**Ans.** 92]
6. Thermal neutrons have an average kinetic energy $\frac{3}{2}kT$, where k is Boltzmann's constant and T is room temperature, 300 K. (i) What is the average energy in eV of a thermal neutrons? (ii) What is the corresponding de Broglie wavelength?

[**Ans.** (i) 0.038 eV, (ii) 1.46×10^{-10} m]

Section 1.4

Solved Problems

1. Calculate the energy of electron at rest.

Solution: Mass of the electron $m = 9.1 \times 10^{-31}$ kg

$$\text{Energy } mc^2 = 9.1 \times 10^{-31} \times (3 \times 10^8)^2 \text{ J}$$

$$= 8.19 \times 10^{-14} \text{ J}$$

$$1.6 \times 10^{-13} \text{ J} = 1 \text{ MeV}$$

$$81.9 \times 10^{-15} \text{ J} = \frac{8.19 \times 10^{-14}}{1.6 \times 10^{-13}} \text{ MeV}$$

$$= 0.511 \text{ MeV}$$

2. Estimate the A value and identify the nucleus if its radius is given to be 3.46 fm.

Solution: Radius of the nucleus $R = 3.46$ fm

We have from Eq. (1.3)

$$r = r_0 A^{1/3}$$

and

$$r_0 = 1.2 \text{ fm}$$

Therefore,

$$\frac{r}{r_0} = A^{1/3}$$

$$A^{1/3} = \frac{r}{r_0}$$

$$= \frac{3.46}{1.2}$$

$$= 2.883$$

$$A = (2.883)^3$$

$$= 23.97$$

$$\approx 24$$

Therefore, $A = 24$, the nucleus is ^{24}Mg.

3. Assume 1 amu = 1.66×10^{-27} kg, estimate the density of nuclear matter.

Solution: Protons and neutrons present in the nucleus constitute nuclear matter. Let mass of a nucleus $A = 40$ amu.

$$\text{Mass of this nucleus } m = 40 \times 1.66 \times 10^{-27} \text{ kg}$$

$$= 6.64 \times 10^{-26} \text{ kg}$$

$$\text{Radius } r = r_0 A^{1/3}, \text{ where } r_0 = 1.2 \text{ fm and } A = 40$$

$$\therefore \quad r = 1.2 \times 40^{1/3} \times 10^{-15} \text{ m}$$

$$= 4.1 \times 10^{-15} \text{ m}$$

$$\text{Volume of nucleon } V = \frac{4}{3}\pi r^3$$

$$= \frac{4}{3} \times 3.1415926 \times (4.1 \times 10^{-15})^3 \text{ m}^3$$

$$= 2.886 \times 10^{-43} \text{ m}^3$$

$$\text{Density} = \frac{\text{Mass}}{\text{Volume}} = \frac{m}{V}$$

$$= \frac{6.64 \times 10^{-26}}{2.886 \times 10^{-43}}$$

$$= 2.3 \times 10^{17} \text{ kg/m}^3$$

$$\therefore \quad \text{Density of nuclear matter} = 2.3 \times 10^{17} \text{ kg/m}^3$$

4. Find the nuclear density of ^{235}U, if $r_0 = 1.2$ fm.

Solution:

$$\text{Mass of 1 nucleon} = 1.66 \times 10^{-27} \text{ kg}$$

$$\text{Mass of 235 nucleons } m = 235 \times 1.66 \times 10^{-27} \text{ kg}$$

$$= 3.901 \times 10^{-25} \text{ kg}$$

$$\text{Radius of } ^{235}\text{U nucleus } r = r_0 A^{1/3}$$

$$= 1.2 \times 10^{-15} \times 235^{1/3}$$

$$= 7.405 \times 10^{-15} \text{ m}$$

$$\text{Volume of } {}^{235}\text{U nucleus } V = \frac{4}{3}\pi r^3$$

$$= \frac{4}{3} \times 3.1415926 \times (7.405 \times 10^{-15})^3 \text{ m}^3$$

$$= 1.701 \times 10^{-42} \text{ m}^3$$

$$\text{Density of } {}^{235}\text{U} = \frac{\text{Mass}}{\text{Volume}}$$

$$= \frac{3.901 \times 10^{-25}}{1.701 \times 10^{-42}} \text{ kg/m}^3$$

$$= 2.29 \times 10^{17} \text{ kg/m}^3$$

5. The radii of oxygen and lead nuclei are found to be 3 fm and 7 fm respectively. Their masses are 2.7×10^{-26} kg and 3.4×10^{-25} kg respectively. Calculate their densities.

Solution: Volume of nucleus with mass number $A = \frac{4}{3}\pi r^3$

$$\text{For oxygen nucleus volume} = \frac{4}{3} \times 3.14159 \times (3 \times 10^{-15})^3$$

$$= 1.131 \times 10^{-43} \text{ m}^3$$

$$\text{Density of oxygen nucleus} = \frac{\text{Mass}}{\text{Volume}}$$

$$= \frac{2.7 \times 10^{-26}}{1.131 \times 10^{-43}}$$

$$= 2.38 \times 10^{17} \text{ kg/m}^3$$

$$\text{For lead nucleus volume} = \frac{4}{3} \times 3.14159 \times (7 \times 10^{-15})^3$$

$$= 1.437 \times 10^{-42} \text{ m}^3$$

$$\text{Density of lead nucleus} = \frac{\text{Mass}}{\text{Volume}}$$

$$= \frac{3.4 \times 10^{-25}}{1.437 \times 10^{-42}}$$

$$= 2.37 \times 10^{17} \text{ kg/m}^3$$

6. If the energy of the α-particle emitted by ^{241}Am is 5.48 MeV, find the closest distance it can approach a Au nucleus.

Solution: The distance of closest approach is given by the relation

$$d = \frac{2Ze^2}{4\pi\varepsilon_0 E}$$

Given:

$$E = 5.48 \text{ MeV}$$
$$= 5.48 \times 1.6 \times 10^{-13} \text{ J}$$

Therefore,

$$d = \frac{8.98 \times 10^9 \times 2 \times 79 \times (1.6 \times 10^{-19})^2}{5.48 \times 1.6 \times 10^{-13}}$$
$$= 4.14 \times 10^{-14} \text{ m}$$

7. Determine the radius of ^{208}Pb.

Solution: For ^{208}Pb, $A = 208$

We have

$$r = r_0 A^{1/3} \qquad \text{and} \qquad r_0 = 1.2 \times 10^{-15} \text{ m}$$

Therefore,

$$\text{Radius } r = 1.2 \times 10^{-15} \times 208^{1/3} = 7.11 \times 10^{-15} \text{ m}$$

or

$$\text{Radius of } ^{208}\text{Pb} = 7.11 \times 10^{-15} \text{ m} = 7.11 \text{ fm}$$

Unsolved Problems

1. It is possible on the basis of shell model that nucleus with $Z = 110$ and $A = 294$ may be exceptionally long-lived. Estimate its nuclear radius.

[**Ans.** 7.98 fm]

2. Ordinary boron is a mixture of $^{10}_{5}$B and $^{11}_{5}$B isotopes and has composite atomic mass of 10.82 amu. What percentage of each isotope is present in ordinary boron? Given $m_{B_{10}}$ = 10.012937 amu and $m_{B_{11}}$ = 11.009305 amu.

[**Ans.** 18.99, 81.01]

3. Find the distance of closest approach, when
(i) 5 MeV protons are bombarded on $^{27}_{13}$Al.
(ii) 5 MeV deuterons are bombarded on $^{27}_{13}$Al.

[**Ans.** (i) 3.5×10^{-15} m, (ii) 3.5×10^{-15} m]

4. Find the distance of closest approach, when
(i) 50 MeV protons are bombarded on $^{208}_{82}$Pb.
(ii) 50 MeV are bombarded on $^{208}_{82}$Pb.

[**Ans.** (i) 2.87×10^{-17} m, (ii) 1.9×10^{-13} m]

5. What is the energy equivalent to the mass of an electron and a proton?

[**Ans.** 0.511 MeV, 939.3 MeV]

6. The density of nuclear matter is 2.294×10^{17} kg/m^3. Estimate the energy equivalent to 1Å^3 of nuclear matter.

[**Ans.** 20.646 kJ]

Section 1.5

Solved Problems

1. Calculate the binding energy of α-particle and express it in MeV and joules. Given $m_p = 1.00758$ amu, $m_n = 1.00897$ amu and $m_{He} = 4.0028$ amu.

Solution: Binding energy of α-particle is given by Eq. (1.18)

$$BE = (Zm_p + Nm_n - m_{He}) \text{ amu}$$
$$= (2 \times 1.00758 + 2 \times 1.00897 - 4.0028) \text{ amu}$$
$$= 0.0293 \text{ amu}$$

Now,

$$1 \text{ amu} = 931.49 \text{ MeV}$$
$$0.0293 \text{ amu} = 0.0293 \times 931 \text{ MeV}$$
$$= 27.29 \text{ MeV}$$

Therefore, the binding energy of α-particle = 27.29 MeV

$$1 \text{ amu} = 1.49239 \times 10^{-10} \text{ J}$$
$$0.0293 \text{ amu} = 0.0293 \times 1.49239 \times 10^{-10}$$
$$= 4.37 \times 10^{-12} \text{ J}$$

2. Find the energy required in joules to break ^{12}C into 3 α-particles. The atomic mass of ^{12}C = 12 amu and $m_{He} = 4.0026$ amu.

Solution: We have the equation

$$^{12}_{6}C \rightarrow 3\,^{4}_{2}He$$

Given:

$$\text{Mass of } ^{12}C = 12 \text{ amu}$$
$$\text{Mass of } \alpha\text{-particle} = 4.0026 \text{ amu}$$
$$\text{Mass of } 3\alpha\text{-particle} = 3 \times 4.0026 \text{ amu}$$
$$= 12.0078 \text{ amu}$$
$$\text{Difference in two masses} = 12 - 12.0078$$
$$= -0.0078 \text{ amu}$$
$$1 \text{ amu} = 1.49239 \times 10^{-10} \text{ J}$$
$$0.0078 \text{ amu} = -0.0078 \times 1.49239 \times 10^{-10}$$
$$= -1.16 \times 10^{-12} \text{ J}$$

Therefore, 1.16×10^{-12} J of energy is required to break up ^{12}C into 3 α-particles.

3. Masses of He nucleus, proton and neutron in amu are 4.0026, 1.007895 and 1.008665. Find the energy required to knock out nucleons from the He nucleus.

Solution: We have the equation

$$^{4}_{2}He \rightarrow 2p + 2n$$

Given:

$$m_{He} = 4.0026 \text{ amu}$$
$$m_p = 1.007895 \text{ amu}$$
$$m_n = 1.008665 \text{ amu}$$

Mass of 2 protons and 2 neutrons

$$= 2 \times 1.007895 + 2 \times 1.008665$$
$$= 4.03312 \text{ amu}$$

Therefore, $\text{mass defect} = 4.0026 - 4.03312$

$$= -0.03052 \text{ amu}$$
$$\text{Equivalent energy} = -0.03052 \times 931.49$$
$$= -28.43 \text{ MeV}$$

Energy required to knock out 2 protons and 2 neutrons from He nucleus = 28.43 MeV.

4. Find the binding energy of $^{56}_{26}Fe$ in MeV. Given: $m_p = 1.007825$ amu, $m_n = 1.008665$ amu and $m_{Fe} = 55.934939$ amu.

Solution: Binding energy (BE) is given as

$$BE = (26m_p + 30m_n - m_{Fe}) \text{ amu}$$
$$= (26 \times 1.007825 + 30 \times 1.008665 - 55.934939) \text{ amu}$$

or

$$\text{Binding energy } BE = 0.5285 \text{ amu}$$

or

$$\text{Binding energy } BE = 0.5285 \times 931.49$$
$$= 492.29 \text{ MeV}$$

5. The mass of a deuteron is 2.014103 amu. If the masses of proton and neutron are respectively 1.007825 amu and 1.008663 amu, find the mass defect and packing fraction.

Solution:

$$\text{Mass defect } \Delta M = m_p + m_n - m_d$$
$$= 1.008825 + 1.008663 - 2.014104 = 0.002385 \text{ amu}$$
$$= 2.22 \text{ MeV}$$
$$\text{Packing fraction } f = \frac{\text{Mass} - A}{A}$$
$$= \frac{2.014103 - 2}{2}$$
$$= 0.00705$$

6. Calculate the average binding energy per nucleon of ^{4}He nucleus. Given $m_{He} = 4.002643$ amu, $m_p = 1.007825$ amu and $m_n = 1.008665$ amu.

Solution: Binding energy is given by

$$BE = (2m_p + 2m_n - m_{He}) \times 931.47 \text{ MeV}$$

$$= (2 \times 1.007825 + 2 \times 1.008665 - 4.002634) \times 931.47 \text{ MeV}$$

$$= 0.03034 \times 931.47 = 28.3 \text{ MeV}$$

$$\text{Binding energy per nucleon} = \frac{28.3}{4} = 7.07 \text{ MeV}$$

Unsolved Problems

1. Find the binding energy of $^{20}_{10}$Ne and of $^{79}_{35}$Br. Given m_p = 1.007825 amu, m_n = 1.008665 amu, m_{Ne} = 19.992436 amu and m_{Hr} = 78.918336 amu.

[**Ans.** 160.64 MeV, 686.29 MeV]

2. Find the energies needed to remove a neutron from $^4_2$He, then to remove a proton, and finally to separate the remaining neutron and proton. Compare with the total binding energy of $^4_2$He, Given m_p = 1.007825 amu, m_n = 1.008665 amu, m_{He4} = 4.002603 amu, m_{He_3} = 3.016029 amu and m_D = 2.014102 amu.

[**Ans.** 20.58 MeV, 5,491 MeV, 2.224 MeV]

3. Which nucleus is more stable $^7_3$Li or $^8_3$Li? Given m_p = 1.007825 amu, m_n = 1.008665 amu, m_{Li_7} = 7.016003 amu and m_{Li_8} = 8.022486 amu.

[**Ans.** $^7_3$Li is more stable as binding energy per nucleon for $^7_3$Li is 5.60 MeV/n and that of $^8_3$Li is 5.16 MeV/n]

4. The binding energy of $^{20}_{10}$Ne is 160.64 MeV. Find its atomic mass. Given m_p = 1.007825 amu and m_n = 1.008665 amu.

[**Ans.** 19.99 amu]

5. Find the average binding energy per nucleon in $^{40}_{20}$Ca. Given m_p = 1.007825 amu, m_n = 1.008665 amu and m_{Ca} = 39.962591 amu.

[**Ans.** 8.55 MeV/n]

6. By referring to Figure 1.4, determine whether energy must be supplied or will be released if we assume that the following transmutations can be accomplished: Fe → Ag, H → He, He → O and Pb → Si.

[**Ans.** supply, released, released, released]

7. Calculate the total binding energy of d and $^{20}_{10}$Ne and also calculate the binding energy per nucleon for these two nuclei. Given m_p = 1.007825 amu, m_n = 1.008665 amu m_d = 2.014102 amu and m_{Ne}= 19.992439 amu.

[**Ans.** 2.2 MeV, 160.6 MeV, 1.1 MeV, 8.0 MeV]

8. What are the atomic and mass numbers of the oxygen isotope with 17 nucleons? Calculate the mass defect, binding energy and binding energy per nucleon for this nuclide, with the assumption that the mass defect is associated with the nucleus. Given m_{17_O} = 16.999133 amu, m_p = 1.007825 amu and m_n= 1.008665 amu.

[**Ans.** 8, 17, 0.141452 amu, 131.76 MeV, 7.75 MeV/nucleons]

9. Natural oxygen contains three isotopes with atomic masses in amu of 15.9949, 16.9991 and 17.9992, and relative abundances of 2500:1:5. Determine to three decimal places the average atomic mass of oxygen.

[**Ans.** 15.999 amu]

Section 1.6

Solved Problems

1. One nucleon is present in *p* state ($\ell_1 = 1$) and another is in *d* state ($\ell_2 = 2$). What will be the resulting orbital angular momenta, if both the nucleons coupled together?

Solution:

$$L = \ell_1 + \ell_2$$

or L values will be between

$$|\ell_1 - \ell_2| \text{ to } |\ell_1 + \ell_2|$$

i.e. L will be 1, 2 or 3.

2. A proton is present in *f* state ($\ell = 3$). What will be the total angular momentum of this proton?

Solution:

$$\left|\ell - \frac{1}{2}\right| \text{ to } \left|\ell + \frac{1}{2}\right|$$

or

$$\frac{5}{2} \text{ and } \frac{7}{2}$$

Unsolved Problems

1. Assume neutron has a radius of 1.5 fm and has uniform density. Calculate its frequency, if its angular momentum is $\hbar\sqrt{s(s+1)}$, where $s = 1/2$, $\hbar = 1.05 \times 10^{-34}$ J s and $m_n = 1.67 \times 10^{-27}$ kg.

[**Ans.** 38.5×10^{21} Hz]

2. Principle quantum number of a nucleus is 3. What are the possible values of the orbital quantum number ℓ?

[**Ans.** 2, 1, 0]

Section 1.11

Solved Problems

1. A singly charged positive ion is accelerated through a potential difference of 1000 V in a mass spectrograph. It then passes through a uniform magnetic field $B = 1500$ gauss, and then deflected into a circular path of radius 0.122 m.

(i) What is the speed of the ion?
(ii) What is the mass of the ion?
(iii) What is the mass number of the ion?

Solution:

(i) Using the relations $\frac{mv^2}{R} = eVB$ and $\frac{1}{2} mv^2 = eV$, where e is ionic charge and V is the potential applied, we get

$$v = \frac{2V}{RB}$$

Now,

$$V = 1000 \text{ V}$$
$$R = 0.122 \text{ m}$$
$$B = 1500 \times 10^{-4} \text{ tesla}$$

Therefore,

$$v = \frac{2 \times 1000}{0.122 \times 1500 \times 10^{-4}} \text{ m}$$

or

$$= 1.093 \times 10^5 \text{ m/s}$$

(ii) Mass of the ion (in amu) $M = \frac{2eV}{v^2}$

$$= \frac{2 \times 1.602 \times 10^{-19} \times 1000}{(1.093 \times 10^5)^2}$$

$$= 2.682 \times 10^{-26} \text{ kg}$$

(iii) Mass number $= \frac{2.682 \times 10^{-26}}{1.673 \times 10^{-27}} \approx 16.03 \approx 16$

$= 16$

2. In a Bainbridge mass spectrograph, singly charged argon ions enter a velocity filter, where the electric field is 5×10^4 V/m and magnetic field is 0.4 tesla. What is the velocity of the ions emerging the velocity filter? If the ions enter the magnetic field of 0.8 tesla, what will be the distances between the ions focused on the photographic plate for the three isotopes: ^{36}Ar, ^{38}Ar and ^{40}Ar?

Solution: Velocity of the ions emerging the velocity filter from Eq. (1.34) $v = \frac{E}{B_1}$

Here,

$$E = 5 \times 10^4 \text{ V/m}$$
$$B_1 = 0.4 \text{ tesla}$$

Therefore,

$$v = \frac{5 \times 10^4}{0.4} = 1.25 \times 10^5 \text{ m/s}$$

When the ions enter the magnetic field, the radius of the path is

$$R = \frac{Mv}{Be}$$

Here,

$$M = \text{mass of the ion in kg}$$
$$B = 0.8 \text{ tesla}$$

For ^{36}Ar ions:

$$R = \frac{36 \times 1.673 \times 10^{-27} \times 1.25 \times 10^5}{0.8 \times 1.602 \times 10^{-19}}$$

$$= 0.0587 \text{ m} = 58.7 \text{ mm}$$

For ^{38}Ar ions:

$$R = \frac{38 \times 1.673 \times 10^{-27} \times 1.25 \times 10^5}{0.8 \times 1.602 \times 10^{-19}}$$
$$= 0.0620 \text{ m} = 62.0 \text{ mm}$$

For ^{40}Ar ions:

$$R = \frac{40 \times 1.673 \times 10^{-27} \times 1.25 \times 10^5}{0.8 \times 1.602 \times 10^{-19}}$$
$$= 0.0653 \text{ m} = 65.3 \text{ mm}$$

Therefore, distance between successive lines due to three different isotopes of Ar

$$= 2 \times (62.0 - 58.7) = 6.6 \text{ mm}$$

or

$$= 2 \times (65.3 - 62.0) = 6.6 \text{ mm}$$

Unsolved Problems

1. In a Bainbridge mass spectrograph, singly charged natural silver ions (^{107}Ag and ^{109}Ag) enter a velocity filter. In the velocity filter electric and magnetic fields are 5×10^4 V/m and 0.4 tesla respectively. What is the speed of the ions emerging the velocity filter? If the ions after leaving velocity filter enter the magnetic field of 0.7 tesla, what is the separation between the two lines due to two isotopes of silver on a photographic plate?

[**Ans.** 1.25×10^5 m/s 7.6 mm]

2. Suppose that singly charged ions with masses close to 12 and 14 amu are accelerated through a potential difference of 1000 V and then travel through a magnetic field of 900 gauss. Where should collector plates for the two different ions be located?

[**Ans.** 0.176 m, 0.190 m]

3. In a Bainbridge mass spectrometer, singly ionized ^{20}Ne passes into the deflection chamber with an initial velocity of 1.2×10^5 m/s. Here it is deflected by a magnetic field of flux density 0.06 tesla. Calculate the radius of ionic path. Also calculate the radius of the path of singly ionized ^{22}Ne having the same initial velocity.

[**Ans.** 0.418 m, 0.460 m]

REVIEW QUESTIONS

Short Answer Type

1. What is electric quadrupole moment?
2. What are thermal neutrons?
3. Explain the terms:
   (i) Nuclear spin.
   (ii) Nuclear magnetic dipole moment.
   (iii) Nuclear electric quadrupole moment.
4. What are the main features of nuclear forces?
5. Discuss:
   (i) Nuclear electric quadrupole moment.
   (ii) Nuclear density.
   (iii) Uncertainty principle forbids electrons in the nucleus.
6. Give any two evidences of neutron–proton model of the nucleus.
7. Does the nucleus of the atom has sharp boundaries? How is the mass distributed in the nucleus?
8. Show that nuclear density is same for all the nuclei.
9. Why the number of neutrons tends to exceed the number of protons in stable nuclei?
10. What are the reasons for assuming spin of electrons?
11. What is nuclear magneton μ_n?
12. What is parity?
13. Why are even–even nuclei most stable?
14. Why does nuclear radius depend on mass number? Calculate its value for ^{27}Al.
15. List different types of magnetic moments that can be associated with nucleons.
16. Predict quadrupole moment for a nucleon whose ground state spin is $\hbar/2$. What is the significance of quadrupole moment?
17. Calculate nuclear density.
18. Explain the non-existence of electrons in the nucleus on the basis of magnetic moments.
19. Explain the concept of nuclear magnetic dipole moment.
20. Explain the concept of nuclear quadrupole moment.
21. What is average binding energy and discuss its variation with A?
22. What do you mean by even–even nuclei?
23. Obtain the ratio of densities of $^{197}_{79}$Au and $^{107}_{47}$Ag.
24. What do you mean by charge independence of nuclear force?
25. Nuclear forces are of short-range. Explain.
26. What is an amu?

27. Explain the terms:
    (i) Binding energy.
    (ii) Mass defect.
28. Find a relation between electron volt and amu.
29. Why do we say that nuclear forces are of short-range?
30. Give one reason as to why electrons cannot exist in nucleus.
31. Why the density of nucleus is large?
32. Which is more, atomic binding energy or nuclear binding energy? Why?
33. What is the unit to measure the size of the nucleus?
34. What is packing fraction?
35. What type of statistics is followed by nucleons?
36. How does the binding energy curve explain fusion?
37. What are amu and electron volt?
38. What is nuclear magnetic moment?
39. What do you mean by odd–odd, odd–even and even–even nuclei? Draw the stability curve.

Long Answer Type

1. Name the forces existing between nucleons. Explain their characteristics.
2. Prove from statistics of angular momentum and wave mechanical considerations that electrons cannot exist in the nucleus.
3. Explain the terms: mass defect, binding energy and amu. Discuss how binding energy varies with A.
4. What is proton–neutron hypothesis? Give a reason for the acceptance of this hypothesis. Why do not electrons exist in the nucleus?
5. Explain the terms: mass defect and binding energy of the nucleus. Give the method of calculation of binding energy for a nucleus. Also give main features of binding energy per nucleon versus mass number graph.
6. Discuss magnetic moment of the nucleus. Explain how neutrons have magnetic moment.
7. Explain, giving two reasons, why the electron–proton hypothesis is not acceptable, while neutron–proton is.
8. What are the forces responsible for holding nucleons together in a nucleus? Give their main characteristics, explaining their origin.
9. Explain the non-existence of electrons inside the nucleus on the basis of wave mechanical considerations.
10. Enumerate the salient features of nuclear forces. How are they different from gravitational and electromagnetic forces?
11. Explain the relationship of nuclear mass with nuclear size. Show how binding energy changes with mass number.
12. Give the reasons for non-existence of electrons in the nucleus.

13. What is binding energy? Explain the stability of the nucleus on the basis of this concept.
14. Define binding energy and explain what one can learn from the variation of binding energy with *A*.
15. Write short notes on:
    (i) Nuclear magnetic moment.
    (ii) Nuclear quadrupole moment.
16. Give one method to determine the size of the nucleus.
17. Calculate nuclear mass density and discuss its variation with distance from the centre of the nucleus.
18. Explain meson or Yukawa theory of nuclear forces.
19. Write notes on:
    (i) Proton–neutron theory.
    (ii) Fermi–Dirac statistics.
    (iii) Nuclear magnetic dipole moment.
20. Discuss four reasons for the failure of proton–electron hypothesis of a nucleus.
21. What are the reasons which lead to the belief that spin is associated with electrons?
22. State the main properties of nuclear forces. Explain the hypothesis of charge and spin independence of nuclear forces.
23. Explain nuclear binding energy and packing fraction. Discuss graphically the variation of average binding energy per nucleon with *A*.
24. Assuming nucleus to be uniformly charged, calculate the work done against electrostatic forces in giving it a charge Ze.
25. Discuss the meson theory of nuclear forces. Give the main properties of these forces.
26. What is electric quadrupole moment of a nucleus? Discuss the shapes of the nucleus on its basis.

Chapter 2

Nuclear Models

2.1 INTRODUCTION

As discussed in Chapter 1, the size of the nucleus is very small and nuclear forces are far more complicated than other well-known forces. In fact, the picture of nuclear forces is still not clear. This picture is different from the case of atom, where the forces are known and atomic model is well established for deducing various properties in atomic domain. Due to the lack of detailed knowledge of nuclear forces, nuclear models, namely liquid drop model, shell model, Nilson model, Fermi gas model, collective model, Bohr Motelson model, interacting boson model, etc. have been developed, each of which is useful in a more or less limited fashion.

In order to understand and predict the properties of the nucleus, we have to know the forces completely. For knowing nuclear forces, we adopt a different approach. In nuclei, we choose an oversimplified theory, the treatment of which is mathematically possible, but the theory should be rich in physics. If this theory is fairly successful in accounting for at least a few properties of the nucleus, we can then improve the model by adding additional terms so that it is capable to account more nuclear properties. In this way, we construct a nuclear model, a simplified view of nuclear structure, which still contains the essentials of the nuclear properties. A good nuclear model must satisfy following two criteria:

- It must reasonably well account for previously measured nuclear properties.
- It must predict additional nuclear properties that can be measured in new experiments.

The development of nuclear models has taken place along the following lines.

In the first type of nuclear models, nucleus has been treated like a drop of liquid, in which nucleons present in the nucleus interact very strongly among themselves. This is like molecules present in a drop of liquid, which interact among themselves very strongly. This treatment gave rise to models like *liquid drop model*, *collective model,* etc. The second type of models is constructed in analogy with the shell model of the atom. In these models, the nucleons are weakly interacting among themselves. This treatment gave rise to *Fermi gas model, shell model, Nilson model,* etc.

In this chapter, we discuss only two models, i.e. liquid drop model and shell model. In the end, a brief description of Fermi gas model is given.

2.2 LIQUID DROP MODEL

Weizsacker in 1935 proposed on the basis of experimental facts that a nucleus resembles a drop of liquid. In 1939, Bohr and Wheeler further developed this model to explain the phenomenon of nuclear fission.

Following are some of the similarities between a drop of liquid and nucleus, which prompted Weizsacker to develop the liquid drop model.

Similarities between Liquid Drop and Nucleus

1. Nuclear forces are analogous to the surface tension of a liquid.
2. The nucleons behave in a manner similar to that of molecules in a liquid drop.
3. The density of the nuclear matter is almost independent of A, showing resemblance to liquid drop where the density of a liquid is independent of the size of the drop.
4. The constant binding energy per nucleon is analogous to the latent heat of vaporization.
5. The disintegration of nuclei by the emission of particles is analogous to the evaporation of molecules from the surface of liquid.
6. The absorption of bombarding particles by a nucleus corresponds to the condensation of drops.
7. The energy of nuclei corresponds to internal thermal vibrations of drop molecules.

Based on these similarities, Weizsacker in 1935 and Bohr and Wheeler in 1939 developed liquid drop model. They ignored the finer features of nuclear forces but strong internucleon attraction is stressed.

Assumptions of the Liquid Drop Model

1. The nucleus consists of incompressible matter.
2. The nuclear force is identical for every nucleon.
3. The nuclear force saturates.
4. In an equilibrium state, the nuclei of atom remain spherically symmetric under the action of strong attractive nuclear forces.

2.2.1 Semiempirical Mass Formula

The analogy between nucleus and liquid drop has been used to set up a semiempirical formula for mass (or binding energy) of a nucleus in its ground state. The formula has been obtained by considering different factors of the nucleus binding.

The mass of the nucleus can be expressed in terms of the total binding energy B and the masses of Z protons and N neutrons as

$$\boxed{M = ZM_p + NM_n - B} \tag{2.1}$$

The binding energy B of a nucleus is given by the sum of five terms as

$$B = B_1 + B_2 + B_3 + B_4 + B_5 \tag{2.2}$$

which are explained as under.

Volume Energy Term (B_1)

The volume term arises from the interaction of the nucleons through the strong force. When a liquid drop evaporates, the energy required for this process is the product of mass of the drop M_m and latent heat of vaporization L. This energy is used to break all the molecular bonds. This is same as the binding energy of the drop B. So

$$B = LM_m N \tag{2.3}$$

where N is the number of molecules in the drop. Equation (2.3) can also be written as

$$\frac{B}{N} = LM_m = \text{constant} \tag{2.4}$$

This means that B/N is independent of the number of molecules present in the liquid drop. As we know that in the liquid drop, a molecule interacts only with its nearest neighbours and number of neighbours is independent of the size of the drop. This characteristics of the system shows that range of interaction among the molecules is much smaller than the dimensions of the drop.

In Chapter 1, we have seen that neutrons and protons are held together in nuclei by short-range attractive forces. These forces reduce the mass of the nucleus below that of its constituents by an amount proportional to the number of nucleons A. Since the volume of the nucleus is proportional to A, hence this term is regarded as a volume binding energy and in analogy to Eq. (2.4) is given by

$$\boxed{B_1 = a_v A} \tag{2.5}$$

where a_v is a proportionality constant and subscript v is for volume.

Surface Energy Term (B_2)

The surface term is a correction to the volume term to take into account that the nucleons at the surface of the nucleus do not have the same level of interactions as nucleons in the interior of the nucleus. In the above discussion, we have assumed that all the molecules are surrounded by its neighbours, while in actual practice the molecules at the surface do not have any neighbours on all the sides. So these molecules are not as tightly bound as the molecules in the interior. Extending this argument to the nuclear case, some nucleons are nearer to the surface, and so they interact with fewer nucleons. Thus, the binding energy is reduced by an amount proportional to the surface area of the nucleus of radius r as the nucleons on the surface are less tightly bound than those in the interior. This term is proportional to the surface area of the nucleus of radius $r(= r_0 A^{1/3})$. Therefore,

$$B_2 \propto -4\pi r^2$$

or

$$B_2 \propto -4\pi r_0^2 A^{2/3}$$

which is usually expressed as

$$\boxed{B_2 = -a_s A^{2/3}} \tag{2.6}$$

where negative sign is for decrease in energy and a_s is constant.

Coulomb Energy Term (B_3)

The Coulomb term represents the energy incorporated in the nucleus as a result of the positive charge present in the nucleus. The only long-range force in the nucleus is the Coulomb repulsion between protons. The total work done in assembling a nucleus consisting of Z protons is given by

$$W = \frac{\frac{3}{5} Z^2 e^2}{4\pi \varepsilon_0 r}$$

where r is the radius of the nucleus.

For a single-proton nucleus

$$w = \frac{\frac{3}{5} e^2}{4\pi \varepsilon_0 r}$$

For a nucleus having Z protons

$$w' = \frac{\frac{3}{5} Ze^2}{4\pi \varepsilon_0 r}$$

For a single-proton nucleus no work is done against Coulomb repulsion in assembling the nucleus. Thus, the true Coulomb energy term for a nucleus containing Z protons is $W - w'$.

$$B_3 = -\left[\frac{\frac{3}{5} Z^2 e^2}{4\pi \varepsilon_0 r} - \frac{Z \frac{3}{5} e^2}{4\pi \varepsilon_0 r}\right]$$

i.e.

$$B_3 = -\frac{3}{5} \frac{Z(Z-1)e^2}{4\pi \varepsilon_0 r} \tag{2.7}$$

The negative sign indicates the repulsive term. As $r = r_0 A^{1/3}$, Eq. (2.7) can be written as

$$\boxed{B_3 = -a_c \frac{Z(Z-1)}{A^{1/3}}} \tag{2.8}$$

where a_c is constant.

Asymmetry Energy Term (B_4)

The asymmetry term reflects the stability of nuclei with the proton and neutron numbers being approximately equal. This is a term, which depends on the neutron excess ($N - Z$) in the nucleus and it decreases with the increasing nuclear binding energy. For very few nuclei of low Z, $N - Z = 0$ and are more stable compared to their neighbours, i.e. their binding energies are maximum. The reduction in binding energy for higher A nuclei is directly proportional to $(N - Z)^2$ or square of excess of neutrons and is inversely proportional to mass number. So, we can write

$$B_4 \propto \frac{(N-Z)^2}{A}$$

$$B_4 = -a_a \frac{(A-2Z)^2}{A} \tag{2.9}$$

As $A = N + Z$ and a_a is constant.

Pairing Energy Term (B_5)

So far we have all the terms in the binding energy have smooth variation with respect to N or Z or A. However, in the actual binding energy versus A curve, there are several discontinuities, particularly when N or Z becomes equal to 2, 4, 8, 20, 28, 50, 82 or 126. These values correspond to shell closure for N or Z. The nuclei having N or Z equal to one of these numbers have large binding energy. This fact did not appear in the liquid drop model, which does not consider intrinsic spin of the nucleons and the shell effects.

It is interesting to classify all the stable nuclei into four groups, first having even Z–even N, second even Z–odd N, third odd Z–even N and last having odd Z–odd N. This classification is shown in Table 2.1.

TABLE 2.1 Number of stable isotopes

Z	N	*Number of stable nuclei*
Even	Even	165
Even	Odd	55
Odd	Even	50
Odd	Odd	5

(the five stable odd Z–odd N nuclei are: ${}^2_1H$, ${}^6_3Li$, ${}^{10}_5B$, ${}^{14}_7N$, ${}^{180}_{73}Ta$)

From Table 2.1, it is clear that even Z–even N nuclei, being most stable, are most abundant. Accordingly, odd Z–odd N nuclei are least abundant and hence least stable. The remaining nuclei have intermediate stability. Therefore, the binding energy also depends upon whether the number of protons and neutrons are odd or even. This pairing effect was incorporated by putting

$$B_5 = a_p A^{-3/4} \tag{2.10}$$

where

a_p = 33.5 MeV for even–even nuclei
= 0 for odd–even (or odd A) nuclei
= –33.5 MeV for odd–odd nuclei

Substituting the values of B_1, B_2, B_3, B_4 and B_5 from Eqs. (2.5), (2.6), (2.8), (2.9) and (2.10) in Eq. (2.2), we get

$$B = a_v A - a_s A^{2/3} - a_c \frac{Z(Z-1)}{A^{1/3}} - a_a \frac{(A-2Z)^2}{A} + a_p A^{-3/4}$$

Substituting the value of B from the above equation in Eq. (2.1), we get the semiempirical mass formula as

$$M = ZM_p + NM_n - a_v A + a_s A^{2/3} + a_c \frac{Z(Z-1)}{A^{1/3}} + a_a \frac{(A-2Z)^2}{A} - a_p A^{-3/4} \tag{2.11}$$

The various constants found are

$$a_v = 15.5 \text{ MeV}$$
$$a_s = 16.8 \text{ MeV}$$
$$a_c = 0.7 \text{ MeV}$$
$$a_a = 23.0 \text{ MeV}$$
$$a_p \begin{cases} = 34 \text{ MeV} & \text{for even–even nuclei} \\ = 0 \text{ MeV} & \text{for odd } A \text{ nuclei} \\ = -34 \text{ MeV} & \text{for odd–odd nuclei} \end{cases}$$

The contribution of various terms for few cases are given (in MeV) in Table 2.2.

TABLE 2.2 Contribution of various terms of semiempirical mass formula for some isotopes

Term	${}^{40}_{20}\text{Ca}$	${}^{120}_{50}\text{Sn}$	${}^{238}_{92}\text{U}$
Volume	+620	+1860	+3689
Surface	−196	−409	−645
Coulomb	−80	−358	−973
Asymmetry	0	−77	−282
Pairing	+2	+1	+0.6
Resultant BE	346	1017	1789.6
BE/n	8.65	8.48	7.52

The semiempirical mass formula reproduces masses of various nuclei quite accurately, but does not account for all the features of the nuclear binding energy.

2.2.2 Mass of Most Stable Isobar

Isobars are nuclides that have same mass number A. The semiempirical mass formula can predict the atomic number Z_0 of most stable isobar for given mass number A.

Neglecting 1 in comparison to Z in the Coulomb term and rewriting Eq. (2.11) as

$$M(Z, A) = aA + bZ + cZ^2 + \delta \tag{2.12}$$

where

$$a = M_n - \left[a_v - a_a - \frac{a_s}{A^{1/3}} \right]$$

$$b = -4a_a - (M_n - M_p)$$

$$c = \left(\frac{4a_a}{A} + \frac{a_c}{A^{1/3}} \right)$$

and

$$\delta = \mp a_p A^{-3/4}$$

Let us find the atomic number of most stable isotope for a given A. This can be calculated by taking the partial derivative of Eq. (2.12) with respect to Z keeping A as constant and equating the resultant equation to zero, i.e.

$$\left(\frac{\partial M}{\partial Z}\right)_A = b + 2cZ = 0$$

which gives

$$Z_0 = Z = -\frac{b}{2c}$$

where Z_0 is the atomic number of most stable isotope for given A. Substituting the values of b and c, we obtain

$$Z_0 = -\frac{-4a_a - (M_n - M_p)}{2\left[\dfrac{4a_a}{A} + \dfrac{a_c}{A^{1/3}}\right]}$$

Since all the quantities in this expression are known, atomic number for most stable isobar can be calculated.

2.2.3 Achievements of Liquid Drop Model

1. It predicts the atomic masses and binding energies of various nuclei accurately.
2. It predicts emission of α- and β-particles in radioactivity.
3. The theory of compound nucleus, which is based on this model, explains the basic features of the fission process.

2.2.4 Failures of Liquid Drop Model

1. It fails to explain the extra stability of certain nuclei, where the numbers of protons or neutrons in the nucleus are 2, 8, 20, 28, 50, 82 or 126 (these numbers are called magic numbers).
2. It fails to explain the measured magnetic moments of many nuclei.
3. It also fails to explain the spin of nuclei.
4. It is also not successful in explaining the excited states in most of the nuclei.
5. The agreement of semiempirical mass formula with experimentally observed masses and binding energies is poor for lighter nuclei compared to the heavy ones.

2.3 SHELL MODEL

Atomic theory based on the shell model has provided remarkable clarification of the complicated details of atomic structure. Nuclear physicist, therefore, attempted to use a similar theory to study nuclear structure. In the atomic shell model, we fill the shells with electrons in order of increasing energy consistent with the requirement of the Pauli principle. When we do so, one obtains an inert core of filled shells, containing 2, 10, 18, 36, 54 and 86 electrons (atomic numbers of inert gases) and some valence electrons; the atomic properties are determined primarily by the valence electrons. When we compare some measured properties of atomic system with the predictions of the model, one finds remarkable agreement. The same kind of effect has been observed in nuclei. Experimentally it was found that nuclei that have 2, 8, 20, 28, 50, 82 and 126 nucleons (protons or neutrons), called magic numbers, are more abundant than other nuclei.

However, there exist several significant differences between atomic and nuclear cases. In the atomic case, the potential is provided by the Coulomb field of the nucleus; the orbits are generated by the external agent i.e. interaction between electrons and nucleus. We can solve the Schrödinger equation for this potential and calculate the energies of the sub-shells into which electrons can then be placed. In case of nucleus, there is no such external agent, the nucleons move in a potential which is not well defined that they themselves create.

Another appealing aspect of atomic shell theory is the existence of spatial orbits. It is often very useful to describe atomic properties in terms of spatial orbits of the electrons. The electrons can move in those orbits relatively free of collisions with other electrons. Nucleons which have a mass about 2000 times larger than that of electrons have a diameter comparable to the size of the nucleus, which is about 10^5 times smaller than that of an atom. How can we regard the nucleons as moving in well-defined orbits when a single nucleon can make many collisions during each orbit?

All these observations tempted nuclear physicists (Barlet, Guggenheimer et al.) to devise an independent particle model formally called the shell model. A shell structure means that nucleons move freely inside the nucleus similar to the electron motion in atom. This approach could explain the existence of first few magic numbers. However, physicist lost interest in this model till 1948 due to its failure to explain higher magic numbers.

In 1948, M.G. Mayer in USA brought together a considerable amount of convincing information showing the evidence for the closed shells, which led to the development of nuclear shell model which could explain all the magic numbers, namely 2, 8, 20, 28, 50, 82 and 126, which apparently represent closed shells in the nucleus. Some of the main aspects of this evidence based on the study of stable nuclei are as under:

1. Binding energy per nucleon vs. A curve. If we plot binding energy per nucleon versus A curve, it shows that binding energy suddenly increases when the number of nucleons is either 2, 8, 20, 28, 50, 82 or 126 indicating that these nuclei are exceptionally stable.
2. Number of stable isotopes. Relative stabilities of different elements are also indicated by the number of stable isotopes per element.

${}_{19}K = 3$	${}_{20}Ca = 6$	${}_{21}Sc = 1$
${}_{49}In = 2$	${}_{50}Sn = 10$	${}_{51}Sb = 2$
${}_{81}Tl = 2$	${}_{82}Pb = 4$	${}_{83}Bi = 1$

   It is clear that number of stable isotopes for $z = 20$, 50 and 82 are much larger compared to neighbouring isotopes.
3. Number of stable isotones. The numbers of stable isotones for $N = 19, 20, 21$; 49, 50, 51 and 81, 82, 83 are shown in Table 2.3.

TABLE 2.3 Number of stable isotones around different magic numbers

N	*Stable isotones*	*N*	*Stable isotones*	*N*	*Stable isotones*
19	0	49	1	81	1
20	5	50	6	82	7
21	1	51	1	83	1

It is clear from the above table that the numbers of stable isotones for $N = 20$, 50 and 82 are much larger as compared to neighbouring stable isotones.

4. A table of relative abundances of nuclei compiled from data on the composition of earth, sun, stars and meteorites shows pronounced peaks at

^{16}O $(N = Z = 8)$
^{40}Ca $(N = Z = 20)$
^{118}Sn $(Z = 50)$
^{88}Sr, ^{89}Y, ^{90}Zr $(N = 50)$
^{138}Ba, ^{139}La, ^{140}Ce $(N = 82)$
^{208}Pb $(Z = 82, N = 126)$

5. Binding energy of next neutron after a magic number is small. The separation energy of the last neutron for $N = 7, 8, 9; 19, 20, 21$ and 27, 28 29 is shown in Table 2.4.

TABLE 2.4 Binding energy of the last neutron around magic numbers

Nucleus	*N*	S_n (MeV)	*Nucleus*	*N*	S_n (MeV)	*Nucleus*	*N*	S_n (MeV)
^{15}O	7	13.2	^{39}Ca	19	13.3	^{47}Ca	27	7.3
^{16}O	8	15.7	^{40}Ca	20	15.7	^{48}Ca	28	9.9
^{17}O	9	4.14	^{41}Ca	21	8.4	^{49}Ca	29	5.1

From the above table it is clear that for neutron numbers = 9, 21 and 29, the separation energy of the last neutron suddenly decreases as compared to the case, when neutron numbers are 8, 20 and 28.

6. It is found that some isotopes are spontaneous neutron emitters. They are:

$${}^{17}_{8}O_9,\ {}^{87}_{36}Kr_{53},\ {}^{137}_{54}Xe_{83},\ {}^{89}_{36}Kr_{53}$$

7. Decay product of radioactive series:

Thorium ${}^{232}_{90}Th_{142}$ 1.4×10^{10} $y(4n$ series) ${}^{208}_{82}Pb_{126}$
Uranium ${}^{238}_{92}U_{146}$ 4.51×10^{9} $y(4n + 2$ series) ${}^{206}_{82}Pb_{124}$
Actinium ${}^{235}_{92}U_{143}$ 7.07×10^{8} $y(4n + 3$ series) ${}^{207}_{82}Pb_{125}$
Neptunium ${}^{237}_{93}Np_{144}$ 2.25×10^{6} $y(4n + 1$ series) ${}^{209}_{83}Bi_{126}$

The end product of each series ends in N or Z equal to either 82 or 126.

8. Neutron absorption cross-section σ, the probability of absorption of neutron by the nucleus is small for nuclides containing magic number of neutrons

N	19	20	21	49	50	51
σ	2.0	0.41	12.0	19.0	0.65	6.4

9. α-decay energies are rather smooth functions of A for a given Z, but it shows striking discontinuities at $N = 126$ or $Z = 82$, the energy of α-particles increases. For example,

$$^{212}_{84}\text{Po}_{128} \rightarrow {}^{208}_{82}\text{Pb}_{126} + \alpha(9\text{ MeV}) \text{ Half-life } t_{1/2} = 3 \times 10^{-7}\text{s}$$

where $N = 126$, $Z = 82$

$$^{210}_{84}\text{Po}_{126} \rightarrow {}^{206}_{82}\text{Pb}_{124} + \alpha(6\text{ MeV}) \qquad t_{1/2} = 3 \times 10^{-7}\text{s}$$

where $N = 126$, $Z = 82$

10. Similar behaviour is exhibited by β-emitters.
11. The electric quadrupole moment measures the departure of nuclear charge distribution from sphericity. This departure is a measure of nuclear deformation. A spherical nucleus has no or nearly zero quadrupole moments. It has been found that nuclei with proton or neutron number equal to one of the magic numbers are spherical in nature, i.e. all the three axes x, y and z are equal like that of a tennis ball. For such nuclei the quadrupole moment is either zero or nearly zero which is also observed experimentally.

However, in some nuclei, out of three, two axes are equal. In the case, where the unequal axis is shorter than the others, the nucleus has somewhat of a pumpkin shape, it is called *oblate*. Extreme case is that of a Hamburger as shown in Figure 1.7. In the other case where the third unequal axis is longer than the other two, the nucleus has somewhat of a football shape, it is called *prolate*. Extreme case is that of a cigar or Hot Dog as shown in Figure 1.6. In general, unequal axis differs in length by about 20%. However, in lighter nuclei, deformations are more. For example, in ^{24}Mg, all the three axes are unequal.

Thus all the facts given above show that magic numbers 2, 8, 20, 28, 50, 82 and 126 correspond to closed shells. The nuclei having any one of these magic number of protons or neutrons or both show more stability than the other nuclei.

Basic Assumptions of the Shell Model

Calculations similar to atoms were also performed for the nucleus also. Following assumptions were made for these calculations:

- Nucleons in a nucleus move independently in a common (mean) potential determined by the average motion of all the other nucleons.
- Protons and neutrons separately fill levels in the nucleus.
- Most of the nucleons are paired and a pair of nucleons contributes zero spin and zero magnetic moment. The paired nucleons thus form an inert core.
- The properties of odd A nuclei are characterized by the unpaired nucleon and odd–odd nuclei by the unpaired proton and neutron.

These assumptions indicate that the nucleus might have a shell structure. It means that nucleons moving in different shells inside the nucleus do not suffer any collisions similar to electrons in different shells in the atom. This assumption is apparently not acceptable as the nucleons have almost the same size as that of the nucleus. So obviously a question arises, why do not so many nucleons moving inside the nucleus suffer any collisions? How can we regard the nucleons as moving in well-defined orbits when a single nucleon can make many collisions during each orbit?

The answer to this question comes from Pauli's exclusion principle. Consider in a heavy nucleus, a collision between two nucleons in a state near the very bottom of the potential well. When the nucleons collide, they transfer energy to one another, but if all of the energy levels are filled up to the level of valence nucleon, there is no way for one of the nucleons to gain energy except to move up to the valence level. The other levels near the original level are filled and cannot accept an additional nucleon. Such a transfer from a low-lying level to the higher-lying level requires more energy than the nucleons are likely to transfer in a collision. Thus, the collisions cannot occur and the nucleons can indeed orbit as if they were transparent to one another!

The first step in developing the shell model is the choice of the potential. Different forms of potential $V(r)$ have been employed in order to obtain the required magic numbers. In the following, we consider two potentials to solve the Schrödinger equation.

2.3.1 The Square Well Potential

The problem can be mathematically simplified, if we assume a potential well with infinite walls as

$$V(r) = -V_0 \quad r < r_0$$
$$= 0 \quad r > r_0$$

The shape of the finite square well potential is shown in Figure 2.1.

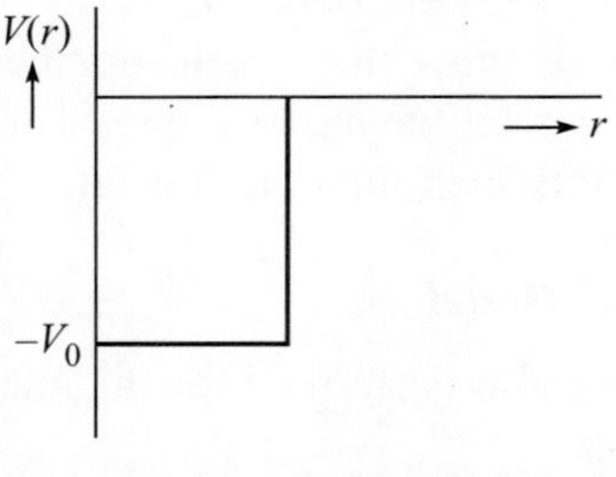

Figure 2.1 Square well potential. $-V_0$ is the depth of the well.

If we solve Schrödinger equation for square well potential, we get the following sequence of levels:

- 1*s*
- 1*p*
- 1*d*

- $2s$
- $1f$
- $2p$
- $1g$
- $2d$
- $1h$
- $3s$
- $2f$
- $1i$
- $3p$
- $2g$

and so on, where $s, p, d, f, g, h, i, \ldots$, etc. stand for usual spectroscopic notation $\ell = 0, 1, 2, 3, 4, 5, 6, \ldots$, respectively.

Because of the two different spin orientation of the nucleon, a level can contain $(2\ell + 1)$ protons or neutrons. For example, number of nucleons in $1s$ ($\ell = 0$) shell will be $2(2 \times 0 + 1) = 2$ and number of nucleons in $1f$ ($\ell = 3$) shell will be $2(2 \times 3 + 1) = 14$. This model predicts the shell closures at nucleon number 2, 8, 18, 20, 34, 40, 58, etc. as shown in Table 2.5.

TABLE 2.5 Nuclear levels and magic numbers predicted by square well potential

Level	*Number of nucleons*	*Magic numbers*
$1s$	2	2
$1p$	6	8
$1d$	10	18
$2s$	2	20
$1f$	14	34
$2p$	6	40
$1g$	18	58
$2d$	10	68
$1h$	22	90
$3s$	2	92
$2f$	14	106
$1i$	26	132
$3p$	6	138

The numbers shown in column 3 of the above table are not the observed magic numbers. The sequence of levels for the square well potential is shown in Figure 2.2.

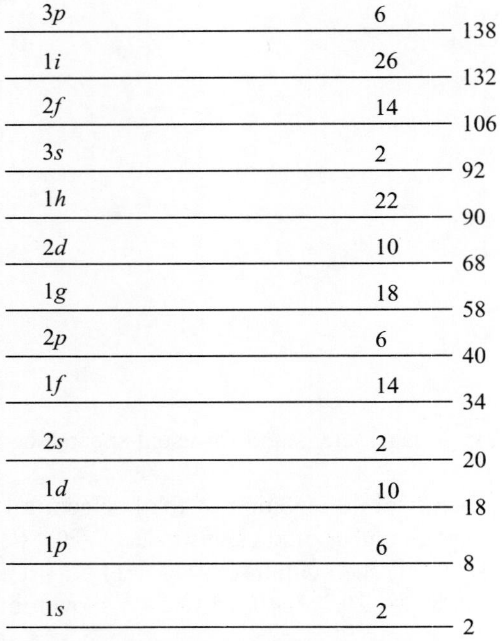

Figure 2.2 Sequence of levels of the square well potential.

The level sequence for square well potential can be remembered in the following way. First in a vertical column write the level sequence 1*s*, 1*p*, 1*d*, 1*f*, 1*g*, 1*h*, 1*i*, etc. as shown in Figure 2.3. Then leave two vertical spaces as blank and again write the level sequence 2*s*, 2*p*, 2*d*, 2*f*, 2*g*, etc. Again leave two vertical blank spaces and write the level sequence 3*s*, 3*p*, 3*d*, 3*f*, etc. as shown in Figure 2.3. In this sequence, 3*s* level will shift between 1*h* and 2*f*, similarly, 1*i* level will shift between 2*f* and 3*p* as indicated by arrow in the Figure 2.3. The resulting sequence will look as shown in Figure 2.4.

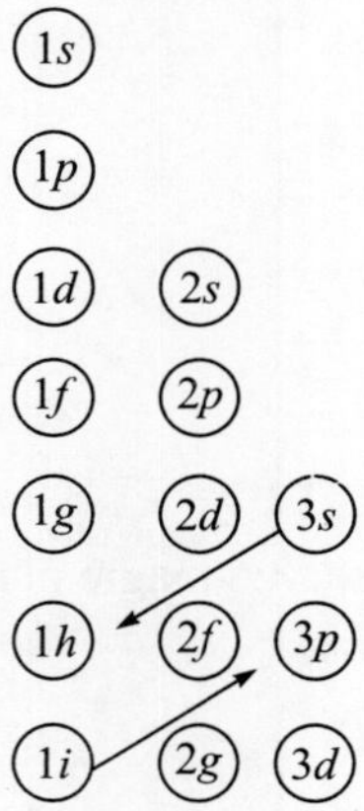

Figure 2.3 Way to remember the square well potential levels.

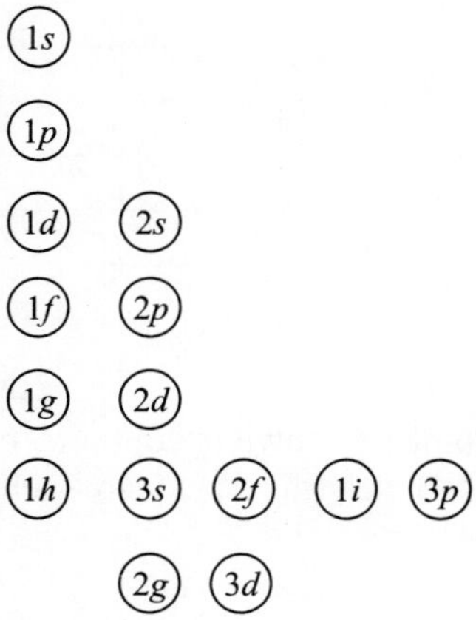

Figure 2.4 The resulting sequence is the sequence of square well potential levels.

Now, starting from the top and move horizontally from left to right, the first level is 1*s*, second 1*p*, third 1*d*, fourth 2*s*, fifth 1*f*, and so on.

2.3.2 The Harmonic Oscillator Potential

The shape of this potential is shown in Figure 2.5.

$$V(r) = -V_0 + \frac{1}{2}Kr^2 \tag{2.13}$$

This potential and the square well potential provide two contrasting view points. The square well has infinite sharp edges. The harmonic oscillator potential diminishes steadily at the edges.

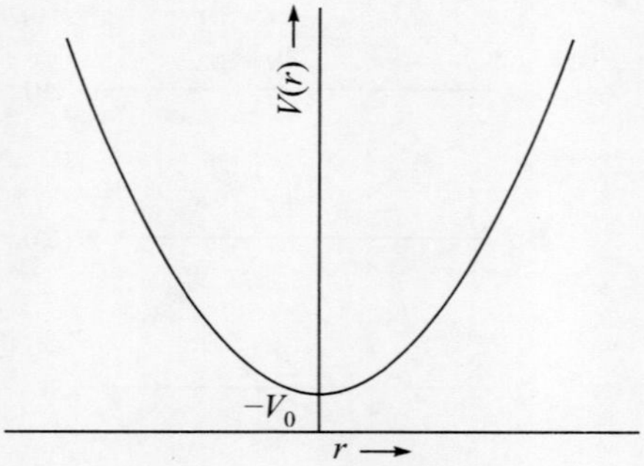

Figure 2.5 Harmonic oscillator potential.

Again solving Schrödinger equation for harmonic oscillator potential, we get the following sequence of levels as shown in first column of Table 2.6. The first level is 1*s*, second is 1*p*, third level contains two sub-shells 2*s*, 1*d*, having same energy, fourth level again contains two sub-shells, 2*p* and 1*f*, and so on.

TABLE 2.6 Nuclear levels and magic numbers predicted by harmonic oscillator potential

Level	*No. of nucleons in various levels*	*Magic number*
1*s*	2	2
1*p*	6	8
2*s*, 1*d*	2 + 10 = 12	20
2*p*, 1*f*	6 + 14 = 20	40
3*s*, 2*d*, 1*g*	2 + 10 + 18 = 30	70
3*p*, 2*f*, 1*h*	6 + 14 + 22 = 42	112
4*s*, 3*d*, 2*g*, 1*i*	2 + 10 + 18 + 26 = 56	168

In this case also each sub-shell contains $2(2\ell + 1)$ protons or neutrons. For example, in fourth shell, we have two sub-shells, 2*p* and 1*f*. For 2*p* ($\ell = 1$), it contains $2(2 \times 1 + 1) = 6$ nucleons and $2f$ ($\ell = 3$), it contains $2(2 \times 3 + 1) = 14$ nucleons. Total number of nucleons in fourth shell = 6 + 14 = 20. These numbers are shown in the third column of the Table 2.6.

This level sequence again does not reproduce experimentally observed magic numbers. The sequence of harmonic oscillator levels is shown in Figure 2.6. The levels are equally spaced.

Harmonic oscillator level sequence can also be remembered in almost similar way as that of square well potential as shown below.

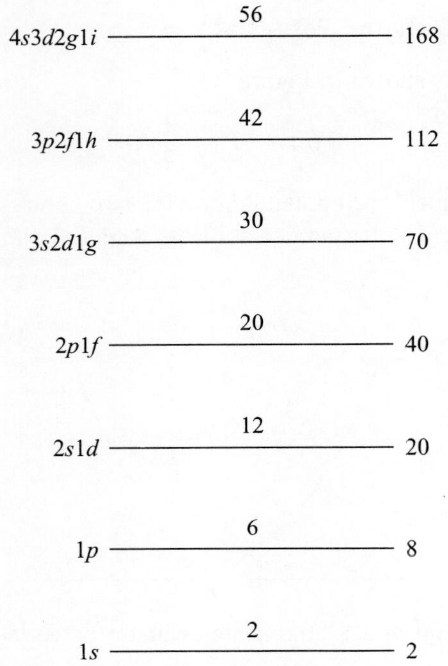

Figure 2.6 Level sequence as obtained for the harmonic oscillator levels.

As shown in Figure 2.7, write in the first vertical column the level sequence 1*s*, 1*p*, 1*d*, 1*f*, 1*g*, 1*h*, 1*i*, etc. Then leave two vertical spaces as blank and again write the level sequence 2*s*, 2*p*, 2*d*, 2*f*, 2*g*, etc. Again leave two vertical blank spaces and write the level sequence 3*s*, 3*p*, 3*d*, 3*f*, etc. Now, these sequences read horizontally are the harmonic oscillator levels. For example, first level is 1*s*, second is 1*p*, third is 1*d*, 2*s* and so on.

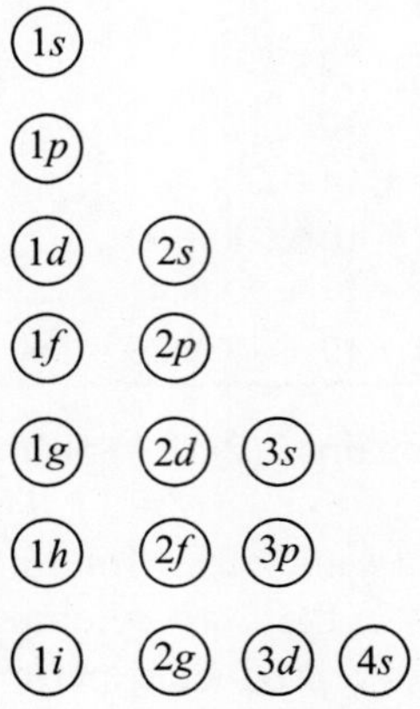

Figure 2.7 Way to remember the harmonic oscillator levels.

The other potential, which is a compromise between square well and harmonic oscillator potential, is

$$V(r) = \frac{-V_0}{1 + e^{\frac{R-r}{d}}} \tag{2.14}$$

This potential is known as Woods–Saxon potential. In this equation $d = 0.524$ fm, R is the mean nuclear radius and $r = r_0 A^{1/3}$. Unlike square well potential, the Woods–Saxon potential does not have any sharp edges at all. The harmonic oscillator potential also does not have any edges. The shape of this potential is shown in Figure 2.8. This potential closely approximates the nuclear charge and matter distribution, falling smoothly to zero beyond the mean radius R. When the Schrödinger equation was solved for this potential, it predicted 2, 8, 20, 40, 58, 92, 112 as magic numbers. We again get the magic numbers 2, 8, and 20, but the higher magic numbers do not emerge from the calculations.

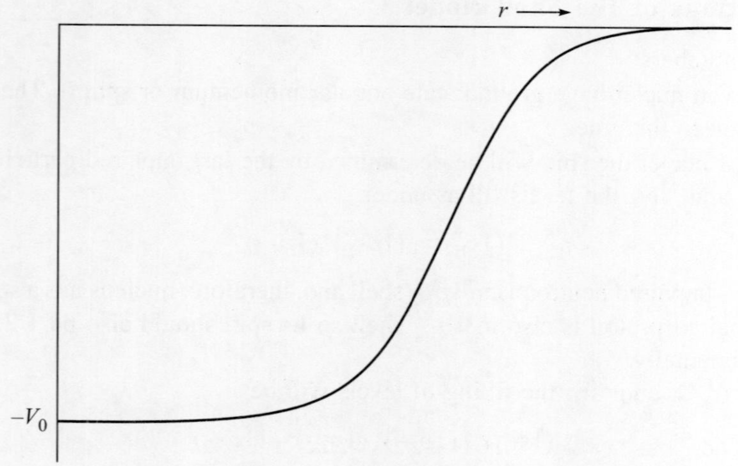

Figure 2.8 Wood–Saxon potential.

2.3.3 Spin-Orbit Coupling

A way out of this difficulty, which proved to be remarkably successful was proposed independently in 1949 by M.G. Mayer in USA and O. Haxel, J.H.D. Jensen and H.E. Suess in Germany. We have seen in Chapter 1 that each nucleon has a spin angular momentum $|\vec{s}| = \sqrt{s(s+1)}\,\hbar$ and orbital angular momentum $|\vec{\ell}| = \sqrt{\ell(\ell+1)}\,\hbar$. It was proposed that there is a strong coupling between the orbital and spin angular momentum of each individual nucleon; referred as spin-orbit coupling. As a result of the spin-orbit coupling, the nucleon energy level for a given value ℓ of the orbital quantum number (except $\ell = 0$) splits into two sub-levels, characterized by total angular momentum quantum number $j = \ell + 1/2$ and $j = \ell - 1/2$ corresponding to spin components of $+1/2$ and $-1/2$ respectively. The sign of this term is chosen in such a way that $\ell + 1/2$ level goes down in energy whereas $\ell - 1/2$ goes up. Further, the total splitting is proportional to ℓ and becomes so large that for a given n, the level with largest ℓ value slides down to energy as low as those of the multiplet with quantum number $n - 1$.

$1s_{1/2}$	2
$1p_{3/2}$, $1p_{1/2}$	8
$1d_{5/2}$, $2s_{1/2}$, $1d_{3/2}$	20
$1f_{7/2}$	28
$2p_{3/2}$, $1f_{5/2}$, $2p_{1/2}$, $1g_{9/2}$	50
$1g_{7/2}$, $2d_{5/2}$, $2d_{3/2}$, $3s_{1/2}$, $1h_{11/2}$	82
$1h_{9/2}$, $2f_{7/2}$, $2f_{5/2}$, $3p_{3/2}$, $3p_{1/2}$, $1i_{13/2}$	126
$2g_{9/2}$, $3d_{5/2}$, $1i_{11/2}$, $2g_{7/2}$, $4s_{1/2}$, $3d_{3/2}$, $1j_{15/2}$	184

The sequence of these levels is shown in Figure 2.9.

2.3.4 Predictions of the Shell Model

1. Magic numbers.
2. Even–even nuclei have ground state angular momentum or spin 0. There is no known exception to this rule.
3. In odd A nuclei the spin will be determined by the last unpaired particle. For example, in ${}^{13}_{6}C_7$ and ${}^{13}_{7}N_6$ the levels fill as under

$$(1s_{1/2})^2 \mid (1p_{3/2})^4 (1p_{1/2})^1$$

Thus, in ${}^{13}_{6}C_7$ unpaired neutron is in $1p_{1/2}$ shell and, therefore, nucleus has a spin 1/2, whereas in ${}^{13}_{7}N_6$ last unpaired proton is also in $1p_{1/2}$ shell so its spin should also be 1/2. This is indeed observed experimentally.

Similarly, in ${}^{17}_{8}O_9$ and ${}^{17}_{9}F_8$ the filling of levels will be

$$(1s_{1/2})^2 \mid (1p_{3/2})^4 (1p_{1/2})^2 \mid (1d_{5/2})^1$$

the predicted spin is 5/2, which is also experimentally observed. Similarly, for ${}^{33}_{16}S_{17}$

$$(1s_{1/2})^2 \mid (1p_{3/2})^4 (1p_{1/2})^2 \mid (1d_{5/2})^6 (2s_{1/2})^2 (1d_{3/2})^1$$

predicted and observed spin is 3/2. However, in ${}^{75}_{33}As_{42}$ and ${}^{61}_{28}Ni_{33}$

$$(1s_{1/2})^2 \mid (1p_{3/2})^4 (1p_{1/2})^2 \mid (1d_{5/2})^6 (2s_{1/2})^2 (1d_{3/2})^4 \mid (1f_{7/2})^8 \mid (2p_{3/2})^4 (1f_{5/2})^1$$

predicted spin is 5/2, whereas observed is 3/2, for both these nuclei.

Similar kind of exceptions has been observed for neutron numbers 57, 59 and 61 also.

One would also expect that for a high atomic mass number A, there will be many stable nuclei with spin 11/2 corresponding to an odd nucleon in the $1h_{11/2}$ state and similarly there should be many stable nuclei with spin 13/2 corresponding to an odd nucleon in the $1i_{13/2}$ state, but not even a single nucleus has ever been observed with ground state spin of 11/2 or 13/2. Many such exceptions have been eliminated by modifying the rules and stating that if the high spin shell (say $1f_{5/2}$) comes after low spin shell ($2p_{3/2}$), the high spin shell fills faster, pairing its particles before the low spin shell can be filled completely. According to this rule, we may write for ${}^{75}As$ and ${}^{61}Ni$.

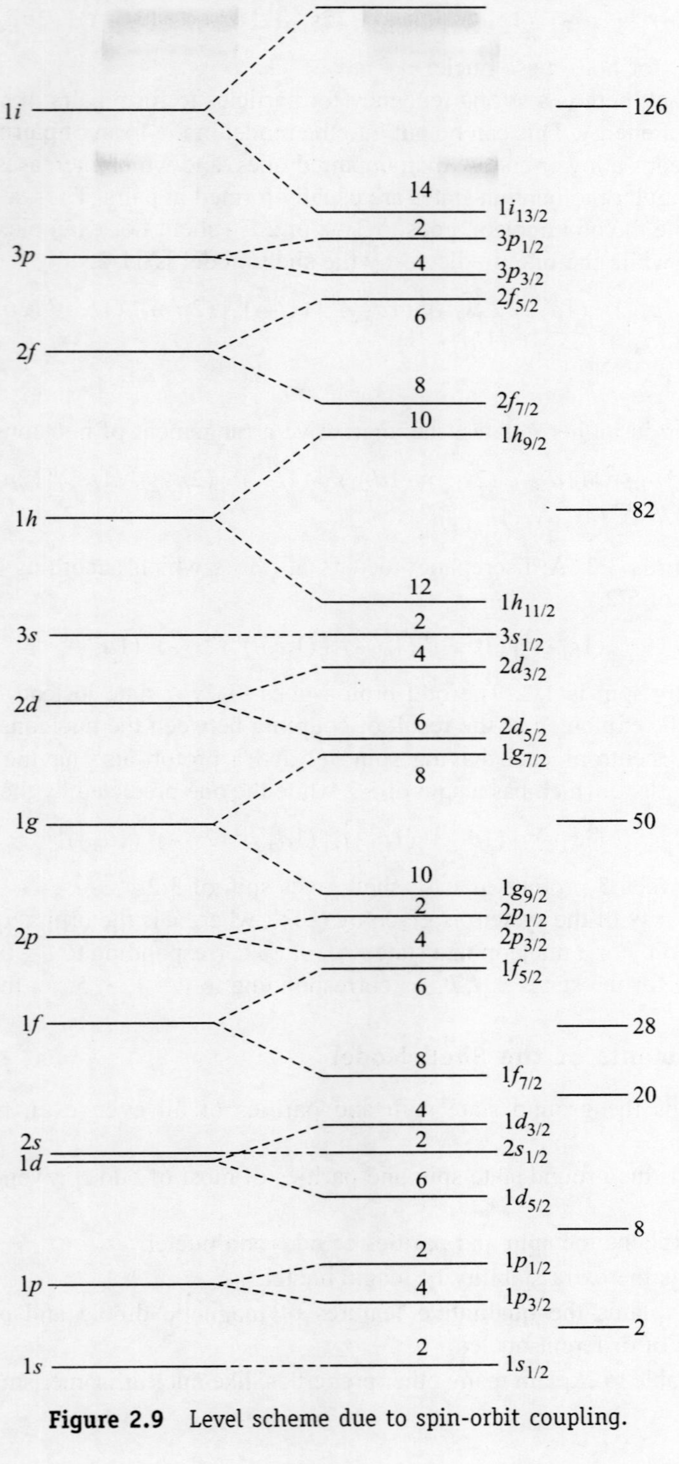

Figure 2.9 Level scheme due to spin-orbit coupling.

$$(1s_{1/2})^2 \mid (1p_{3/2})^4\,(1p_{1/2})^2 \mid (1d_{5/2})^6\,(2s_{1/2})^2\,(1d_{3/2})^4 \mid (1f_{7/2})^8 \mid (2p_{3/2})^3\,(1f_{5/2})^2$$

giving spin as 3/2 for both these nuclei.

We can say that there is a strong tendency for particles to form pairs in higher ℓ states even at some expense of energy. This can be put into the model in the form of pairing potential, which gives paired nucleons a lower energy than unpaired ones, and which increases with increasing ℓ.

The higher angular momentum states are usually formed in pairs. Thus, a level $1h_{11/2}$ may be filled in pairs while the odd nucleon goes to $3s_{1/2}$ or $2d_{3/2}$ shell. For example, the measured spin of ${}^{137}_{56}Ba_{81}$ is 3/2, while the one predicted by the shell model is 11/2.

$$(1s_{1/2})^2 \mid (1p_{3/2})^4\,(1p_{1/2})^2 \mid (1d_{5/2})^6\,(2s_{1/2})^2\,(1d_{3/2})^4 \mid (1f_{7/2})^8 \mid (2p_{3/2})^4\,(1f_{5/2})^6\,(2p_{1/2})^2\,(1g_{9/2})^{10}$$
$$\mid (1g_{7/2})^8\,(2d_{5/2})^6\,(2d_{3/2})^4\,(3s_{1/2})^2\,(1h_{11/2})^{11} \mid$$

predicts spin as 11/2.

Due to pairing in higher l states, the alternative arrangement of nucleons is as under

$$(1s_{1/2})^2 \mid (1p_{3/2})^4\,(1p_{1/2})^2 \mid (1d_{5/2})^6\,(2s_{1/2})^2\,(1d_{3/2})^4 \mid (1f_{7/2})^8 \mid (2p_{3/2})^4\,(1f_{5/2})^6\,(2p_{1/2})^2\,(1g_{9/2})^{10} \mid$$
$$(1g_{7/2})^8\,(2d_{5/2})^6\,(2d_{3/2})^3\,(3s_{1/2})^2\,(1h_{11/2})^{12} \mid$$

This predicts spin as 3/2. A discrepancy occurs at ${}^{19}_{9}F_{10}$, which according to the shell model should have spin of 5/2.

$$(1s_{1/2})^2 \mid (1p_{3/2})^4\,(1p_{1/2})^2 \mid (1d_{5/2})^1\,(2s_{1/2})^0\,(1d_{3/2})^0$$

Experimentally spin is 1/2. The odd proton goes to $2s_{1/2}$ state instead of $1d_{5/2}$ state. This discrepancy may be explained as the result of coupling between the nucleons outside the closed shell, i.e. the two neutrons each having spin 5/2 and a proton also having spin 5/2. Another discrepancy is for ${}^{23}Na$, which has a spin of 3/2 while the one predicted by the shell model is 5/2.

$$(1s_{1/2})^2 \mid (1p_{3/2})^4\,(1p_{1/2})^2 \mid (1d_{5/2})^3\,(2s_{1/2})^0\,(1d_{3/2})^0$$

Coupling between 3 protons in $1d_{5/2}$ shell gives spin of 3/2.

Finally, the parity of the system is given by $(-1)^{\ell}$, where ℓ is the orbital quantum number of the last odd nucleon. For a nucleon in a state *s*, *d*, *g*, ... corresponding to $\ell = 0, 2, 4, \ldots$ the parity is even (+), while for the states *p*, *f*, *h*, ... corresponding to $\ell = 1, 3, 5, \ldots$ the parity is odd (–).

2.3.5 Achievements of the Shell Model

1. It explains the ground state spin and parities of all even–even nuclei without any exception.
2. It explains the ground state spin and parities of most of odd *A* (even–odd or odd–even) nuclei.
3. It also explains the spin and parities of odd–odd nuclei.
4. It explains the extra stability of magic nuclei.
5. It also explains the qualitative features of magnetic dipole and electric quadrupole moments of different nuclei.
6. It is also able to explain many other properties, like nuclear isomerism of different nuclei.

2.3.6 Failures of Shell Model

1. Shell model fails to explain spin values for certain nuclei.
2. Shell model is unable to explain the energy of first excited states in even–even nuclei.
3. It is unable to explain magnetic moments of some nuclei.
4. This model is also unable to explain quadrupole moments of many nuclei.
5. Shell model is also unable to explain the ground states of odd A nuclei in the mass region $150 \le A \le 190$ and $A \ge 220$.

2.4 FERMI GAS MODEL

The semiempirical binding energy formula is based on treating the nucleus like a liquid drop. Such an analogy is an oversimplification and nucleus has many properties that can be explained more simply in terms of independent particle behaviour rather than in terms of the strong interaction picture implied by the liquid drop model. The most primitive independent particle model is obtained if the nucleus is treated as a degenerate Fermi gas of nucleons. The nucleons are assumed to move freely except for effects of the exclusion principle, throughout a sphere of radius $R = R_0 A^{1/3}$, $R_0 = 1.2$ fm. The situation is represented in Figure 2.10 by two wells, one for neutrons and other for protons. Free neutrons and free protons, far away from the walls, have the same energy, and the zero level for the two wells is the same. The two potential wells though have slightly different shapes, mainly because of the Coulomb part, the well for protons is less deep because of the Coulomb potential by an amount E_c and externally the $\frac{1}{r}$ dependence of the Coulomb potential extends the range. Protons trying to enter the nucleus from the outside are repelled by the nuclear charge; they must either tunnel through the barrier explained in Chapter 3 or have enough energy to pass over the barrier.

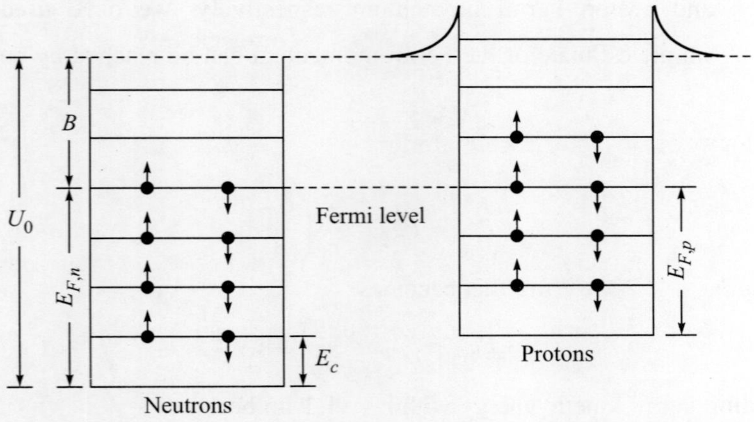

Figure 2.10 Protons–neutrons levels in Fermi model of the nucleus. The two potential wells have slightly different shapes because of the Coulomb repulsion of protons.

The wells contain a finite number of levels; each level can be occupied by two nucleons, one with spin up and other with spin down. It is assumed that the nuclear temperature is so low that the nucleons occupy the lowest states available to them. The term degenerate Fermi gas describes such a situation. The nucleons populate all states up to a maximum kinetic energy equal to the Fermi energy E_F.

From the density of the states obtained from a free Fermi gas confined to stay within a volume V, the nuclear volume, the number of particle states are given by

$$n = \frac{2V}{(2\pi\hbar)^3}\int_0^{P_F} d^3P \tag{2.15}$$

or

$$n = \frac{VP_F^3}{3\pi^2\hbar^3}$$

and the Fermi momentum P_F is given by

$$P_F = \hbar\left(3\pi^2\frac{n}{V}\right)^{1/3} = h\left(\frac{3}{8\pi}\frac{n}{V}\right)^{1/3} \tag{2.16}$$

Let us calculate the depth of the potential well for Z protons and N neutrons. We can write

$$P_{F,n} = \frac{\hbar}{r_0}\left(\frac{9\pi N}{4A}\right)^{1/3}$$

$$P_{F,p} = \frac{\hbar}{r_0}\left(\frac{9\pi Z}{4A}\right)^{1/3}$$

for the neutron and proton Fermi momentum respectively. We have used the relations $V = 4/3\,\pi r_0^3 A$. A simple estimate of the Fermi momentum can be reached by considering

$$N = Z = \frac{A}{2},$$

resulting in

$$P_{F,n} \approx P_{F,p} \approx \frac{\hbar}{r_0}\left(\frac{9\pi}{8}\right)^{1/3}$$

Using the value $\hbar c$ = 197 MeV fm, this becomes

$$P_{F,n} \approx P_{F,p} \approx \frac{297}{r_0}\text{ MeV/c} \tag{2.17}$$

The corresponding Fermi kinetic energy with r_0 = 1.2 fm becomes

$$\epsilon_F = \frac{P_{F,n}^2}{2m} \approx \frac{P_{F,p}^2}{2m} \approx 33\text{ MeV} \tag{2.18}$$

This energy corresponds to the kinetic energy of the highest occupied orbit (smallest binding energy). Given average binding energy ≈8 MeV, we can make a good estimate of the nuclear well depth ≈33 + 8 = 41 MeV, which agrees well with the experimental value.

The Fermi gas model gave some information about nuclear excited states even at low excitation energies observed in radioactive decay.

2.5 COLLECTIVE MODEL

The successes of the these models presented here led to a serious dilemma, whereas the liquid drop model can account for the behaviour of nucleus as a whole, shell model indicates the individual and nearly independent behaviour of nucleons. Combining the two aspects together, a new model called Collective Model was developed. In addition, this new model gave good results for magnetic moments, excited states, etc. of some nuclei. Giving more details of this model is beyond the scope of this book.

NUMERICAL PROBLEMS

Section 2.2

Solved Problems

1. Use the semiempirical mass formula to calculate the binding energy of ${}^{40}_{20}\text{Ca}$. What is the percentage discrepancy between this value and the actual value?

Solution: Binding energy = $[M(Z, N) - Zm_p - Nm_n] \times 931.49$ MeV

For ${}^{40}_{20}\text{Ca}$, actual binding energy

$$= [M(20, 20) - 20m_p - 20m_n] \times 931.49 \text{ MeV}$$

$$= (39.962591 - 20 \times 1.007825 - 20 \times 1.008665) \times 931.49 \text{ MeV}$$

$$= -342.05 \text{ MeV}$$

Binding energy as per semiempirical mass formula

$$B = a_v - a_s A^{2/3} - a_c \frac{Z(Z-1)}{A^{1/3}} - a_a \frac{(A-2Z)^2}{A} \pm a_p A^{-3/4}$$

Here, $A = 40$, $Z = 20$, $N = 20$, $a_v = 15.5$ MeV, $a_s = 16.8$ MeV, $a_c = 0.7$ MeV, $a_a = 23.0$ MeV and $a_p = 34.0$ MeV

Substituting various values, we get

$$B = 40 \times 15.5 - 16.8 \times 40^{2/3} - 0.7 \times \frac{20 \times 19}{40^{1/3}} - 23.0 \frac{40 - 2 \times 20}{40} - 30 \times 40^{-3/4}$$

$$= 620 - 196.4939 - 77.77887 - 1.8862$$

$$= 343.84 \text{ MeV}$$

Therefore, percentage discrepancy between the actual and semiempirical mass formula value

$$= \frac{343.84 - 342.05}{342.05}$$

$$\approx 0.5\%$$

2. Coulomb energy is given as

$$E_c = \frac{3}{5}\frac{Z(Z-1)\,e^2}{4\pi\varepsilon_0 R}$$

For mirror nuclei such as $^{15}_{7}N$ and $^{15}_{8}O$, the difference in masses Δm is due to difference in Coulomb energy and difference between proton and neutron mass. Show whether this value of Δm agrees with the actual value.

Solution: Given:

$$E_c = \frac{3}{5}\frac{Z(Z-1)\,e^2}{4\pi\varepsilon_0 R}$$

or

$$E_c = a_c\frac{Z(Z-1)}{A^{1/3}},$$

where

$$a_c = \frac{3e^2}{5\times 4\pi\varepsilon_0 r_0^{1/3}}$$

$$= 0.7 \text{ MeV}$$

For $^{15}_{7}N$, Coulomb energy contribution is

$$E_c = \frac{0.7\times 7\times 6}{15^{1/3}}$$

$$= 11.92 \text{ MeV}$$

For $^{15}_{8}O$, Coulomb energy contribution is

$$E_c = \frac{0.7\times 8\times 7}{15^{1/3}}$$

$$= 15.895 \text{ MeV}$$

Difference in both the Coulomb energy contribution = 15.895 – 11.92 = 3.975 MeV

Difference in the masses of $^{15}_{8}O$ and $^{15}_{7}N$ is simply the difference in the masses of hydrogen and neutron. Therefore,

$$m_O - m_N = (m_H - m_n)\times 931.49 \text{ MeV}$$

$$= (1.007276 - 1.008665)\times 931.49 \text{ MeV}$$

$$= -1.2938 \text{ MeV}$$

Therefore, according to the problem, difference in Coulomb energy and difference between proton and neutron mass is given by

$$= -1.2938 + 3.975 = 2.681 \text{ MeV}$$

Actual difference in the masses of ${}^{15}_{8}O$ and ${}^{15}_{7}N$

$$= 15.003070 - 15.000108$$

$$= 0.002962 \text{ amu}$$

$$= 0.002962 \times 931.49$$

$$= 2.759 \text{ MeV}$$

Therefore, for mirror nuclei, the difference in masses Δm is due to difference in Coulomb energy and difference between proton and neutron mass.

3. Find the energy needed to remove a neutron from ${}^{81}Kr$, ${}^{82}Kr$ and ${}^{83}Kr$.

Solution: Separation energy of neutron for ${}^{81}Kr$ = [BE of ${}^{81}Kr$ – BE of ${}^{80}Kr$]

$$\text{Binding energy} = [Zm_p + Nm_n - M(Z, N)] \times 931.49 \text{ MeV}$$

For ${}^{81}Kr$, binding energy $= (36 \times 1.007825 + 45 \times 1.008665 - 80.91661) \times 931.49$ MeV

$$= 703.28 \text{ MeV}$$

For ${}^{80}Kr$, binding energy $= (36 \times 1.007825 + 44 \times 1.008665 - 80.91661) \times 931.49$ MeV

$$= 695.50 \text{ MeV}$$

Therefore, separation energy of neutron for ${}^{81}Kr = 703.27 - 695.49 = 7.78$ MeV

Separation energy of neutron for ${}^{82}Kr$ = [BE of ${}^{82}Kr$ – BE of ${}^{81}Kr$]

For ${}^{81}Kr$, binding energy $= (36 \times 1.007825 + 46 \times 1.008665 - 81.913482) \times 931.49$ MeV

$$= 714.28 \text{ MeV}$$

Therefore, separation energy of neutron for ${}^{82}Kr = 714.28 - 703.28 = 11.00$ MeV

Separation energy of neutron for ${}^{83}Kr$ = [BE of ${}^{83}Kr$ – BE of ${}^{82}Kr$]

For ${}^{83}Kr$, binding energy $= (36 \times 1.007825 + 47 \times 1.008665 - 82.914134) \times 931.49$ MeV

$$= 721.74 \text{ MeV}$$

Therefore, separation energy of neutron for ${}^{83}Kr = 721.74 - 714.28 = 7.46$ MeV

4. Which isobars of $A = 75$ does the liquid drop model suggest to be the most stable nucleus?

Solution: According to liquid model, binding energy is given by

$$B = a_v - a_s A^{2/3} - a_c \frac{Z(Z-1)}{A^{1/3}} - a_a \frac{(A-2Z)^2}{A} \pm a_p A^{-3/4}$$

Here, $a_v = 15.5$ MeV, $a_s = 16.8$ MeV, $a_c = 0.7$ MeV, $a_a = 23.0$ MeV and $a_p = 34.0$ MeV

In order to find most stable isotope for $A = 75$, we take the partial derivative of B with respect to Z and equate that to zero.

$$\frac{\partial B}{\partial Z} = -a_c \frac{2Z-1}{A^{1/3}} + 4a_a \frac{A-2Z}{A} = 0$$

Substituting various values

$$= -\frac{0.7}{75^{1/3}}(2Z-1) + \frac{4 \times 23}{75}(75 - 2Z) = 0$$

which gives

$$Z = 32.89 \approx 33$$

Therefore, according to liquid drop model, most stable isotope for $A = 75$ is with $Z = 33$.

5. Using the liquid drop model, find the most stable isobars for $A = 27$, $A = 118$ and $A = 238$.

Solution: Binding energy is given by

$$B = a_v A - a_s A^{2/3} - a_c \frac{Z(Z-1)}{A^{1/3}} - a_a \frac{(A-2Z)^2}{A} \pm a_p A^{-3/4}$$

For most stable isobar, $\dfrac{\partial B}{\partial Z} = 0$

or
$$\frac{\partial B}{\partial Z} = \frac{a_c}{A^{1/3}}(2Z-1) + \frac{4a_a}{A}(A-2Z) = 0$$

which gives

$$Z = \frac{4a_a + \dfrac{a_c}{A^{1/3}}}{\dfrac{2a_c}{A^{1/3}} + \dfrac{8a_a}{A}}$$

Substituting $a_c = 0.7$ MeV and $a_a = 23.0$ MeV and $A = 27$, we get

$$Z = \frac{4 \times 23 + \dfrac{0.7}{27^{1/3}}}{\dfrac{2 \times 0.7}{27^{1/3}} + \dfrac{8 \times 23}{27}} \approx 12.6$$

or
$$Z = 13$$

Similarly, for $A = 118$, we get $Z = 49.8$ or $Z = 50$

And for $A = 238$, we get $Z = 91.98$ or $Z = 92$

6. For the mirror nuclei $^{23}_{11}\text{Na}_{12}$ and $^{23}_{12}\text{Mg}_{11}$, calculate the Coulomb coefficient a_c and hence estimate r_0. Given:

$$M(^{23}\text{Mg}) = 22.994124 \text{ amu}$$

$$M(^{23}\text{Na}) = 22.989768 \text{ amu}$$

$$M(n) = 1.008665 \text{ amu}$$

$$M(p) = 1.007276 \text{ amu}$$

Solution: We have Eq. (2.11)

$$M(A, Z) = ZM_p + (Z-1)\,M_n - a_v A + a_s A^{2/3} + a_c\,\frac{Z(Z-1)}{A^{1/3}} + a_a\,\frac{(A-2Z)^2}{A} \mp a_p A^{-3/4}$$

For $^{23}_{12}\mathrm{Mg}_{11}$, $(A-2Z)^2 = 1$ and for odd A nucleus $a_p = 0$. Taking the approximation $Z(Z-1) \approx Z^2$, this equation reduces to

$$M(A, Z) = ZM_p + (Z-1)\,M_n - a_v A + a_s A^{2/3} + a_c\frac{Z^2}{A^{1/3}} + \frac{a_a}{A}$$

And for $^{23}_{11}\mathrm{Na}_{12}$ nucleus, $(A-2Z)^2 = 1$ and for odd nuclei $a_p = 0$, we get

$$M(A, Z-1) = (Z-1)\,M_p + ZM_n - a_v A + a_s A^{2/3} + a_c\frac{(Z-1)^2}{A^{1/3}} + \frac{a_a}{A} \mp a_p A^{-3/4}$$

Subtracting these two equations, we get

$$M(A, Z) - M(A, Z-1) = -(M_n - M_p) + \frac{a_c}{A^{1/3}}(2Z-1)$$

Therefore,
$$a_c = \frac{A^{1/3}}{(2Z-1)}\,[M(A, Z) - M(A, Z-1) + (M_n - M_p)]$$

Substituting various values, we get

$$a_c = \frac{23^{1/3}}{(2\times 12-1)}\,[M(^{23}\mathrm{Mg}) - M(^{23}\mathrm{Na}) + (M_n - M_p)]$$

or

$$a_c = \frac{23^{1/3}}{23}\,[22.994124 - 22.989768 + 1.008665 - 1.007276]$$

$$= 0.00071 \text{ amu}$$

Thus,

$$a_c = 0.00071 \times 931.47 = 0.66 \text{ MeV}$$

Now,

$$a_c = -\frac{3}{5}\,\frac{e^2}{4\pi 4\varepsilon_0 r_0}$$

$$0.66 \times 1.6 \times 10^{-13} = -\frac{3}{5}\,\frac{8.98\times 10^9 \times (1.6\times 10^{-19})^2}{r_0}$$

which gives

$$r_0 = 1.3 \times 10^{-15} \text{ m}$$

7. Calculate the contribution of Coulomb energy and surface energy terms for $^{236}_{92}\mathrm{U}$ nucleus.

Solution: Contribution due to Coulomb energy term for $^{236}_{92}U$ nucleus

$$= -\frac{3}{5}\frac{Z(Z-1)\,e^2}{4\pi\varepsilon_0 r}$$

$$= -\frac{3}{5}\frac{Z(Z-1)\,e^2}{4\pi\varepsilon_0 r_0 A^{1/3}}$$

Substituting various values, we get

$$= -\frac{3}{5} \times 8.98 \times 10^9 \times \frac{(92\times 91)\,(1.6\times 10^{-19})^2}{1.2\times 10^{-15}\times (236)^{1/3}}$$

$$= -1.557 \times 10^{-10}\text{ J}$$

$$= -972.0\text{ MeV}$$

Contribution due to surface energy term for $^{236}_{92}U$ nucleus

$$= a_s A^{2/3}$$

$$= -16.8 \times 236^{2/3}$$

$$= -641.6\text{ MeV}$$

Unsolved Problems

1. Calculate the maximum energy required to remove a neutron from $^{91}_{40}Zr$, $^{92}_{40}Zr$ and $^{93}_{40}Zr$. Given:

$$M_H = 1.007825\text{ amu}$$
$$M_n = 1.008665\text{ amu}$$
$$M(^{90}Zr) = 89.904703\text{ amu}$$
$$M(^{91}Zr) = 90.905644\text{ amu}$$
$$M(^{92}Zr) = 91.905039\text{ amu}$$
$$M(^{93}Zr) = 92.906474\text{ amu}$$

[**Ans.** 7.19 MeV, 8.635 MeV and 6.739 MeV]

2. Use liquid drop model to find the most stable isobar for $A = 27$, 64 and 216.

[**Ans.** 12, 27, and 73]

3. When thermal neutrons (zero kinetic energy) is captured by a nucleus of Z protons and N neutrons, the excitation energy of the compound nucleus is given by

$$S_n = M(Z, N) + M_n - M(Z, N+1)$$

Calculate the difference in excitation energy for the case

$$^{235}_{92}U + n \rightarrow {}^{236}_{92}U \text{ and } {}^{238}_{92}U + n \rightarrow {}^{239}_{92}U$$

Given:

$$M(^{235}_{92}U) = 235.04924\text{ amu}$$
$$M(^{236}_{92}U = 236.045563\text{ amu}$$

$$M({}^{238}_{92}\text{U}) = 238.050785 \text{ amu}$$

$$M({}^{239}_{92}\text{U}) = 239.054290 \text{ amu}$$

[Ans. 1.74 MeV]

4. Show that average binding energy per nucleon is given as

$$\frac{B}{A} = a_v - \frac{a_s}{A^{1/3}} - a_c\frac{Z^2}{A^{4/3}} - a_a\frac{(A-2Z)^2}{A} \pm a_p A^{-3/4}$$

5. Calculate the binding energy per nucleon in case of ^{124}Xe, using problem 4 and compare with result of Chapter 1. Given:

$$m({}^{124}\text{Xe}) = 123.9061 \text{ amu}$$

[Ans. 8.45 MeV, 8.44 MeV]

6. Calculate the surface and Coulomb energy of ${}^{238}_{92}$U.

[Ans. 459.17 MeV, 812.05 MeV]

7. For the mirror nuclei ${}^{23}_{11}\text{Na}_{12}$ and ${}^{23}_{12}\text{Mg}_{11}$, calculate the Coulomb term coefficient a_c. Given:

$$\text{M(Na)} = 22.9897 \text{ amu}$$

$$\text{M(Mg)} = 22.9941 \text{ amu}$$

[Ans. 0.602 MeV]

8. For the isobaric family with $A = 39$, estimate the nuclear charge of the most stable isobar.

[Ans. 18]

9. Estimate the separation of the $1p_{1/2}$ and $1d_{5/2}$ energy levels for nuclei with mass number $A \sim 16$. Given the following information:

Total binding energy of ^{15}O = 111.96 MeV

Total binding energy of ^{16}O = 127.62 MeV

Total binding energy of ^{17}O = 131.76 MeV

The sequence of levels is $1s_{1/2}$; $1p_{3/2}$; $1p_{1/2}$; $1d_{5/2}$; $1d_{3/2}$.

[Ans. 11.52 MeV]

Section 2.3

Solved Problems

1. Using the shell model, predict the characteristics of ground state of ${}^{15}_{8}$O, ${}^{16}_{8}$O and ${}^{17}_{8}$O.

Solution: **${}^{15}_{8}$O:**

According to shell model, 8 protons pair to give no contribution to the spin. Seven neutrons will be distributed in different shells as

$$(1s_{1/2})^2 \mid (1p_{3/2})^4 \, (1p_{1/2})^1 \mid$$

Last unpaired neutron is in $p_{1/2}$ shell. Therefore, spin of ${}^{15}_{8}$O is 1/2. Since the last unpaired neutron is in p shell for which $\ell = 1$, therefore, parity is odd. Therefore, spin parity of ${}^{15}_{8}$O is $1^-/2$.

$^{16}_{8}$O:

According to shell model, 8 protons pair to give no contribution to the spin. Similarly, 8 neutrons also pair together giving no contribution to spin. Therefore, spin parity of $^{16}_{8}$O is 0^+.

$^{17}_{8}$O:

According to shell model, 8 protons pair to give no contribution to the spin. Nine neutrons will be distributed in different shells as

$$(1s_{1/2})^2 \mid (1p_{3/2})^4 (1p_{1/2})^2 \mid (1d_{5/2})^1 (2s_{1/2})^0 (1d_{3/2})^0$$

Last unpaired neutron is in $d_{5/2}$ shell. Therefore, spin of $^{17}_{8}$O is 5/2. Since the last unpaired neutron is in d shell for which $\ell = 2$, therefore, parity is even. Therefore, spin parity of $^{17}_{8}$O is $5^+/2$.

2. Use single particle shell model to predict the ground state spins and parities of $^{63}_{29}$Cu and $^{40}_{18}$Ar.

Solution: **$^{63}_{29}$Cu$_{34}$:**

According to shell model, 34 neutrons pair to give no contribution to the spin. Twenty-nine protons will be distributed in different shells as

$$(1s_{1/2})^2 \mid (1p_{3/2})^4 (1p_{1/2})^2 \mid (1d_{5/2})^6 (2s_{1/2})^2 (1d_{3/2})^4 \mid (1f_{7/2})^8 \mid (2p_{3/2})^1 (1f_{5/2})^0$$

Last unpaired proton is in $p_{3/2}$ shell. Therefore, spin of $^{63}_{29}$Cu$_{34}$ is 3/2. Since the last unpaired proton is in p shell for which $\ell = 1$, therefore, parity is odd. Therefore, spin parity of $^{63}_{29}$Cu$_{34}$ is $3^-/2$.

$^{40}_{18}$Ar$_{22}$:

According to shell model, 18 protons pair to give no contribution to the spin. Similarly, 22 neutrons also pair together giving no contribution to spin. Therefore, spin parity of $^{40}_{18}$Ar is 0^+.

3. Use single particle shell model to predict the ground state spin and parities of $^{19}_{10}$Ne, $^{20}_{10}$Ne and $^{21}_{10}$Ne.

Solution: **$^{19}_{10}$Ne$_9$:**

According to shell model, 10 protons pair to give no contribution to the spin. Nine neutrons will be distributed in different shells as

$$(1s_{1/2})^2 \mid (1p_{3/2})^4 (1p_{1/2})^2 \mid (1d_{5/2})^1 (2s_{1/2})^0 (1d_{3/2})^0$$

Last unpaired neutron is in $d_{5/2}$ shell. Therefore, spin of $^{19}_{10}$Ne$_9$ is 5/2. Since the last unpaired neutron is in d shell for which $\ell = 2$, therefore, parity is even. Therefore, spin parity of $^{19}_{10}$Ne$_9$ is $5^+/2$.

$^{20}_{10}$Ne$_{10}$:

According to shell model, 10 protons pair to give no contribution to the spin. Similarly, 10 neutrons also pair together giving no contribution to spin. Therefore, spin parity of $^{20}_{10}$Ne is 0^+.

$^{21}_{10}$Ne$_{11}$:

According to shell model, 10 protons pair to give no contribution to the spin. Eleven neutrons will be distributed in different shells as

$$(1s_{1/2})^2 \mid (1p_{3/2})^4 (1p_{1/2})^2 \mid (1d_{5/2})^3 (2s_{1/2})^0 (1d_{3/2})^0$$

Last unpaired neutron is in $d_{5/2}$ shell. Therefore, spin of ${}^{21}_{10}Ne$ is 5/2. Since the last unpaired neutron is in d shell for which $\ell = 2$, therefore, parity is even. Therefore, spin parity of ${}^{21}_{10}Ne$ is $5^+/2$.

4. Calculate the magnetic moments of nuclei in problem 3.

Solution: $\mathbf{{}^{19}_{10}Ne_9}$:

$$J = \frac{5}{2}$$

The magnetic moment $\mu = J + 2.29$

or

$$\mu = \frac{5}{2} + 2.29 = 4.79 \text{ nuclear magneton}$$

$\mathbf{{}^{20}_{10}Ne_{10}}$:

$$J = 0$$

The magnetic moment $\mu = J + 2.29$

or

$$\mu = 2.29 \text{ nuclear magneton}$$

$\mathbf{{}^{21}_{10}Ne_{11}}$:

$$J = \frac{5}{2}$$

The magnetic moment $\mu = J + 2.29$

or

$$\mu = \frac{5}{2} + 2.29 = 4.79 \text{ nuclear magneton}$$

Unsolved Problems

1. Predict the shell model state for the odd nucleon in the undermentioned cases:

(i) ${}^{63}_{29}Cu$ (ii) ${}^{25}_{12}Mg$

2. Draw schematic shell model diagram for the following nuclei, showing the filling-up of the levels by protons and neutrons.

$${}^{7}_{3}Li,\ {}^{27}_{13}Al \text{ and } {}^{43}_{20}Ca$$

3. The ground state spin of ${}^{19}_{9}F$ as predicted by the shell model is 5/2, while the experimentally observed spin is 1/2. How can this discrepancy be explained by taking into consideration the order of shell model levels?

4. Predict the characteristics of the ground states of

(i) ${}^{58}_{27}Co$, ${}^{59}_{27}Co$ and ${}^{60}_{27}Co$, and (ii) ${}^{207}_{82}Pb$, ${}^{208}_{82}Pb$ and ${}^{209}_{82}Pb$

5. Predict the shell model states of the odd nucleon in the following cases:

$${}^{63}_{29}Cu,\ {}^{25}_{12}Mg,\ {}^{17}_{9}F,\ {}^{127}_{53}I$$

$$\left[\textbf{Ans. } \frac{3^-}{2}, \frac{5^+}{2}, \frac{5^+}{2}, \frac{7^+}{2}\right]$$

6. Ground state spin and parity are given for the following nuclei:

$$^{3}\text{He}(1/2^{+}),\ ^{21}\text{Ne}(3/2^{-}) \text{ and } ^{209}\text{Bi}(9/2^{-})$$

Justify these observed values on the basis of the shell model.

7. Given that the ordering of the nuclear levels is

$$1s_{1/2};\ 1p_{3/2};\ 1p_{1/2};\ 1d_{5/2};\ 1d_{3/2};\ 2s_{1/2};\ 1f_{7/2};\ 2p_{3/2};\ 1f_{5/2};$$

Assign spin and parity to the following nuclei $^{3}_{2}\text{He}$, $^{20}_{10}\text{Ne}$, $^{27}_{13}\text{Al}$, $^{41}_{21}\text{Sc}$ and $^{69}_{31}\text{Ga}$.

$$\left[\textbf{Ans.}\ \frac{1^{+}}{2}, 0^{+}, \frac{5^{+}}{2}, \frac{7^{-}}{2}, \frac{3^{-}}{2}\right]$$

8. Assume that the level sequence according to shell model is

$$1s_{1/2};\ 1p_{3/2};\ 1p_{1/2};\ 1d_{5/2};\ 2s_{1/2};\ 1d_{3/2};\ 1f_{7/2};\ 2p_{3/2}.$$

Find the ground state spin and parity assignments for the following nuclei:

$$^{7}_{3}\text{Li},\ ^{23}_{11}\text{Na} \text{ and } ^{33}_{16}\text{S}$$

$$\left[\textbf{Ans.}\ \frac{3^{-}}{2}, \frac{5^{+}}{2}, \frac{3^{+}}{2}\right]$$

REVIEW QUESTIONS

Short Answer Type

1. What are the drawbacks of liquid drop model?
2. What are magic numbers?
3. What are the main achievements and limitations of shell model?
4. What are the similarities between a nucleus and liquid drop?
5. Give two achievements of nuclear shell model.
6. What are magic numbers? What is their significance?
7. What are magic numbers, singly magic and doubly magic nuclei? Which nuclear model explains the stability of magic nuclei?
8. What is the importance of asymmetric term in semiempirical mass formula?
9. State predictions of shell model.
10. Name the various contributions to the mass of a nucleus as taken in semiempirical mass formula.
11. Which numbers out of these are magic numbers 6, 8, 12, 14, 20, 40?
12. List various nuclear models.
13. What are magic numbers? Which numbers out of these are magic numbers 6, 8, 10, 12, 14, 16, 18, 20, 25, 30?

14. What was the need of various nuclear models? What are the main successes of liquid drop model?
15. Give two experimental evidences of magic numbers.
16. Discuss various factors in favour of nuclear shell model.

Long Answer Type

1. Discuss the shell model of the nucleus. What are its merits and demerits?
2. Give experimental evidence of magic numbers.
3. Explain the limitations of shell model.
4. Derive the expression for semiempirical mass formula.
5. Discuss the basic assumptions of liquid drop model of nucleus. Explain how the model is used to estimate the semiempirical mass formula.
6. What are assumptions and drawbacks of liquid drop model?
7. What are magic numbers? Give experimental evidences of nuclear magic numbers. What are the achievements of shell model?
8. What do you understand by magic numbers? What are the experimental evidences of nuclear magic numbers?
9. Assuming nucleus to be uniformly charged shell of radius r, calculate the work done against electrostatic force in giving it a charge Ze.
10. Starting from assumptions discuss the achievements and failures of liquid drop model.
11. What is semiempirical mass formula? Explain the significance of various terms.
12. Discuss relative merits and demerits of various nuclear models. What is the importance of magic numbers? Explain shell model in detail.
13. State assumptions of liquid drop model.
14. Derive asymmetric energy term of semiempirical mass formula.
15. Explain the existence of magic numbers on the basis of single particle shell model.
16. Write down the major predictions of shell model.
17. Discuss the stability of nuclei on the basis of semiempirical mass formula.
18. Derive the nuclear energy level scheme on the basis of extreme single particle shell model.
19. List some of the experimental evidences for shell model.
20. Derive the Coulomb energy term of semiempirical mass formula.
21. Give the elementary account of shell model and explain how this model predicts the existence of magic numbers.
22. Give the elementary account of single particle shell model and state its predictions.
23. Explain various predictions of the liquid drop model. Give a brief description of semiempirical mass formula.
24. Derive various factors which contribute to the binding energy of the nucleus and derive semiempirical mass formula based on these factors.

25. State various assumptions of liquid drop model. Discuss the semiempirical mass formula of the liquid drop model. Discuss the utility of this formula.
26. In what ways liquid drop model is different from the shell model? What are magic numbers? How does the shell model explain the existence of magic numbers and other nuclear properties?
27. What are the basic differences between liquid drop model and shell model of the nucleus? What is the evidence of shell structure in the nuclei? Explain the main assumptions of the shell model of the nucleus. Discuss its achievements, failures and limitations.
28. Explain liquid drop model. Obtain semiempirical mass formula. Give any two achievements of the model.

Chapter 3

Radioactivity

3.1 INTRODUCTION

In the periodic table there are nuclei, which are called stable, i.e. the properties of these nuclei do not change with the passage of time, unless the nuclei are otherwise disturbed. Yet there is another class of nuclei, whose number is much more, which are unstable. They disintegrate spontaneously by emitting either electromagnetic radiations or some particles.

The phenomenon of spontaneous decay of a nucleus accompanied by the emission of alpha-particles, beta-particles or gamma-rays is known as radioactivity.

It could be either natural or artificial. Radioactivity was first discovered by Henry Becquerel in 1896 accidentally in the same way as X-rays were discovered by Röentgen in 1895. French scientist Henry Becquerel in 1896 accidentally discovered the phenomenon of radioactivity, when he took a photographic plate wrapped in black paper and placed a uranium salt on it and exposed it to sunlight. On developing the photographic plate he found it to be foggy or darkened indicating that the uranium salt has emitted radiations in the presence of sunlight, which could penetrate into the black paper and affect the photographic plate. In those days it was thought that this phenomenon was possible only in sunlight. Once again Becquerel got ready to repeat this experiment, but the day was cloudy. So he put the whole set-up into his table drawer. After two days, he developed these plates without exposing to sun. To his surprise, the photographic plates were still foggy. It became clear to him that the uranium salt always gives invisible radiations irrespective of illumination. He was able to show that these invisible penetrating radiations are able to ionize gases and some part of the radiations consisted of fast moving charged particles. This was the beginning of the era of radioactivity.

Soon afterwards, Pierre Curie and Marie Curie working in the Becquerel laboratory, found two elements other than Uranium which were also radioactive. They named one as Polonium and the other as Radium.

The experiments of Rutherford and his co-workers found out three components in these penetrating radiations. They called these components as α-, β- and γ-particles. Later on α-particles were identified as ^4He nuclei, β-particles as electrons and γ-particles as high-energy photons. Later studies discovered three modes of β-decay. They are electron emission (β^-), positron emission (β^+) and electron capture. Similarly, γ-decay also has three different decay modes. They are γ-emission, internal conversion and internal pair conversion. The various decay modes of a radioactive nucleus are summarized in Figure 3.1. These modes are given below:

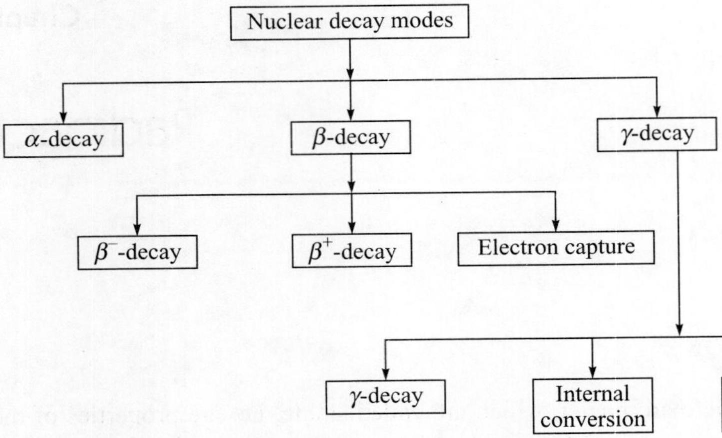

Figure 3.1 Various decay modes of radioactive nuclei.

Alpha Decay

In this process, parent nucleus disintegrates to give a daughter nucleus and helium nucleus or an α-particle. Mass number of the daughter nucleus decreases by four units and atomic number decreases by two units. A typical example of this decay mode is

$$^{238}_{92}\text{U} \rightarrow {}^{234}_{90}\text{Th} + {}^{4}_{2}\text{He}$$

Beta Decay

In this process, parent nucleus disintegrates to give a daughter nucleus and a β-particle. As shown in Figure 3.1, β-decay is classified into three categories.

(i) **Electron emission or β^--decay:** Here, parent nucleus disintegrates to give a daughter nucleus, a β^--particle (or an electron) and a new particle called *antineutrino* ($\overline{\nu}$). Mass number of the daughter nucleus remains the same and atomic number increases by one unit. One such example is

$$^{14}_{6}\text{C} \rightarrow {}^{14}_{7}\text{N} + \beta^- + \overline{\nu}$$

(ii) **Positron emission or β^+-decay:** In this process, parent nucleus disintegrates to give a daughter nucleus, a β^+-particle (or a positron) and a new particle called *neutrino* (ν). Mass number of the daughter nucleus remains the same and atomic number decreases by one unit. One example is

$$^{22}_{11}\text{Na} \rightarrow {}^{22}_{10}\text{Ne} + \beta^+ + \nu$$

(iii) **Electron capture:** Here the parent nucleus captures one of the orbital electrons with the emission of a neutrino (ν). Mass number of the daughter nucleus remains the same and atomic number decreases by one unit. For example,

$$^{54}_{25}\text{Mn} + \beta^- \rightarrow {}^{54}_{24}\text{Cr} + \nu$$

Gamma Decay

If the excitation energy available with a nucleus is not sufficient for particle emission, it loses its energy by following three processes, as shown in Figure 3.1.

(i) **Gamma decay:** Alpha and beta decays of a radioactive nucleus usually leave the daughter nucleus in an excited state. If the excitation energy available with the daughter nucleus is not sufficient for further particle emission, it loses its energy by emitting electromagnetic radiations, also known as γ-rays. Mass and charge of the daughter nucleus remains the same as before the emission of γ-rays. One such example is

$$^{137}_{56}\text{Ba}^* \rightarrow {}^{137}_{56}\text{Ba} + \gamma$$

The star (*) on ^{137}Ba indicates that it is in excited state.

(ii) **Internal conversion:** In internal conversion process, an excited nucleus instead of emitting γ-rays directly transfers its excitation energy to one of the orbital electrons and the orbital electron is ejected as conversion electron. Mass and charge of the daughter nucleus remains the same as before the emission of the orbital electron. This is so because electron is ejected from the electronic orbits.

(iii) **Internal pair conversion:** If the excitation energy of the nucleus > 1.022 MeV, the nucleus can de-excite by emitting directly an electron and positron pair in its own Coulomb field. This process is known as *internal pair conversion.*

More details of these processes will be given later on.

3.2 LAWS OF DISINTEGRATION

The laws of radioactive disintegration are:

1. There is an equal probability for all nuclei of a radioactive element to decay.
2. The rate of spontaneous disintegration of a radioactive element is proportional to the number of nuclei present at that time.

Mathematically, it can be written as

$$\frac{dN}{dt} \propto N \tag{3.1}$$

where N is the number of atoms present at time t. Removing proportionality sign, we get

$$\frac{dN}{dt} = -\lambda N \tag{3.2}$$

where λ is a constant of proportionality and is known as decay constant of the element. Negative sign indicates that as t increases N decreases. Rewriting Eq. (3.2) as

$$\frac{dN}{N} = -\lambda dt \tag{3.3}$$

Integrating both sides, we have

$$\int \frac{dN}{N} = -\lambda \int dt$$

$$\ln(N) = -\lambda t + C \tag{3.4}$$

where C is constant of integration and is evaluated by the fact that at $t = 0$, number of atoms of the radioactive element is N_0. Using this condition, we get

$$C = \ln(N_0) \tag{3.5}$$

Substituting this value of C in Eq. (3.4), we get

$$\ln(N) = -\lambda t + \ln(N_0)$$

or

$$\ln \frac{N}{N_0} = -\lambda t$$

Thus,

$$\boxed{N = N_0 e^{-\lambda t}} \tag{3.6}$$

The exponential nature of this equation shows that it takes infinite time for whole of the radioactive material to disintegrate. N vs. t has been plotted for ^{24}Na ($t_{1/2} = 15$ h) in Figure 3.2. A plot of $\log(N)$ vs. t would be a straight line.

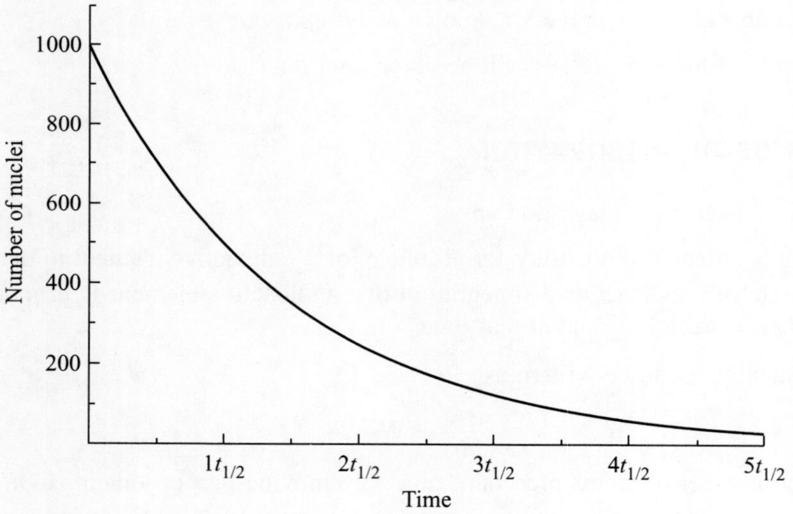

Figure 3.2 Exponential decay curve for ^{24}Na.

3.2.1 Activity and its Units

Activity

The activity (A) of a radioactive nuclide is defined as the number of disintegration per unit time.

$$A = -\frac{dN}{dt} = \lambda N$$

Substituting for N from Eq. (3.6), we get

$$A = \lambda N_0 e^{-\lambda t}$$

Substituting $A_0 = \lambda N_0$, the exponential relation for activity is

$$\boxed{A = A_0 e^{-\lambda t}} \tag{3.7}$$

In this relation A_0 is the activity at $t = 0$. The exponential factor shows that activity is decreasing with time in the same fashion as N.

Unit

- **The SI unit of activity is becquerel (Bq):** 1 becquerel = 1 Bq = 1 disintegration/s (dps or disintegrations per second).

  In day-to-day life, we encounter activities of the order of kilobecquerel (kBq), megabecquerel (MBq) or gigabecquerel (GBq).

$$1 \text{ kBq} = 10^3 \text{ Bq}$$
$$1 \text{ MBq} = 10^6 \text{ Bq}$$
$$1 \text{ GBq} = 10^9 \text{ Bq}$$

- **Traditional unit of activity is curie (Ci) and rutherford (R):**

  1 curie (Ci) = 3.7×10^{10} disintegration/s

$$= 3.7 \times 10^{10} \text{ Bq}$$
$$= 37 \text{ GBq}$$
$$1 \text{ rutherford} = 10^6 \text{ disintegration/s}$$
$$= 1 \text{ MBq}$$

  Curie is a big unit, normal activities of radioactive sources are defined in terms of millicurie (1 mCi = 10^{-3} Ci) and microcurie (1 μCi = 10^{-6} Ci).

3.2.2 Half-life

Rutherford in 1904 introduced the term *half-life* ($t_{1/2}$) for a radioactive material. It is the characteristic property of a radioactive nuclide.

Half-life is the time required for number of radioactive nuclides to reduce to half by disintegration.

If N_0 is the initial number of radioactive nuclei present, then in one half-life this number will reduce to $N_0/2$. Therefore, after 1 half-life

$$N = \frac{N_0}{2}$$

Substituting N in Eq. (3.6), we get

$$\frac{N_0}{2} = N_0 e^{-\lambda t_{1/2}}$$

or

$$\frac{1}{2} = e^{-\lambda t_{1/2}}$$

or

$$e^{\lambda t_{1/2}} = 2$$

Taking natural log on both sides

$$\lambda t_{1/2} = \ln 2$$

$$\boxed{t_{1/2} = \frac{\ln 2}{\lambda} = \frac{0.693}{\lambda}} \tag{3.8}$$

Therefore, half-life of a radioactive material depends upon the disintegration constant λ. The larger the value of disintegration constant, smaller is the half-life, i.e. nuclei of radioactive material will decay faster and vice versa. Half-life of a radioactive material is its intrinsic property and it cannot be altered by any physical or chemical means.

After each half-life, the number of radioactive atoms reduces to half. For example, if we take on 1st Jan. 2006, 10,000 atoms of a radioactive material whose half-life is 1 year, then on 1st Jan. 2007, the number of radioactive atoms left becomes 5000. On 1st Jan. 2008, the number will be 2500 and on 1st Jan. 2009, it will be 1250 and so on. It is clear that after infinite half-lives, the number of radioactive nuclei will reduce to zero.

3.2.3 Average (Mean) Life

As we have seen (Eq. 3.6) that infinite time is required for the complete disappearance of the radioactive material, one does not know which atom of the material is going to decay at a given instant of time, i.e. individual atoms have lifetime from zero to infinity. Because of this Soddy in 1904 introduced the concept of average or mean life (τ).

The average or mean life is defined as the average life expectancy of the radioactive species.

Therefore, the average life is calculated by summing the lives of all the nuclei and dividing by the total number of nuclei. Suppose dN_1 nuclei decay in time t_1, dN_2 nuclei decay in time t_2, dN_3 nuclei decay in time t_3 and so on, then the average life or mean life will be

$$\tau = \frac{t_1 dN_1 + t_2 dN_2 + t_3 dN_3 + \ldots}{dN_1 + dN_2 + dN_3 + \ldots}$$

This equation can be written in integral form as

$$\tau = \frac{\int_0^{N_0} t\,dN}{\int_0^{N_0} dN}$$

Since

$$\int_0^{N_0} dN = N_0$$

Therefore,

$$\tau = \frac{1}{N_0} \int_0^{N_0} t\,dN$$

where N_0 is the total number of radioactive atoms present initially.

Differentiating Eq. (3.6) with respect to t, we have

$$dN = -\lambda N_0 e^{-\lambda t} dt$$

Now for $t = \infty$, $N = 0$ and for $t = 0$, $N = N_0$
Therefore,

$$\tau = -\frac{1}{N_0}\int_{\infty}^{0} \lambda t N_0 e^{-\lambda t} dt$$

$$\tau = \lambda \int_{0}^{\infty} t e^{-\lambda t} dt \qquad (3.9)$$

The integral in Eq. (3.9) can be evaluated by parts

$$\int_{0}^{\infty} t e^{-\lambda t} dt = \left[t \frac{e^{-\lambda t}}{-\lambda} \right]_0^{\infty} - \int_{0}^{\infty} 1 \frac{e^{-\lambda t}}{-\lambda} dt$$

$$= (0 - 0) - \left[\frac{1}{\lambda} \times \frac{e^{-\lambda t}}{-\lambda} \right]_0^{\infty}$$

$$= -\frac{1}{\lambda^2}(0 - 1) = \frac{1}{\lambda^2}$$

Therefore, the value of average life from Eq. (3.9) is

$$\boxed{\tau = \frac{1}{\lambda} = \frac{t_{1/2}}{0.693}} \qquad (3.10)$$

where we have used the value of λ from Eq. (3.8).

3.3 RADIOACTIVE SERIES

There are about 65 naturally occurring radioisotopes. Out of these about 42 radioactive nuclides have $Z \geq 82$. These radioactive nuclides decay by α- or β-emission, changing Z and/or A until a stable nucleus is reached. α-decay changes A by 4 amu while β-decay does not change A, therefore, we have four independent decay chains with mass numbers $4n$, $4n + 1$, $4n + 2$ and $4n + 3$, where n is an integer. Any other change of mass is repetition of these four series, i.e. $4n + 4$ will be a member of $4n$ series, and so on. This decay process tends to terminate to the stable member of the series. These four series are listed in Table 3.1.

TABLE 3.1 Radioactive series

Series	*Type*	*Parent nucleus*	*Final nucleus*	*Half-life of parent nucleus years*
Thorium	$4n$	^{232}Th	^{208}Pb	1.41×10^{10}
Neptunium	$4n + 1$	^{237}Np	^{209}Bi	2.14×10^{6}
Uranium	$4n + 2$	^{238}U	^{206}Pb	4.47×10^{9}
Actinium	$4n + 3$	^{235}U	^{207}Pb	7.04×10^{8}

Notice that the neptunium series is not observed in nature as the longest-lived member of the neptunium series has far too short a lifetime (2.14×10^6 years) *to have survived since the formation of the earth* ($\sim 4.6 \times 10^9$ years). *Further, unlike other series, the parent nucleus of actinium series is* ^{235}U *and not actinium. This is due to the fact that actinium was first discovered and it was thought that actinium is the parent nucleus of this series. So, this series was called actinium series. But later on when* ^{235}U *was discovered, it was found that not actinium but long-lived* ^{235}U (7.04×10^8 years) *is the parent nucleus of this series. However, the name actinium series was not changed and still continuing.*

The four series corresponding to $4n$, $4n + 1$, $4n + 2$ and $4n + 3$ are displayed in Figures 3.3 to 3.6 respectively.

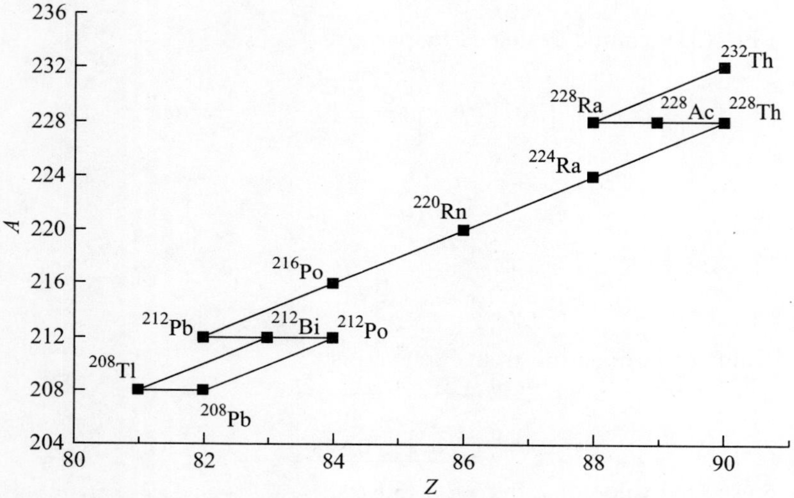

Figure 3.3 $A = 4n$, Thorium series. End product is ^{208}Pb.

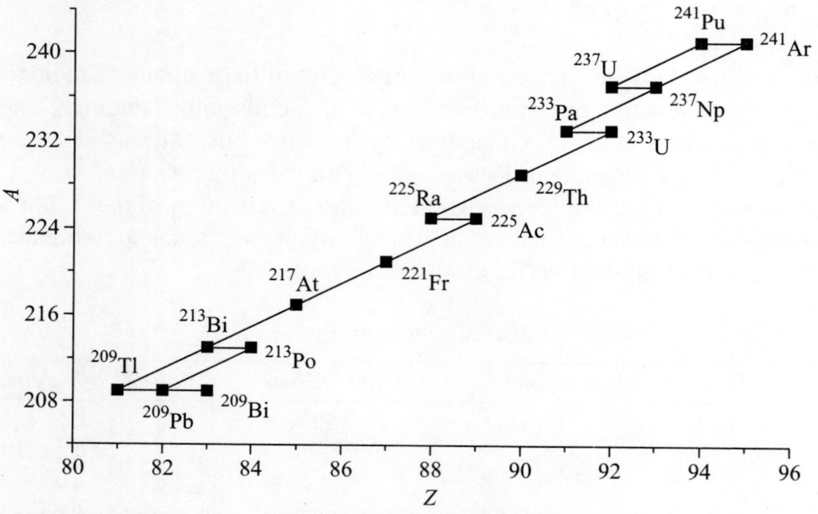

Figure 3.4 $A = 4n + 1$, Neptunium series. End product is ^{209}Bi.

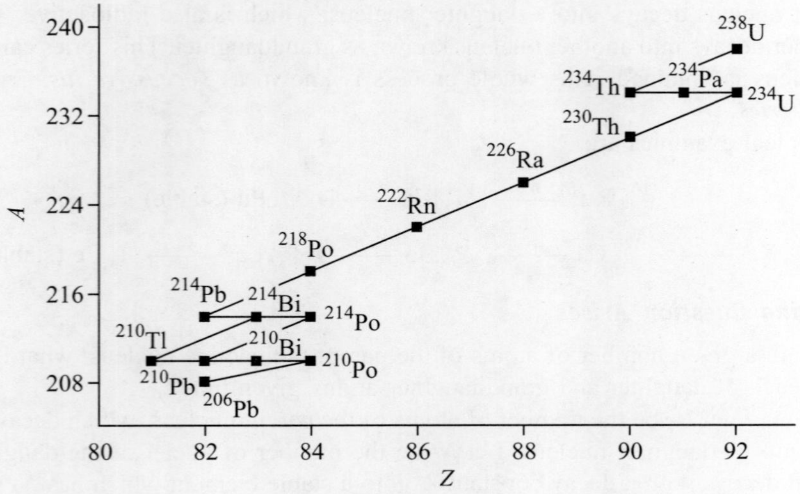

Figure 3.5 $A = 4n + 2$, Uranium series. End product is ^{206}Pb.

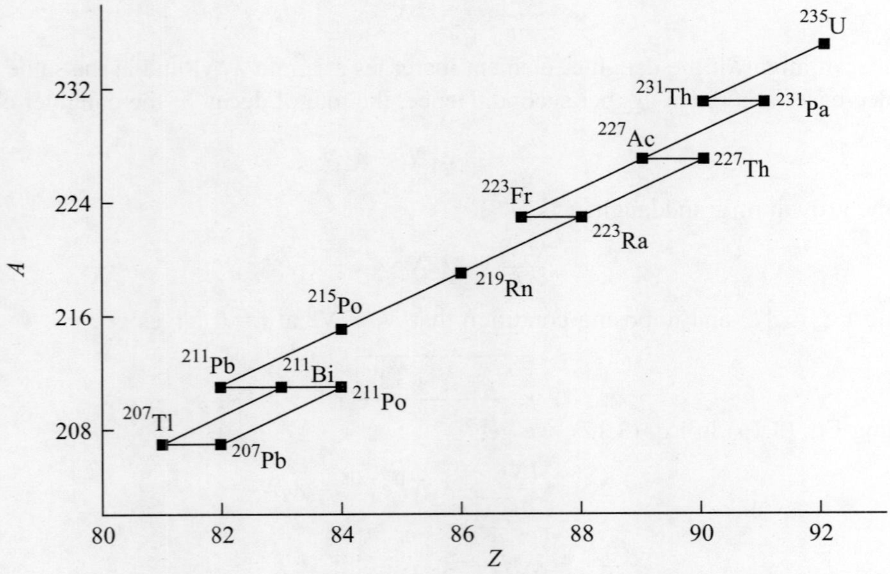

Figure 3.6 $A = 4n + 3$, Actinium series. End product is ^{207}Pb.

3.4 LAW OF SUCCESSIVE DISINTEGRATION

In all the examples considered so far, radioactive nucleus also known as parent nucleus decays to a stable nucleus, called daughter nucleus. There can be other cases too.

A parent nucleus decays into a daughter nucleus, which is also radioactive. The daughter nucleus further decays into another nucleus known as granddaughter. This series can continue till a stable nucleus is obtained. This whole process is known as *successive disintegrations* or a *radioactive series*.

Some typical examples are

$$^{105}_{44}\text{Ru} \xrightarrow{\beta^-} {}^{105}_{45}\text{Rh} \xrightarrow{\beta^-} {}^{105}_{46}\text{Pd (stable)}$$

$$^{140}_{54}\text{Xe} \xrightarrow{\beta^-} {}^{140}_{55}\text{Cs} \xrightarrow{\beta^-} {}^{140}_{56}\text{Ba} \xrightarrow{\beta^-} {}^{140}_{57}\text{La} \xrightarrow{\beta^-} {}^{140}_{58}\text{Ce (stable)}$$

An Interesting Question Arises

If we start with a given number of atoms of the parent radioactive nucleus, what is the number of nuclei in each of daughter and granddaughter at any given time?

At any time t, let N_1 be the number of atoms of the parent nucleus, which decays with decay constant λ_1 into its daughter nucleus. Let N_2 be the number of nuclei of the daughter element, which in turn decays with a decay constant λ_2 into a stable element which has N_3 stable atoms. Also assume that at time $t = 0$, $N_1 = N_1^0$, $N_2 = N_2^0$ and $N_3 = N_3^0$.

Now, the rate of decay of the parent nucleus is

$$\frac{dN_1}{dt} = -\lambda_1 N_1 \tag{3.11}$$

The number of atoms of the daughter element increases at a rate $\lambda_1 N_1$ and at the same time this number decreases at a rate $\lambda_2 N_2$ per second. Hence, the rate of decay of the daughter element is

$$\frac{dN_2}{dt} = \lambda_1 N_1 - \lambda_2 N_2 \tag{3.12}$$

Finally, the growth of granddaughter is

$$\frac{dN_3}{dt} = \lambda_2 N_2 \tag{3.13}$$

Integrating Eq. (3.11) and imposing condition that $N_1 = N_1^0$ at $t = 0$, gives

$$\boxed{N_1 = N_1^0 e^{-\lambda_1 t}} \tag{3.14}$$

Substituting Eq. (3.14) in Eq. (3.12), we get

$$\frac{dN_2}{dt} = \lambda_1 N_1^0 e^{-\lambda_1 t} - \lambda_2 N_2$$

$$\frac{dN_2}{dt} + \lambda_2 N_2 = \lambda_1 N_1^0 e^{-\lambda_1 t} \tag{3.15}$$

Multiplying Eq. (3.15) by $e^{\lambda_2 t}$, we have

$$e^{\lambda_2 t}\frac{dN_2}{dt} + \lambda_2 N_2 e^{\lambda_2 t} = \lambda_1 N_1^0 e^{-\lambda_1 t} e^{\lambda_2 t}$$

or

$$\frac{d}{dt}[N_2 e^{\lambda_2 t}] = \lambda_1 N_1^0 e^{(\lambda_2 - \lambda_1)t} \tag{3.16}$$

Integrating Eq. (3.16), we get

$$N_2 e^{\lambda_2 t} = \frac{\lambda_1}{\lambda_2 - \lambda_1} N_1^0 e^{(\lambda_2 - \lambda_1)t} + C \tag{3.17}$$

where C is a constant of integration. At $t = 0$, $N_2 = N_2^0$

$$C = -\frac{\lambda_1}{\lambda_2 - \lambda_1} N_1^0 + N_2^0$$

Substituting C in Eq. (3.17), we get

$$N_2 e^{\lambda_2 t} = \frac{\lambda_1}{\lambda_2 - \lambda_1} N_1^0 e^{(\lambda_2 - \lambda_1)t} - \frac{\lambda_1}{\lambda_2 - \lambda_1} N_1^0 + N_2^0$$

Dividing by $e^{\lambda_2 t}$, we get

$$N_2 = \frac{\lambda_1}{\lambda_2 - \lambda_1} N_1^0 e^{-\lambda_1 t} - \frac{\lambda_1}{\lambda_2 - \lambda_1} N_1^0 e^{-\lambda_2 t} + N_2^0 e^{-\lambda_2 t}$$

$$\boxed{N_2 = \frac{\lambda_1}{\lambda_2 - \lambda_1} N_1^0 (e^{-\lambda_1 t} - e^{-\lambda_2 t}) + N_2^0 e^{-\lambda_2 t}} \tag{3.18}$$

Substituting for N_2 in Eq. (3.13) and integrating with respect to t, we get

$$N_3 = \lambda_2 \int \left\{ \frac{\lambda_1}{\lambda_2 - \lambda_1} N_1^0 [e^{-\lambda_1 t} - e^{-\lambda_2 t}] + N_2^0 e^{-\lambda_2 t} \right\} dt$$

$$N_3 = \frac{\lambda_1 \lambda_2}{\lambda_2 - \lambda_1} N_1^0 \left[\frac{e^{-\lambda_1 t}}{-\lambda_1} - \frac{e^{-\lambda_2 t}}{-\lambda_2} \right] + \lambda_2 N_2^0 \frac{e^{-\lambda_2 t}}{-\lambda_2} + D$$

where D is a constant of integration. Simplifying, we get

$$N_3 = \frac{\lambda_1 \lambda_2}{\lambda_2 - \lambda_1} N_1^0 \left[\frac{e^{-\lambda_2 t}}{\lambda_2} - \frac{e^{-\lambda_1 t}}{\lambda_1} \right] - N_2^0 e^{-\lambda_2 t} + D \tag{3.19}$$

At $t = 0$ $N_3 = N_3^0$, we get

$$D = N_1^0 + N_2^0 + N_3^0$$

Substituting D in Eq. (3.19), we get

$$N_3 = \frac{\lambda_1 \lambda_2}{\lambda_2 - \lambda_1} N_1^0 \left[\frac{e^{-\lambda_2 t}}{\lambda_2} - \frac{e^{-\lambda_1 t}}{\lambda_1} \right] - N_2^0 e^{-\lambda_2 t} + N_1^0 + N_2^0 + N_3^0$$

$$\boxed{N_3 = N_3^0 + N_2^0 (1 - e^{-\lambda_2 t}) + N_1^0 \left(1 + \frac{\lambda_1}{\lambda_2 - \lambda_1} e^{-\lambda_2 t} - \frac{\lambda_2}{\lambda_2 - \lambda_1} e^{-\lambda_1 t} \right)} \tag{3.20}$$

Number of nuclei of parent N_1, daughter N_2 and granddaughter N_3 at any given instant of time t can be calculated using Eqs. (3.14), (3.18) and (3.20) respectively.

Special Case

When only parent nuclei are present at $t = 0$, i.e. $N_1 = N_1^0$ and $N_2^0 = N_3^0 = 0$. Substituting these values, Eqs. (3.14), (3.18) and (3.20) for $t > 0$ become

$$\boxed{N_1 = N_1^0 e^{-\lambda_1 t}} \tag{3.21}$$

$$\boxed{N_2 = \frac{\lambda_1}{\lambda_2 - \lambda_1} N_1^0 (e^{-\lambda_1 t} - e^{-\lambda_2 t})} \tag{3.22}$$

and

$$\boxed{N_3 = N_1^0 \left(1 + \frac{\lambda_1}{\lambda_2 - \lambda_1} e^{-\lambda_2 t} - \frac{\lambda_2}{\lambda_2 - \lambda_1} e^{-\lambda_1 t}\right)} \tag{3.23}$$

With the help of Eqs. (3.21), (3.22) and (3.23), we can compute the number of nuclei of parent, daughter and granddaughter (assuming granddaughter to be stable) at any instant of time. In Figure 3.7, we have plotted the number of nuclei of parent, daughter and granddaughter versus t assuming initially at $t = 0$, only parent nuclei exist.

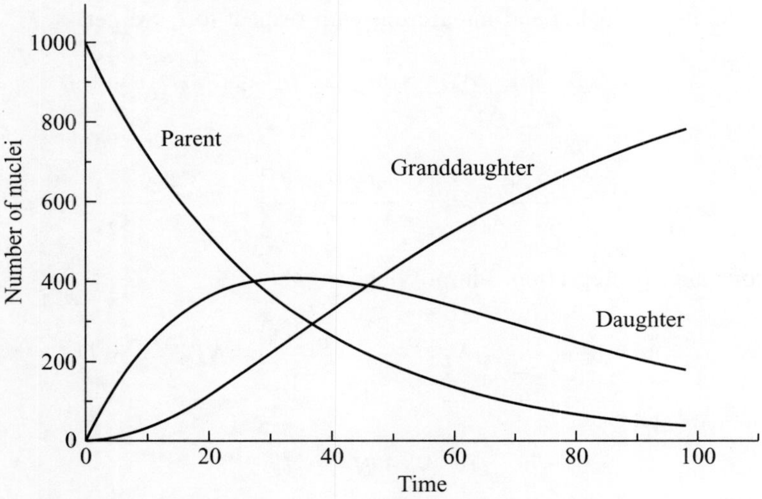

Figure 3.7 Decay of parent, decay and growth of daughter and growth of granddaughter nuclei.

3.4.1 Radioactive Equilibrium

Using Eqs. (3.21), (3.22) and (3.23), one can define two types of equilibria depending upon the decay constants. First case is when $\lambda_1 << \lambda_2$, i.e. when decay constant of the parent nucleus is very small compared to the daughter nucleus. Second is when $\lambda_1 < \lambda_2$. Other possible cases like $\lambda_1 \approx \lambda_2$, $\lambda_1 > \lambda_2$ and $\lambda_1 >> \lambda_2$ are also there. The first two cases lead to secular and transient equilibrium respectively. While the other possible cases do not approach equilibrium. We discuss these cases in the following sections.

Permanent or Secular Equilibrium

For $\lambda_1 << \lambda_2$ or $t_1 >> t_2$, where t_1 and t_2 are half-lives of parent and daughter nuclei.

$$\lambda_2 - \lambda_1 \approx \lambda_2$$

and $$e^{-\lambda_1 t} \approx 1$$

Using these values, Eq. (3.22) becomes

$$N_2 = \frac{\lambda_1}{\lambda_2} N_1^0 (1 - e^{-\lambda_2 t}) \tag{3.24}$$

If t is large compared to half-life of the daughter, i.e. $t >> t_2$, the exponential factor in Eq. (3.24) becomes negligible compared to 1. In this case Eq. (3.24) reduces to

$$N_2 = \frac{\lambda_1}{\lambda_2} N_1^0 \tag{3.25}$$

Since the half-life of the parent nuclei is much larger than the time t, so N_1^0 almost remains constant, i.e. $N_1^0 = N_1$, therefore, Eq. (3.25) becomes

$$\boxed{N_2 \lambda_2 = N_1 \lambda_1 = \text{constant}} \tag{3.26}$$

From this equation it is clear that the amount of daughter N_2 is always constant because N_1, λ_1 and λ_2 are constants. Since $\frac{N_2}{N_1}$ is constant, the daughter is said to be in permanent or secular or long-term equilibrium with the parent. Equation (3.26) can also be written in the following form using Eq. (3.8)

$$\boxed{\frac{N_2}{N_1} = \frac{\lambda_1}{\lambda_2} = \frac{t_2}{t_1}} \tag{3.27}$$

This equation shows that the number of parent nuclei is inversely proportional to its decay constant or directly proportional to its half-life. Similar argument is true for daughter nucleus. Similarly, if we consider that the daughter is unstable, and it decays into granddaughter, then it can be shown that

$$\boxed{N_1 \lambda_1 = N_2 \lambda_2 = N_3 \lambda_3 = \text{constant}} \tag{3.28}$$

From this equation it is clear that the product of number of parent nuclei and its decay constant is always constant. Some examples of secular equilibrium are

$$^{90}_{38}\text{Sr} \xrightarrow{\beta^-} {}^{90}_{39}\text{Y} \xrightarrow{\beta^-} {}^{90}_{40}\text{Zr}$$

$$^{226}_{88}\text{Ra} \xrightarrow{\alpha} {}^{222}_{86}\text{Rn} \xrightarrow{\alpha} {}^{218}_{84}\text{Po}$$

In Figure 3.8 we have plotted time versus number of atoms of the parent, daughter and granddaughter nuclei for ^{90}Sr decay. In Figure 3.9 we have plotted the activities of parent, daughter and total activity, i.e. parent and daughter activity versus time.

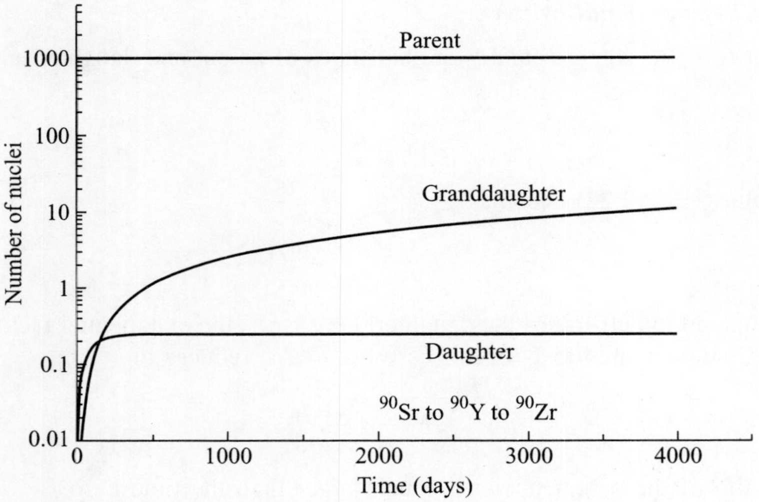

Figure 3.8 A plot of time versus number of atoms of the parent, daughter and granddaughter for ^{90}Sr decay.

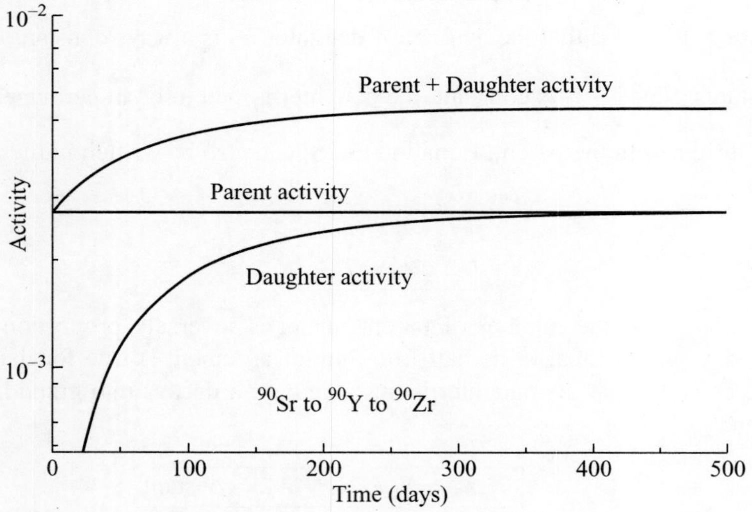

Figure 3.9 A plot of activities of parent, daughter and parent + daughter versus time for ^{90}Sr decay.

Transient Equilibrium

In case $\lambda_1 < \lambda_2$ or $t_1 > t_2$, the number of daughter nuclei must be calculated using Eq. (3.22) for small t. For this case where t becomes sufficiently large, the factor $e^{-\lambda_2 t}$ becomes very small compared to $e^{-\lambda_1 t}$. After this time a state of equilibrium known as transient equilibrium is reached. Neglecting $e^{-\lambda_2 t}$ in comparison to $e^{-\lambda_1 t}$ Eq. (3.22) becomes

$$N_2 = \frac{\lambda_1}{\lambda_2 - \lambda_1} N_1^0 e^{-\lambda_1 t} \tag{3.29}$$

Rearranging Eq. (3.29) becomes

$$N_2 = N_1 \frac{\lambda_1}{\lambda_2 - \lambda_1}$$

or

$$\boxed{\frac{N_2}{N_1} = \frac{\lambda_1}{\lambda_2 - \lambda_1} = \text{constant}} \qquad (3.30)$$

Some examples of this case are:

$$^{140}\text{Ba} \xrightarrow{\beta^-} {}^{140}\text{La} \xrightarrow{\beta^-} {}^{140}\text{Ce}$$

$$^{132}\text{Te} \xrightarrow{\beta^-} {}^{132}\text{I} \xrightarrow{\beta^-} {}^{132}\text{Xe}$$

In Figure 3.10 we have plotted time versus number of atoms of the parent, daughter and granddaughter nuclei for ^{140}Ba decay. In Figure 3.11 we have plotted the activities of parent, daughter and total activity, i.e. parent and daughter activity versus time.

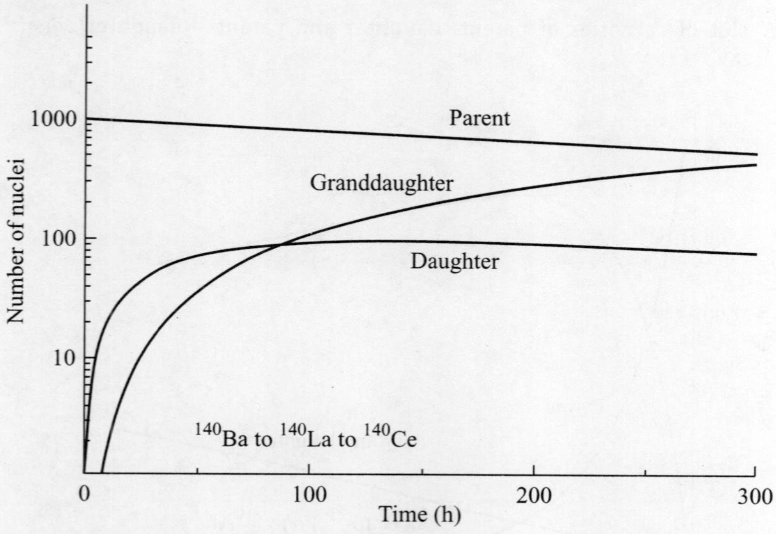

Figure 3.10 Transient equilibrium for ^{140}Ba to ^{140}La. Initially, we start with ^{140}Ba.

Nonequilibrium Case

In case the above conditions of secular and transient equilibria are not met, then we do not approach equilibrium. Some examples of these cases are

$$^{210}_{83}\text{Bi} \xrightarrow{\beta^-} {}^{210}_{84}\text{Po} \xrightarrow{\beta^-} {}^{210}_{85}\text{At}$$

This decay is shown in Figure 3.12.

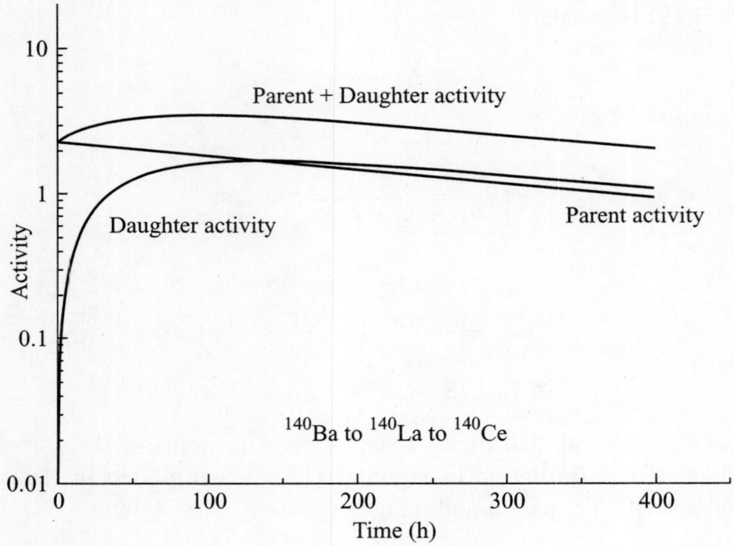

Figure 3.11 A plot of activities of parent, daughter and parent + daughter versus time for ^{140}Ba decay.

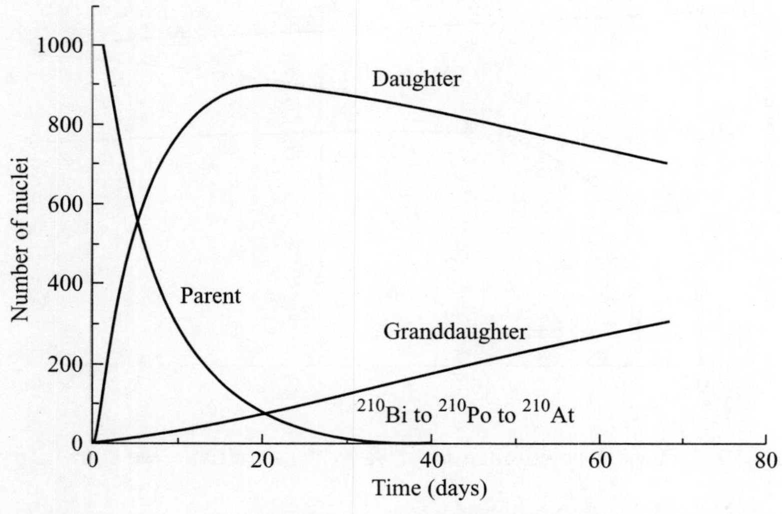

Figure 3.12 A plot of number of nuclei of parent, daughter and granddaughter versus time depicting nonequilibrium for ^{210}Bi decay.

3.5 ALPHA EMISSION

Alpha emission is a process in which parent nucleus spontaneously disintegrates into a daughter nucleus and an α-particle.

Rutherford showed that α-particles are ${}^{4}_{2}$He nuclei. Therefore, in this process parent nucleus loses 4 nucleons, its mass decreases by 4 amu and charge decreases by 2. The equation for this decay is

$$\boxed{{}^{A}_{Z}X \rightarrow {}^{A-4}_{Z-2}Y + {}^{4}_{2}\text{He} + Q} \tag{3.31}$$

where Q is the total energy released in this decay process and is called the *disintegration energy*. In order to evaluate Q let us assume that M_p is the mass of parent nucleus M_d is the mass of daughter nucleus and m_α is the mass of α-particle. The parent nucleus is at rest before the emission of α-particle, i.e. its kinetic energy is zero. When α-particle is emitted with velocity v_α, to conserve momentum, the daughter nucleus must recoil with velocity V_d. Therefore, by applying law of conservation of momentum

$$\text{Initial momentum} = \text{Final momentum}$$

$$0 = M_d V_d - m_\alpha v_\alpha$$

Negative sign is taken as daughter nucleus and α-particles move in opposite directions.

or

$$V_d = \frac{m_\alpha v_\alpha}{M_d}$$

Energy evolved or Q-value of the decay is defined as

$$Q = \text{Final kinetic energy} - \text{Initial kinetic energy}$$

$$= \left(\frac{1}{2} M_d V_d^2 + \frac{1}{2} m_\alpha v_\alpha^2\right) - 0$$

Therefore,

$$Q = \frac{1}{2} M_d V_d^2 + \frac{1}{2} m_\alpha v_\alpha^2$$

Substituting for V_d

$$Q = \frac{1}{2} M_d \left(\frac{m_\alpha v_\alpha}{M_d}\right)^2 + \frac{1}{2} m_\alpha v_\alpha^2$$

or

$$= \frac{1}{2} m_\alpha v_\alpha^2 \left(\frac{m_\alpha}{M_d} + 1\right)$$

If K_α is the kinetic neergy of α-particles or

$$K_\alpha = \frac{1}{2} m_\alpha v_\alpha^2$$

Therefore,

$$Q = K_\alpha \left(\frac{m_\alpha}{M_d} + 1\right)$$

$$\frac{m_\alpha}{M_d} \approx \frac{4}{A-4}$$

Therefore,

$$Q = K_\alpha \frac{A}{A-4}$$

or

$$\boxed{K_\alpha = \frac{Q(A-4)}{A}} \tag{3.32}$$

Some common examples of α-emitters are:

$$^{241}_{94}\text{Am} \rightarrow {}^{237}_{92}\text{U} + {}^{4}_{2}\text{He}$$

$$^{226}_{88}\text{Ra} \rightarrow {}^{222}_{86}\text{Rn} + {}^{4}_{2}\text{He}$$

3.5.1 Properties of Alpha Particles

1. Alpha particles are ${}^{4}_{2}$He nuclei with 2 units of positive charge and mass of 4 amu.
2. Because they are charged particles so they can be deflected by electric and magnetic fields.
3. They ionize the medium through which they pass. Their ionizing power is much higher than β- and γ-rays.
4. Because of the high-ionizing power, they can be easily absorbed by a few centimetres of air or fraction of millimetre thick aluminium.
5. Naturally occurring α-emitters emit α-particles with energy in the range of 5 MeV to 10 MeV.
6. Alpha particles when fall on certain materials like ZnS, barium platinocynide, etc. emit flashes of light called scintillations. This property of scintillation was initially used to detect α-particles.
7. Long exposures to α-emitters produce harmful effects on the human body.

3.5.2 Alpha Spectrum

There are only two products in the decay of parent nucleus—daughter nucleus and α-particle. Therefore, energy evolved in α-decay or Q-value is shared by these two particles in the inverse proportion to their masses. Thus, α-particle carries more energy as compared to the daughter nucleus. These values of the energy of α-particle and daughter nucleus are unique. As an example, suppose disintegration energy in the decay of a nucleus is 5.085 MeV, the energy carried by the α-particle is 5.000 MeV, while the daughter nucleus carries 0.085 MeV energy. All the α-particles emitted in this decay have the same energy. If we plot a curve between the energy of α-particle and their number, it would be a straight line as shown in Figure 3.13. This curve is known as *energy spectrum* or *simply spectrum*.

The energy spectrum of α-particles is also known as line spectrum or discrete spectrum. In actual experiment instead of straight line we observe a broad peak with a maxima at 5.000 MeV (in the example taken above). This broadening is explained on the fact that detection of α-particles in a detector is a statistical process. This means that there is a finite probability that all the α-particles may not produce same number of electron–ion pairs in the detector medium. It is due to this variation in the number of electron–ion pair produced that the straight line gets broadened.

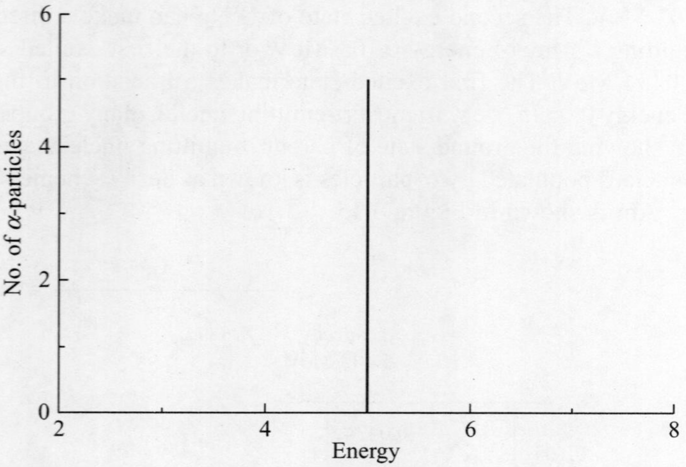

Figure 3.13 Alpha spectrum, a plot between number of α-particles versus energy of α-particles as recorded with an ideal detector.

The actual experimental α-particle spectrum in the decay of ^{241}Am as recorded by a semiconductor detector is shown in Figure 3.14. In this spectrum, there are two peaks, first at 5.48 MeV and second at 5.44 MeV. These two peaks correspond to two groups of α-particles emitted by $^{241}_{95}$Am source. The α-decay of ^{241}Am is represented as

$$^{241}_{95}\text{Am} \rightarrow \, ^{237}_{93}\text{Np} + \, ^{4}_{2}\text{He}$$

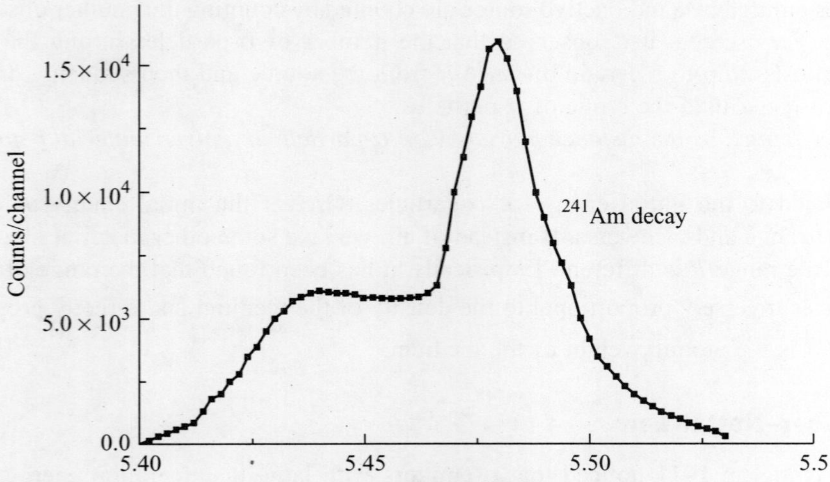

Figure 3.14 Actual alpha spectrum as observed in the decay of ^{241}Am. The two peaks at 5.44 and 5.48 MeV correspond to two groups of α-particles emitted by ^{241}Am.

When 5.48 MeV α-particle is emitted, ^{237}Np is left in the first excited state having energy 0.0595 MeV. When 5.44 MeV α-particle is emitted, ^{237}Np is left in the second excited state having energy 0.103 MeV. The second excited state of ^{237}Np can make a direct transition to the ground state by emitting a γ-ray of energy 0.103 MeV, or to the first excited state by emitting a γ-ray of energy 0.0434 MeV. The first excited state makes a transition to the ground state by emitting a γ-ray of energy 0.0595 MeV. In most α-emitting nuclei, many groups of α-particles are emitted. A diagram showing the ground state of parent α-emitting nucleus and different excited states of daughter nucleus populated by α-particles is known as decay scheme of the nucleus. The decay scheme of ^{241}Am is shown in Figure 3.15.

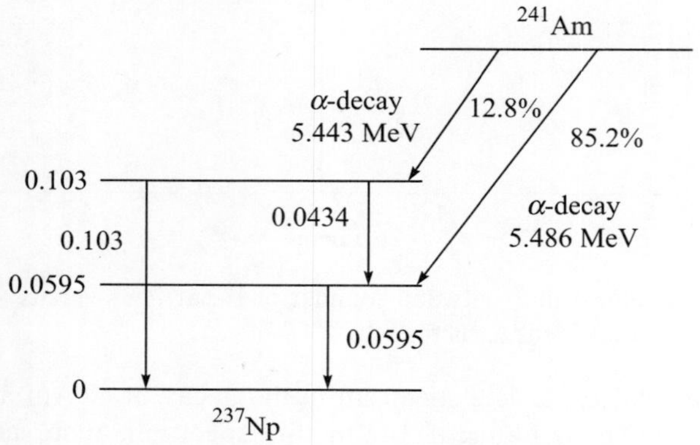

Figure 3.15 Decay scheme of ^{241}Am.

3.5.3 Range of Alpha Particles

If α-particles emitted by a radioactive source are counted by counting the number of scintillations produced on ZnS screen, it is observed that the number of α-particles hitting the screen are practically constant up to a certain distance R from the source and then suddenly drop to zero. This distance R is called the range of α-particles.

Range is defined as the distance moved by an α-particle in a given material before it comes to rest.

It is related to the initial energy of α-particles. Greater the initial energy of α-particles, greater is the range and vice versa. If instead of air, we take some other material like aluminium or iron, etc. the range R is different. Empirically, it has been found that the range of α-particles in a medium is inversely proportional to the density of the medium and directly proportional to $\sqrt{A}$, where A is the atomic weight of the medium.

3.5.4 Geiger–Nuttal Law

Geiger and Nuttal in 1911 noticed that α-emitters with large disintegration energy have short half-lives and vice versa. They observed that a change of factor 2 in energy scale changes the half-life by a factor of 10^{24}. As an example ^{232}Th has a half-life of 1.4×10^{10} years (4.4×10^{17} s) and energy of α-particles emitted by it is 4.08 MeV. ^{218}Th has a half-life of 10^{-7} s while energy of the emitted α-particle is 9.85 MeV. In other words, disintegration constant varies extremely rapidly with small changes in energy. The longest-lived α-emitters emit least energetic α-particles and vice versa.

These empirical facts were correlated by Geiger–Nuttal in 1911 when $\log(\lambda)$ was plotted against $\log(R)$. Plots for members of the three naturally occurring radioactive series (discussed in Section 3.3) were found to fall along three parallel straight lines, one line for each series as shown in Figure 3.16. These straight lines are expressed by Geiger–Nuttal law which is as under

$$\boxed{\log(\lambda) = A \times \log(R) + B} \tag{3.33}$$

where A and B are constants. Value of A is nearly the same for all the three lines but B has different values.

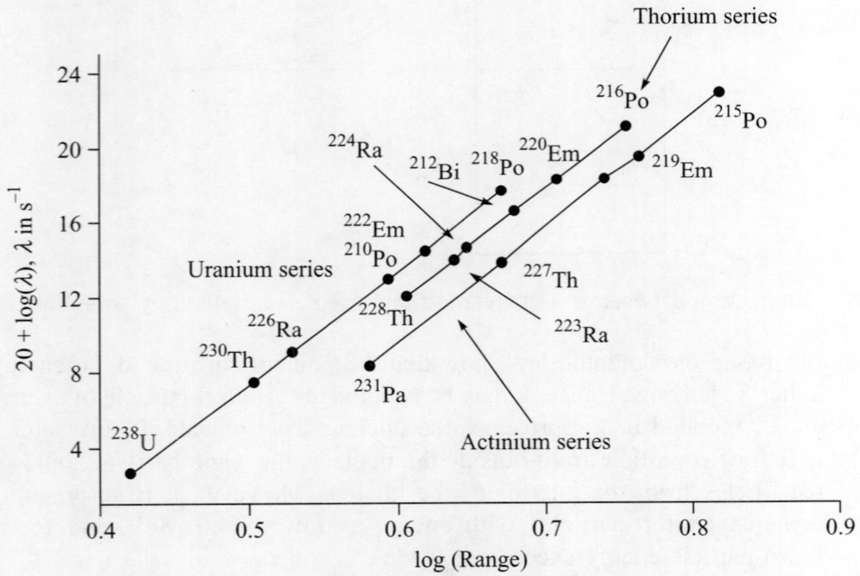

Figure 3.16 Plot of $\log(\lambda)$ versus log (Range) for uranium, thorium and actinium series.

In terms of energy, Geiger–Nuttal law is

$$\boxed{\log(\lambda) = A' \times \log(E) + B'} \tag{3.34}$$

The constants A and B of Eq. (3.33) become A' and B' when the range is converted to energy.

3.5.5 Gamow Theory of Alpha Decay

Let us discuss the scattering of α-particles from a nucleus. In scattering experiments as α-particle from infinity comes in close proximity of scattering nucleus, the Coulomb repulsion between the two increases. When the nucleus and α-particle are just touching each other, the repulsion is maximum. If the kinetic energy of α-particle is larger than the Coulomb repulsion, the α-particle will overcome the repulsion and will enter the nucleus. At this point, suddenly the α-particle enters the region of nuclear force, where it is attracted by the nucleons in the nucleus. This interaction between the nucleus and an α-particle can be represented pictorially by a potential energy curve as shown in Figure 3.17. In this figure, the rising portion from A to B indicates increasing electrostatic repulsion of an α-particle as it approaches the nucleus. The sharp fall from B to C is essentially the region within the nucleus, implies attraction of the α-particle by the nucleons present in the nucleus.

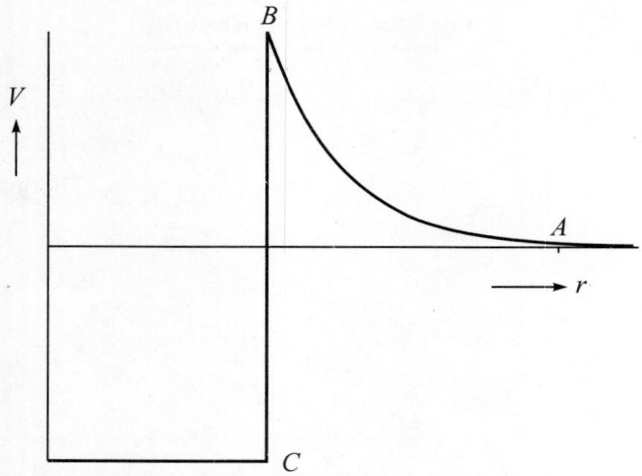

Figure 3.17 Interaction between α-particle and the nucleus represented by potential energy curve.

Calculations based on Coulomb law show that the point B corresponds to energy of about 25 MeV for higher Z elements. Hence, it may be concluded that an α-particle of energy less than 25 MeV should be repelled if it approaches the nucleus from outside. If this potential barrier prevents the access of α-particle from outside the nucleus, the same barrier should prevent the emission of α-particles from the interior of the nucleus. However, it is surprising that some radioactive elements emit α-particles with energy as low as 4.0 MeV, and for no natural radioisotope the α-particle energy exceeds 11.0 MeV.

Classical mechanics has no explanation to α-decay process. In 1928, Gurney and Condon of U.S.A., and Gamow of Russia independently developed a theory known as Gamow theory of α-decay to tackle the problem of α-decay using quantum mechanics.

According to classical mechanics an α-particle having energy 5–10 MeV inside the nucleus cannot cross the potential barrier and escape, if its energy is less than the maximum height of the Coulomb barrier which is about 25 MeV, but quantum mechanics predicts that there is a definite but small probability that such a particle from the interior can be found outside the nucleus. In other words, there is a definite probability that the α-particle can escape from the nucleus even when its energy is less than that of the top of the potential barrier. Thus, quantum mechanically, there is finite probability that α-particle leaks through the potential barrier, as if there are holes in the potential barrier. This effect is called *quantum mechanical tunnelling* or *simply tunnelling*.

In order to explain the phenomenon of α-decay, Gamow made the following assumptions:

- An α-particle may exist as an α-particle within a heavy nucleus.
- Such an α-particle is in constant motion and is held in the nucleus by a potential barrier.
- There is a small but definite likelihood that the α-particle may tunnel through the barrier (despite the height of the barrier) each time it collides with the barrier.

Based on these assumptions, Gamow derived the relation for the probability that α-particle may tunnel through the potential barrier as under.

In order to make the calculations simple, we consider only one-dimensional rectangular potential barrier of width a and height V, which is greater than the kinetic energy of α-particle. This rectangular potential barrier is shown in Figure 3.18. There are three regions of interest marked as I, II, and III. In region I, incident and reflected α-particles or waves associated with α-particles are present. In region II, both forward moving transmitted waves and reflected waves from the other side of the barrier are present. In region III, only forward moving transmitted waves are present. Mathematically, the rectangular potential is defined as

$$\begin{aligned} V(x) &= 0 \qquad && \text{for } x < 0 \\ V(x) &= V && \text{for } 0 \le x \le a \\ V(x) &= 0 && \text{for } x > a \end{aligned}$$

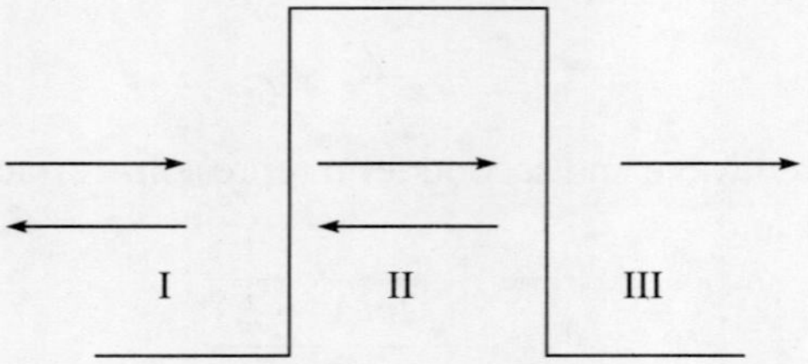

Figure 3.18 The rectangular potential barrier $V(x)$ versus x.

One dimensional time independent Schrödinger equation is given as

$$\frac{d^2\psi}{dx^2} + \frac{2m}{\hbar^2}(E - V)\psi = 0$$

In regions I and III, $V = 0$, so one-dimensional time-independent Schrödinger equation becomes

$$\frac{d^2\psi}{dx^2} + \frac{2m}{\hbar^2}E\psi = 0 \tag{3.35}$$

where $m = \dfrac{m_\alpha m_d}{m_\alpha + m_d}$ is the reduced mass of the α-particle and daughter nucleus; $\hbar = \dfrac{h}{2\pi}$, h being the Planck constant. As both incident and reflected α-waves are present in region I, so the solution of Eq. (3.35) in region I is

$$\psi_I = Ae^{ik_1x} + Be^{-ik_1x} \tag{3.36}$$

where

$$k_1 = \frac{\sqrt{2mE}}{\hbar} = \frac{p}{\hbar} = \frac{2\pi}{\lambda} \tag{3.37}$$

Here p is the momentum and λ is the de Broglie wavelength associated with α-particles of energy E. A and B are constants to be evaluated later on. The first term in the solution (Eq. 3.36) represents the incident wave and the second term represents the reflected wave.

In region III, only the forward moving transmitted wave is present, hence the solution of Eq. (3.35) is

$$\psi_{III} = Ce^{ik_1x} \tag{3.38}$$

Here C is constant.

The one-dimensional time-independent Schrödinger equation for the region II, where $E < V$ is

$$\frac{d^2\psi}{dx^2} + \frac{2m}{\hbar^2}(E - V)\psi = 0$$

or

$$\frac{d^2\psi}{dx^2} - \frac{2m}{\hbar^2}(V - E)\psi = 0 \tag{3.39}$$

The solution of this equation in region II is

$$\psi_{II} = Fe^{-k_2x} + Ge^{k_2x} \tag{3.40}$$

First term represents reflected wave and second term represents refracted wave. Here F and G are constants to be determined and

$$k_2 = \frac{\sqrt{2m(V - E)}}{\hbar} \tag{3.41}$$

In order to evaluate constants A, B, C, F and G, we apply certain conditions known as boundary conditions. These conditions are:

1. Wave functions of the two regions must be same or match at the boundary of the two regions, i.e. at $x = 0$ and at $x = a$.
2. First derivative of wave functions of the two regions must be same or match at the boundary of the two regions, i.e. at $x = 0$ and at $x = a$.

Combining these two conditions, we get a new set of conditions i.e.,

Condition 1

$$\psi_I = \psi_{II} \text{ at } x = 0 \qquad \text{and} \qquad \frac{d\psi_I}{dx} = \frac{d\psi_{II}}{dx} \text{ at } x = 0$$

Condition 2

$$\psi_{II} = \psi_{III} \text{ at } x = a \qquad \text{and} \qquad \frac{d\psi_{II}}{dx} = \frac{d\psi_{III}}{dx} \text{ at } x = a$$

Applying condition 1 to Eqs. (3.36) and (3.40), we get

$$A + B = F + G \tag{3.42}$$

and

$$ik_1(A - B) = -k_2(F - G) \tag{3.43}$$

Similarly, applying condition 2 to Eqs. (3.38) and (3.40), we get

$$Ce^{ik_1a} = Fe^{-k_2a} + Ge^{k_2a} \tag{3.44}$$

and

$$ik_1 Ce^{ik_1 a} = -Fk_2 e^{-k_2 a} + Gk_2 e^{k_2 a} \tag{3.45}$$

Dividing Eq. (3.45) by k_2 and adding to Eq. (3.44), we get

$$Ce^{ik_1 a}\left(\frac{ik_1}{k_2}+1\right) = 2Ge^{k_2 a}$$

or

$$G = \frac{Ce^{ik_1 a}}{2}\left(\frac{ik_1}{k_2}+1\right)e^{-k_2 a} \tag{3.46}$$

Similarly, dividing Eq. (3.45) by k_2 and subtracting from Eq. (3.44), we get

$$F = \frac{Ce^{ik_1 a}}{2}\left(1-\frac{ik_1}{k_2}\right)e^{k_2 a} \tag{3.47}$$

Substituting G and F from Eqs. (3.46) and (3.47) in Eq. (3.42), we get

$$A + B = \frac{Ce^{ik_1 a}}{2}\left(1-\frac{ik_1}{k_2}\right)e^{k_2 a} + \frac{Ce^{ik_1 a}}{2}\left(\frac{ik_1}{k_2}+1\right)e^{-k_2 a}$$

Rearranging this equation, we get

$$A + B = \frac{C}{2}e^{ik_1 a}(e^{-k_2 a} + e^{k_2 a}) + \frac{C}{2}\frac{ik_1}{k_2}e^{ik_1 a}(e^{-k_2 a} - e^{k_2 a}) \tag{3.48}$$

We have

$$2\cos hx = e^x + e^{-x} \text{ and } 2\sin hx = e^x - e^{-x}$$

Therefore,

$$A + B = Ce^{ik_1 a}\left(\cos hk_2 a - \frac{ik_1}{k_2}\sin hk_2 a\right) \tag{3.49}$$

Similarly, substituting G and F from Eqs. (3.46) and (3.47) in Eq. (3.43), we get

$$A - B = \frac{-k_2}{ik_1}\frac{C}{2}e^{ik_1 a}\left[(e^{k_2 a} - e^{-k_2 a}) - \frac{ik_1}{k_2}(e^{k_2 a} + e^{-k_2 a})\right]$$

or

$$A - B = \frac{-k_2}{ik_1}Ce^{ik_1 a}\left[\frac{e^{k_2 a} - e^{-k_2 a}}{2} - \frac{ik_1}{k_2}\frac{e^{k_2 a} + e^{-k_2 a}}{2}\right]$$

or

$$A - B = \frac{-k_2}{ik_1}Ce^{ik_1 a}\left(\sin h(k_2 a) - \frac{ik_1}{k_2}\cos h(k_2 a)\right)$$

or

$$A - B = Ce^{ik_1a}\left(\cos h(k_2a) + \frac{ik_2}{k_1}\sin h(k_2a)\right) \tag{3.50}$$

Adding Eqs. (3.49) and (3.50), we get

$$\frac{A}{C} = e^{ik_1a}\left[\cos h(k_2a) + \frac{i}{2}\left(\frac{k_2}{k_1} - \frac{k_1}{k_2}\right)\sin h(k_2a)\right] \tag{3.51}$$

Subtracting Eqs. (3.49) and (3.50), we get

$$\frac{B}{C} = -\frac{i}{2}e^{ik_1a}\left[\frac{k_1}{k_2} + \frac{k_2}{k_1}\right]\sin h(k_2a) \tag{3.52}$$

Let v_1 be the velocity of incident flux of α-particles in region I and v_3 the velocity of transmitted flux of α-particles in region III. Since regions I and III are force free, so $v_3 = v_1$. Transmission coefficient T is given by

$$T = \frac{CC^*}{AA^*}$$

$$T = \frac{1}{\cos h(k_2a) + \frac{i}{2}\left(\frac{k_2}{k_1} - \frac{k_1}{k_2}\right)\sin h(k_2a)} \times \frac{1}{\cos h(k_2a) - \frac{i}{2}\left(\frac{k_2}{k_1} - \frac{k_1}{k_2}\right)\sin h(k_2a)}$$

or

$$T = \frac{1}{\cos h^2(k_2a) + \frac{1}{4}\left(\frac{k_2}{k_1} - \frac{k_1}{k_2}\right)^2 \sin h^2(k_2a)} \tag{3.53}$$

Now,

$$\cos h^2(x) - \sin h^2(x) = 1$$

or

$$\cos h^2(x) = 1 + \sin h^2(x)$$

Therefore,

$$T = \frac{1}{1 + \frac{1}{4}\frac{(k_2^2 + k_1^2)^2}{k_1^2 k_2^2}\sin h^2(k_2a)} \tag{3.54}$$

From Eqs. (3.37) and (3.41), we have

$$k_1 = \frac{\sqrt{2mE}}{\hbar}$$

$$k_2 = \frac{\sqrt{2m(V - E)}}{\hbar}$$

$$k_1^2 + k_2^2 = \frac{2mV}{\hbar^2} \quad \text{and} \quad k_1k_2 = \frac{2m}{\hbar^2}\sqrt{E(V - E)}$$

Therefore, Eq. (3.54) becomes

$$T = \frac{1}{1+\frac{1}{4}\frac{V^2}{E(V-E)}\sinh^2(k_2a)}$$

$$T = \left[1+\frac{1}{4}\frac{V^2}{E(V-E)}\sinh^2(k_2a)\right]^{-1} \tag{3.55}$$

Since $k_2a >> 1$ and $\sinh^2(x) = \frac{1}{4}(e^{2x} + e^{-2x} - 2)$

So, $\sinh^2(k_2a) = \frac{1}{4}e^{2k_2a}$, where we have neglected $e^{-2k_2a} - 2$, since this term is much smaller than e^{2k_2a}. Thus, Eq. (3.55) becomes

$$T = \left[1+\frac{1}{16}\frac{V^2}{E(V-E)}e^{2k_2a}\right]^{-1}$$

Now, $V > E$ and $e^{2k_2a} >> 1$, so we can neglect 1 in comparison to the second term of the above equation. So, we have

$$T = \left[\frac{1}{16}\frac{V^2}{E(V-E)}e^{2k_2a}\right]^{-1} = 16\frac{E(V-E)}{V^2}e^{-2k_2a} \tag{3.56}$$

When $2k_2a >>1$ the exponential factor in the above equation becomes extremely small, while the factor before the exponential is of the order of 1. For the order of magnitude calculations

$$T = e^{-2k_2a} \tag{3.57}$$

or

$$\log_e(T) = -2k_2a \tag{3.58}$$

Equation (3.57) represents the fraction of α-particles that will penetrate the barrier of width a and of height $V > E$. Inside the nucleus the α-particle is faced with a barrier of varying height. Because of the varying height Eq. (3.58) can be expressed as an integral, i.e.

$$\log_e(T) = -2\int_{r_0}^{r} k_2x\,dx$$

where r_0 is the distance from the centre of the nucleus for which potential is constant. r is the distance from the centre of the nucleus where the potential energy becomes equal to E. For $x > r$, $E > V$, i.e. α-particle can escape permanently from the nucleus.
To calculate the integral, we define the potential energy $V(x)$ as

$$V(x) = \frac{2(Z-2)e^2}{4\pi\varepsilon_0 x}$$

where $(Z-2)e$ is the charge of the daughter nucleus and $2e$ is the charge of α-particle.

Now,

$$k_2 = \frac{\sqrt{2m(V-E)}}{\hbar} = \left(\frac{2m}{\hbar^2}\right)^{1/2}\left(\frac{2(Z-2)e^2}{4\pi\varepsilon_0 x} - E\right)^{1/2}$$

Since $V = E$ at $x = r$, so

$$E = \frac{2(Z-2)e^2}{4\pi\varepsilon_0 r} \tag{3.59}$$

Thus,

$$k_2 = \left(\frac{2mE}{\hbar^2}\right)^{1/2}\left(\frac{r}{x} - 1\right)^{1/2}$$

Now,

$$\log_e(T) = -2\int_{r_0}^{r}\left(\frac{2mE}{\hbar^2}\right)^{1/2}\left(\frac{r}{x} - 1\right)^{1/2} dx$$

Solving this integral, we get

$$\log_e(T) = -2\left(\frac{2mE}{\hbar^2}\right)^{1/2} r\left\{\cos^{-1}\left(\frac{r_0}{r}\right)^{1/2} - \left(\frac{r_0}{r}\right)^{1/2}\left(1 - \frac{r_0}{r}\right)^{1/2}\right\}$$

The potential barrier is relatively wide, i.e. $r >> r_0$. For this approximation, we can write

$$\cos^{-1}\left(\frac{r_0}{r}\right)^{1/2} = \frac{\pi}{2} - \left(\frac{r_0}{r}\right)^{1/2}$$

and

$$\left(1 - \frac{r_0}{r}\right)^{1/2} = 1$$

Thus,

$$\log_e(T) = -2\left(\frac{2mE}{\hbar^2}\right)^{1/2} r\left\{\frac{\pi}{2} - 2\left(\frac{r_0}{2}\right)^{1/2}\right\} \tag{3.60}$$

We have from Eq. (3.59)

$$r = \frac{2(Z-2)e^2}{4\pi\varepsilon_0 E} \tag{3.61}$$

Substituting r from Eq. (3.61) in Eq. (3.60), we get

$$\log_e(T) = \frac{4e}{\hbar}\left(\frac{m}{\pi\varepsilon_0}\right)^{1/2}(Z-2)^{1/2} r_0^{1/2} - \frac{e^2}{\hbar\varepsilon_0}\left(\frac{m}{2}\right)^{1/2}(Z-2)E^{-1/2} \tag{3.62}$$

Substituting the values of various constants, we get

$$\log_e(T) = 2.97(Z-2)^{1/2} r_0^{1/2} - 3.95(Z-2)E^{-1/2} \tag{3.63}$$

where r_0 is expressed in fermi. We also find

$$\text{Decay constant } \lambda = vT$$

or

$$\log_e(\lambda) = \log_e(v) + \log_e(T) \tag{3.64}$$

where ν is the number of times the α-particle makes collisions with the potential barrier per unit time. If v is the velocity of α-particle in the nucleus, then the collision frequency is approximately given by

$$\nu = \frac{v}{2r_0} \tag{3.65}$$

Combining Eqs. (3.63), (3.64) and (3.65), we get

$$\log_e(\lambda) = \log_e\left(\frac{v}{2r_0}\right) + \log_e[2.97(Z-2)^{1/2}r_0^{1/2} - 3.95(Z-2)E^{-1/2}] \tag{3.66}$$

which is the required equation of Gamow theory.

3.6 BETA DECAY

We have already seen (Figure 3.1) one of the modes of radioactive decay is by β-emission. In this mode three types of decay processes are encountered.

1. The nucleus decays by an electron emission. This process is known as β^--emission, e.g. ${}^{14}_{6}\text{C} \rightarrow {}^{14}_{7}\text{N} + \beta^- + \overline{\nu}$.
2. The nucleus decays by positron emission. This decay is known as β^+-decay, e.g. ${}^{22}_{11}\text{Na} \rightarrow {}^{22}_{10}\text{Ne} + \beta^+ + \nu$.
3. The nucleus captures one of its orbital electrons. This process is called electron capture, e.g. ${}^{54}_{25}\text{Mn} + \beta^- \rightarrow {}^{54}_{24}\text{Cr} + \nu$.

These disintegrations are also called *isobaric transformations* as there is no change in the mass of the nucleus but there is a change of nuclear charge by one unit.

In Chapter 1, we have seen that nucleus consists of protons and neutrons. Electrons cannot exist inside the nucleus. Therefore, in β^--decay a neutron must be converted into a proton at the time of β^--emission. Thus, in this process the charge of the decaying nucleus increases by 1. In β^+-decay a proton is converted into a neutron. Also in electron capture a proton is converted into a neutron. In both these processes, charge of the nucleus decreases by 1.

3.6.1 Conditions for Spontaneous Emission of β^--Particles

β^--decay is represented by the general equation

$$\boxed{{}^{A}_{Z}X \rightarrow {}_{Z+1}^{A}Y + {}^{0}_{-1}e + \overline{\nu}}$$

or

$$ {}^{1}_{0}n \rightarrow {}^{1}_{1}\text{H} + {}^{0}_{-1}e + \overline{\nu}$$

As is well known now, two types of neutrinos exist, one emitted in the decay of nucleon which is denoted as ν_e and the second in the decay of muon, which is denoted as ν_μ. For the sake of simplicity, in this discussion of β-decay, we denote the neutrino emitted in nucleon decay as ν.

Let M_X, M_Y and m_e be the rest masses of the parent nucleus, daughter nucleus and electron respectively. Assume that the parent nucleus is at rest. Let K_Y and K_e be the kinetic energies of the daughter nucleus and the electron respectively. Using the law of conservation of energy, we have

$$M_X c^2 = M_Y c^2 + K_Y + m_e c^2 + K_e$$

In this equation, we have neglected the energy carried by neutrino K_ν. In β-decay, total energy evolved is shared by nucleon, β^--particle and neutrino in all possible proportions. Here we consider a special case, when neutrino carries zero kinetic energy.

or

$$K_Y + K_e = Q = (M_X - M_Y - m_e)c^2 \tag{3.67}$$

For spontaneous decay Q must be positive, i.e.

$$M_X > (M_Y + m_e)$$

Thus, for β^--emission rest mass of the parent nucleus must be greater than the sum of rest masses of the daughter nucleus and the electron.

If $M(Z)$ and $M(Z + 1)$ are the rest masses of the parent and daughter atom, then

$$M(Z) = M_X + m_e Z$$
$$M(Z + 1) = M_Y + m_e(Z + 1)$$

In writing these equations, we have neglected the binding energy of the electrons.

Rewriting the above equations in the following form

$$M_X = M(Z) - m_e Z$$
$$M_Y = M(Z + 1) - m_e(Z + 1)$$

Substituting these values in (3.67) equation, we get

$$Q = [M(Z) - M(Z + 1)]c^2 \tag{3.68}$$

Therefore, β^--decay occurs whenever the mass of the parent atom is greater than that of the daughter atom. In other words, nucleus decays by β^--emission, if it is heavier than another atom with Z greater by one but having same A.

3.6.2 Conditions for Spontaneous Emission of β^+-Particles

β^+-decay is represented by the equation

$$^{A}_{Z}X \rightarrow {}^{A}_{Z-1}Y + {}^{0}_{1}e + \nu$$

or

$$^{1}_{1}\mathrm{H} \rightarrow {}^{1}_{0}n + {}^{0}_{+1}e + \nu$$

As in β^--decay, the disintegration energy Q for β^+-decay is given by the relation

$$Q = K_Y + K_e = (M_X - M_Y - m_e)c^2$$

If $M(Z)$ and $M(Z - 1)$ are the rest masses of the parent and daughter atom, then

$$M(Z) = M_X + m_e Z$$
$$M(Z - 1) = M_Y + m_e(Z - 1)$$

We get

$$Q = [M(Z) - M(Z + 1) - 2m_e]c^2 \tag{3.69}$$

For β^+-decay Q must be positive.

Therefore, β^+-decay of atom occurs only if its mass is greater than the sum of the rest mass of the atom with same A but Z less by one and twice the rest mass of an electron.

3.6.3 Beta-Particle Spectrum

In Section 3.6, we have discussed that β-decay is a three body problem, i.e. there are three products in the decay of parent nucleus—daughter nucleus, β-particle and a neutrino. If we look at the energy carried by β-particle, it can carry any amount of energy starting from zero to Q. If β-particle carries zero energy, the remaining energy (Q) will be shared by daughter nucleus and neutrino. If β-particle carries energy Q, then the other two decay products will carry zero energy. If it carries energy between zero and Q, the remaining energy will be shared by the other two decay products. Therefore, the β-particle spectrum is a continuous spectrum unlike α-particle spectrum, which is a line spectrum or discrete spectrum.

Figures 3.19, 3.20 and 3.21 show the β^--, β^+-spectra from $^{64}_{27}$Cu and β^--spectrum from $^{210}_{83}$Bi. All the three figures indicate that the β's emitted in nuclear decay have a continuous energy distribution and energy of β's ranges from zero to certain maximum value called the end point energy. There are certain characteristics of these spectra.

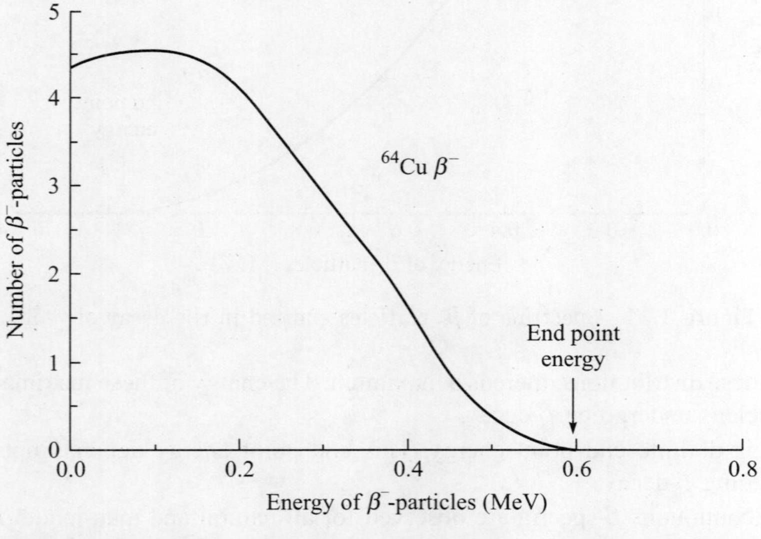

Figure 3.19 Spectrum of β^--particles emitted in the decay of ^{64}Cu.

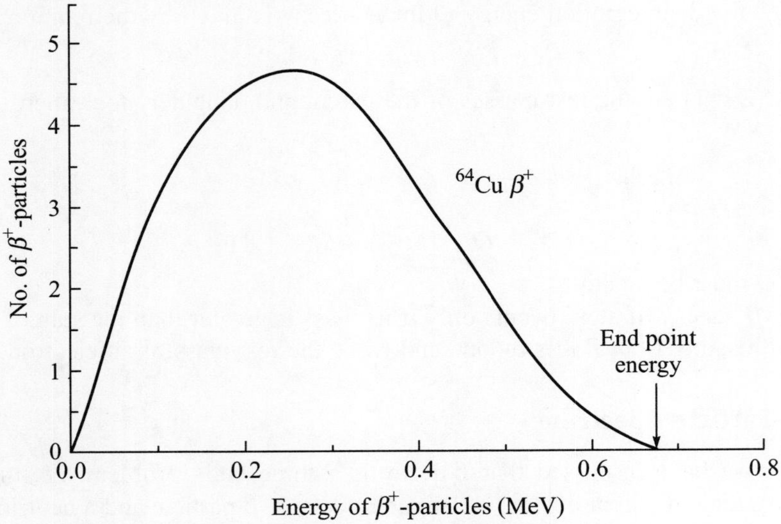

Figure 3.20 Spectrum of β^+-particles emitted in the decay of ^{64}Cu.

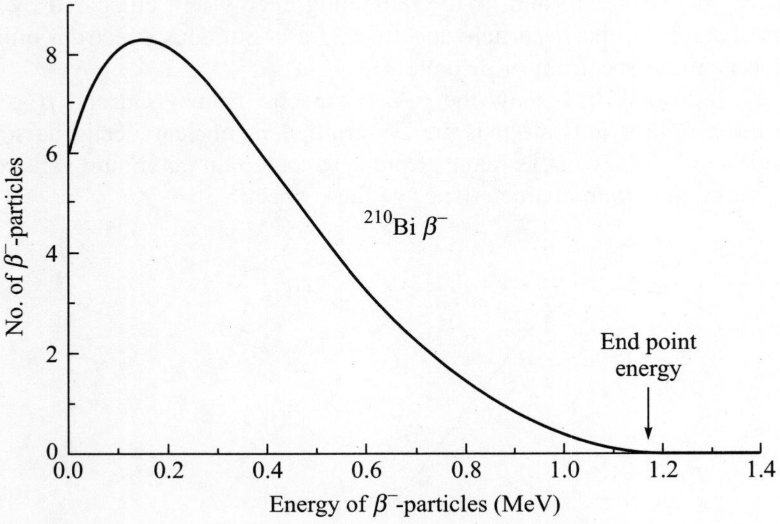

Figure 3.21 Spectrum of β^--particles emitted in the decay of ^{210}Bi.

1. In all these distributions, there is a maximum. The energy of these maxima depends upon the nucleus undergoing β-decay.
2. There is definite end point energy. This end point energy depends upon the nucleus undergoing β-decay.
3. These continuous β-spectra are observed for all natural and man-made β-emitters.

4. Figures 3.19 and 3.20 are the spectra of β^-- and β^+-decay. They have been plotted for the same nucleus (^{64}Cu) and have nearly same end point energy. The difference in both the curves is

- That the β^--spectrum is shifted towards to lower energy side.
- That the β^+-spectrum is shifted towards to higher energy side.

This is expected from the fact that the β^--particles are slowed down and β^+-particles are speeded by the Coulomb force of the nucleus. Theoretically, it has been shown that if Coulomb force effect is not taken into account, then both the curves are identical.

Figure 3.22 shows the β^--spectrum of ^{137}Cs. This spectrum again is continuous but there are some sharp peaks or lines superimposed on the continuous distribution. These lines are due to internal conversion, which can be explained as under. In ^{137}Cs β^--decay does not take place to ground state of ^{137}Ba but to an excited state of ^{137}Ba. ^{137}Ba can return to the ground state either by emitting a γ-ray or it may directly transfer the excess energy to one of the orbital electron, e.g. *K, L, M,* ..., shall etc. electrons and that electron is ejected from the atom. These electrons are known as conversion electrons. The energy of conversion electron is the total excess energy available with the nucleus minus the binding energy of the electron in the respective shell. So, the conversion electrons have unique energy and hence show in the spectrum as peaks superimposed on the continuous distribution.

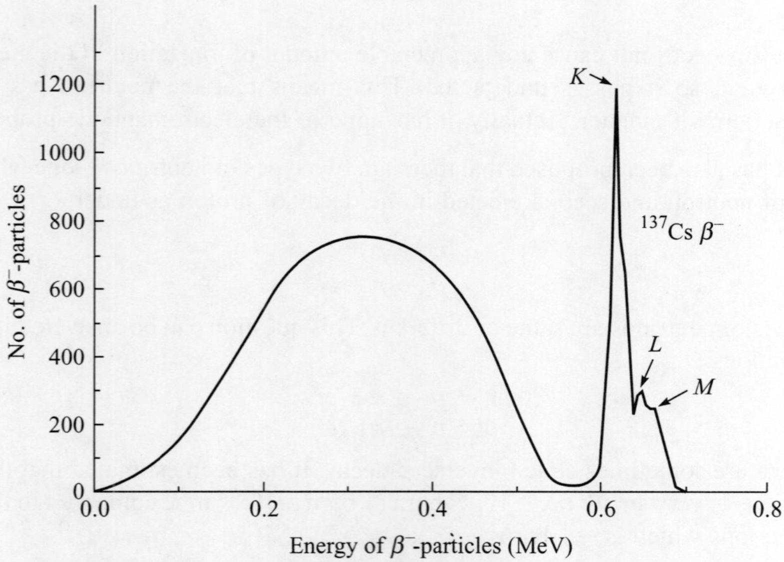

Figure 3.22 Spectrum of β^--particles emitted in the decay of ^{137}Cs.

It is important to mention that atomic number is defined as the number of protons in the nucleus. In internal conversion, the electron is ejected from the electronic orbits, therefore, the atomic number of the nucleus remains the same. This is unlike β-decay, where a proton/neutron inside the nucleus decays to neutron/proton.

3.6.4 Electron Capture

This process is represented by the equation

$$ {}^{A}_{Z}X + {}^{0}_{-1}e \rightarrow {}^{A}_{Z-1}Y + \nu $$

Proceeding in the same way as that β^--decay, the disintegration energy for this process is

$$ Q = [M(Z) - M(Z-1)]c^2 \qquad (3.70) $$

Therefore, for electron capture to occur, the mass of the parent atom must be greater than the mass of a nucleus with same A but with Z less by one.

3.6.5 Neutrino (ν) and Antineutrino ($\overline{\nu}$)

As already discussed in β^--decay, during β-decay a new particle neutrino or antineutrino is emitted. Here are some properties of neutrino[1].

1. The ν must have zero charge as the charge is already conserved without neutrino.
2. Because the maximum energy carried away by electrons is equal to the maximum available energy or end point energy, the neutrino must have zero or almost zero rest mass.
3. The conservation of angular momentum requires the neutrino to have intrinsic spin as $\frac{1}{2}\hbar$.
4. A neutrino does not cause any appreciable amount of ionization in the medium when it is moving, so it passes undetected. This means that the neutrino has a very weak interaction with matter. Actually, it has none of the electromagnetic properties.

Further, it has also been proposed that there are two types of neutrinos—one which is ejected in the decay of neutron and second ejected in the decay of proton as under:

$$ n \rightarrow p + e^- + \overline{\nu} $$
$$ p \rightarrow n + e^+ + \nu $$

These two types of neutrinos are same or different. This question can be answered if we study the following reactions:

$$ \overline{\nu} + p \rightarrow n + e^+ $$
$$ \nu + n \rightarrow p + e^- $$

These reactions are sometimes called inverse β-decay. It has been estimated that the probability of these reactions is very small, i.e. ~ 10^{-46} barn (1 barn = 10^{-28} m^2), compared to the probability of normal reactions which is ~ 1 barn.

3.6.6 Detection of Antineutrino ($\overline{\nu}$)

To show the existence of $\overline{\nu}$, experiments were started by Reines and Cowan in 1952 and completed in 1960. They used nuclear reactor as a source of $\overline{\nu}$. In nuclear reactors fission of

1. Precisely speaking, these neutrinos are called e-neutrinos or ν_e and $\overline{\nu}_e$. For details see Chapter 8 (section 8.5.1)

^{235}U produces neutron-rich fission products, which undergo β^--decay and emit $\overline{\nu}$. The flux of $\overline{\nu}$ was estimated to be ~ 10^{13} per cm^2 per second around the reactor. If the $\overline{\nu}$ are real particles, then they should interact with protons as under:

$$\overline{\nu} + p \rightarrow n + e^+$$

For proton target Reines and Cowan took large quantities of water (~ 0.3 m^3) in which small quantity of $CdCl_2$ was dissolved. When $\overline{\nu}$ interact with proton (from water) a n and β^+ are produced. The positron β^+, will quickly encounter an electron and will annihilates giving two γ- rays of 511 keV each moving in opposite directions to each other to conserve momentum. This annihilation process takes 10^{-14} to 10^{-15} seconds. The neutrons produced collide with mostly hydrogen present in water and keep on losing energy and gradually slow down. This slowing down process takes about 8–10 μs. These slow neutrons are captured by Cd, forming a compound nucleus, which quickly decays emitting a γ-ray of 9.1 MeV. Both these γ-rays of 511 keV and 9.1 MeV are detected by scintillation detector giving light flashes. Electronic circuit picks up these light flashes and gives electrical pulses. Thus, the characteristic signal for the presence of a $\overline{\nu}$ will be an electrical signal due to 511 keV γ-ray followed about 10 μs later by electrical signal due to 9.1 MeV γ-ray. With the experimental set-up developed by Reines and Cowan, they were able to observe about two such events per hour. After many years of ingenious and painstaking series of experiments, Reines and Cowan in 1960 were able to confirm directly the existence of $\overline{\nu}$.

3.6.7 Detection of Neutrino (ν)

Another series of experiments were performed by Davis in 1955 to show that ν and $\overline{\nu}$ are different particles. He demonstrated that $\overline{\nu}$ capture by neutron is not possible using the reaction

$$\overline{\nu} + {}^{37}_{17}\text{Cl} \rightarrow {}^{37}_{18}\text{Ar} + e^- \text{ or } \overline{\nu} + n \rightarrow p + e^-$$

If ν are different from $\overline{\nu}$, then this reaction is not possible.

To prove that this reaction is not possible, Davis placed ~ 1000 gallons of CCl_4 in front of large flux of $\overline{\nu}$. The source of $\overline{\nu}$ in this experiment is again a nuclear reactor. If this capture of $\overline{\nu}$ by n is possible, then ${}^{37}_{18}Ar$ will be formed which is radioactive. By periodically searching for ${}^{37}_{18}Ar$ produced, he found no traces of ${}^{37}_{18}Ar$. He concluded that reaction was not observed indicating that ν and $\overline{\nu}$ are different particles.

Experiments for direct detection of ν were carried out by Davis in 1955 and they lasted for about 10 years. These experiments were done in a mine which was 1500 metres deep. He used neutrinos emitted by sun as sun is a strong source of neutrinos because a large number of proton rich nuclei are formed during fusion process in sun. We know that proton-rich nuclei decay by β^+ and ν. These experiments were based on the reaction

$$\nu + {}^{37}_{17}\text{Cl} \rightarrow {}^{37}_{18}\text{Ar} + e^- \text{ or } \nu + n \rightarrow p + e^-$$

He was able to detect 1 atom of ${}^{37}_{18}Ar$ per day. This ${}^{37}_{18}Ar$ formation confirms the presence of ν.

3.7 GAMMA DECAY

Alpha and beta decays of a radioactive nucleus usually leave the daughter nucleus in an excited state. If the excitation energy available with the daughter nucleus is not sufficient for further particle emission, it loses its excess energy by the following processes:

1. Gamma decay.
2. Internal conversion.
3. Internal pair conversion.

Gamma decay takes place more often than the other two processes.

3.7.1 Gamma Decay

Suppose a nucleus is in higher excited state with energy E_i, it makes a transition to lower energy state E_f or loses energy ΔE

$$h\nu = \Delta E = E_i - E_f \tag{3.71}$$

If E_f is ground state of the nucleus, no further emission of γ-ray photons will take place, otherwise the nucleus will emit one or more γ-ray photons before reaching to ground state. During γ-decay mass and charge of the decaying nucleus does not change. For example, consider ^{137}Cs nucleus. It emits β^--particles and the resulting nucleus ^{137}Ba is left in excited state, which emits a γ-ray of 662 keV energy. This total decay is represented by a diagram called the decay scheme of ^{137}Cs which is shown in Figure 3.23.

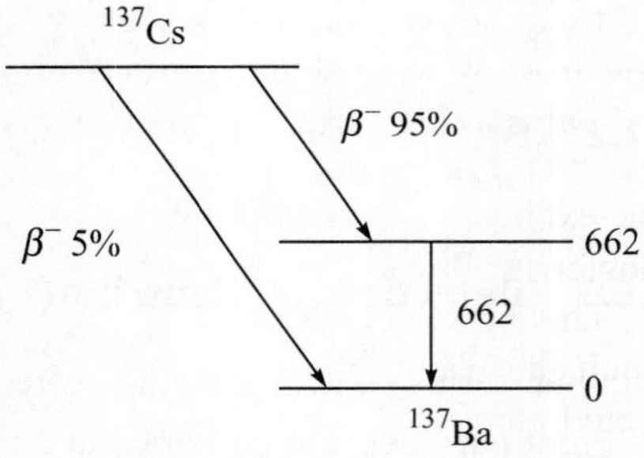

Figure 3.23 Decay scheme of ^{137}Cs.

About 95% of ^{137}Cs nuclei decay by β^--emission to the first excited state with energy 662 keV of ^{137}Ba. Remaining 5% of ^{137}Cs nuclei decays by β^--emission to the ground state of ^{137}Ba. The excited state of ^{137}Ba which is denoted by putting * on Ba as ^{137}Ba* decays to ground state by emission of 662 keV γ-ray. This γ-decay is also written as

$$^{137}\text{Ba}^* \rightarrow {}^{137}\text{Ba} + \gamma$$

Similarly, the decay scheme of ^{60}Co is as shown in Figure 3.24.

$$^{60}\text{Ni}^* \rightarrow {}^{60}\text{Ni} + \gamma$$

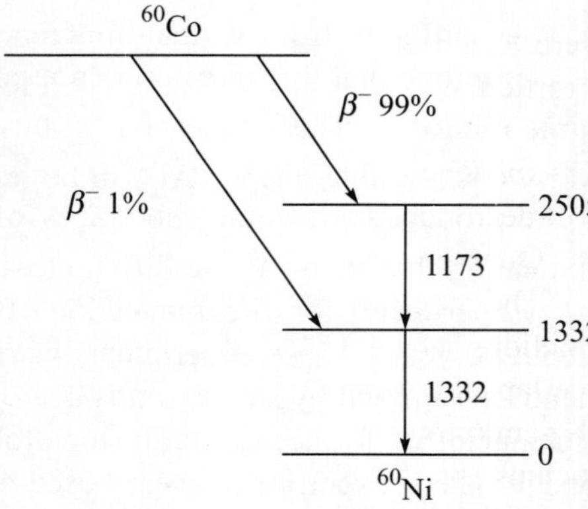

Figure 3.24 Decay scheme of ^{60}Co.

Remember γ-decay is a two-body problem, i.e. products are the decayed nucleus and a γ- ray. Therefore, the γ-ray spectrum is a line spectrum or discrete spectrum.

Properties of Gamma Rays

- Gamma rays are electromagnetic waves just like X-rays or visible light.
- They always travel with velocity of light irrespective the energy of γ-rays.
- They do not cause any appreciable ionization and are not deflected by electric or magnetic fields.
- Gamma-rays have high penetrating power, much larger than that of α- or β-particles.

It is interesting to mention here that both X- and γ-rays are electromagnetic radiations and are identical. The differences between the two are shown in Table 3.2.

TABLE 3.2 Differences between X-rays and γ-rays

S.No.	*X-rays*	*γ-rays*
1.	They originate from outside the nucleus.	They originate from inside the nucleus.
2.	X-rays are emitted when electrons make a transition from higher shell to lower shell.	γ-rays are emitted when nucleons inside the nucleus makes a transition from higher shell to lower shell.
3.	Generally, energy of X-rays is low, of the order of keV.	Energy of γ-ray is high, of the order of MeV.

3.7.2 Internal Conversion

The excited nucleus can perform transition to a lower state either by emitting a γ-ray or by transferring its energy directly to electrons surrounding the nucleus.

This process in which nucleus directly transfers its energy to one of the orbital electrons, by which the orbital electron is ejected from the atom, is known as *Internal Conversion.* These ejected electrons are known as conversion electrons.

The kinetic energy E_e of the conversion electron is given by

$$E_e = E_\gamma - I_{be}$$

Here E_γ is the energy available with the nucleus and I_{be} is the binding energy of the electron in a particular shell. Thus, the kinetic energy of the conversion electrons is unique. *Since the energy of the conversion electrons is unique, therefore, the energy spectrum of conversion electrons is line spectrum.* This line spectrum is superimposed on the continuous electron spectrum ejected in β-decay as shown in Figure 3.22.

Quantum mechanically this effect can be explained as under:

The nucleus and the orbital electrons being quantum mechanical systems have wave functions associated with them. The wave function of electron for a certain fraction of time overlaps the wave function of the nucleus. Thus, there is a finite probability that nucleus instead of γ-emission transfers its energy to the orbital electron. The *K*-shell electron being nearest to the nucleus, so the probability of energy transfer to *K*-shell electron will be maximum and next comes the *L*-shell electron, and so on. Because internal conversion and γ-decay are competing processes and if λ_γ and λ_e are the transition probabilities for γ-decay and internal conversion, then the total transition probability λ is given by

$$\lambda = \lambda_\gamma + \lambda_e$$

If the conversion electrons are ejected from K, L and M shells, then

$$\lambda = \lambda_\gamma + \lambda_K + \lambda_L + \lambda_M$$

If a given radioactive material emits N_γ rays in a given interval of time and N_e conversion electrons are also ejected in the same time, then we define

$$\alpha = \frac{N_e}{N_\gamma} = \frac{\lambda_e}{\lambda_\gamma}$$

Here α is called Internal Conversion Coefficient (ICC). If we record all the electrons (from K, L, M, etc. shells), the α is denoted by α_T (total internal conversion coefficient). However, if N_K, N_L, N_M, etc. are electrons ejected from K, L, M, etc. shells, then

$$\alpha_T = \frac{N_K + N_L + N_N + \cdots}{N_\gamma} = \frac{N_K}{N_\gamma} + \frac{N_L}{N_\gamma} + \frac{N_M}{N_\gamma} + \cdots$$

Therefore,

$$\alpha_T = \alpha_K + \alpha_L + \alpha_M + \cdots$$

Here α_K, α_L, α_M, etc. are called K, L, M, etc. conversion coefficients. The value of α lies between 0 and ∞.

3.7.3 Internal Pair Conversion

In addition to γ-emission and internal conversion, there is another competing process—Internal Pair Conversion for the de-excitation of the nucleus. We have already seen that if the excitation energy of γ-ray is greater than 1.022 MeV, this γ-ray is absorbed in the medium and an electron–positron pair is created. It has been shown by Oppenheimer and Nedelsky in 1933 that if the excitation energy of the nucleus is greater than 1.022 MeV ($> 2m_0c^2$), the nucleus may de-excite by the creation of electron–positron pair somewhere in its own Coulomb field. Though internal pair conversion can take place anywhere in the Coulomb field of the nucleus, but its probability is greatest at a distance from the nucleus which is of the order of $\left(\frac{Z}{137}\right)^2$ times the radius of the K shell of the atomic electrons.

Such a process in which nucleus de-excites by ejecting an electron–positron pair in its own Coulomb field is known as *Internal Pair Conversion*. This process is a competing process with γ-emission and internal conversion.

Just like γ-emission and internal conversion, internal pair conversion is due to electromagnetic interaction. Total excitation energy E_0 available with the nucleus is consumed as

$$E_0 = 2m_0c^2 + E_+ + E_-$$

where E_+ and E_- are the kinetic energies carried by positron and electron respectively and $2m_0c^2$ is the sum of the rest masses of the positron and electron. This process was explained theoretically by Rose in 1950. He found that

1. the probability of internal pair conversion increases with increasing the energy of excitation of the nucleus.

2. the probability of internal pair conversion decreases with the increase in multipolarity L (orbital angular momentum) of the transition.
3. the internal pair creation very slowly decreases with increasing Z.
4. if the excitation energy of the nucleus is very high ($\geq$ 10 MeV), internal pair conversion becomes independent of multipolarity L of the transition.

These theoretical facts of Rose were nicely verified by experiments.

Differences between Pair Production and Internal Pair Conversion

- In pair production the excited nucleus emits a γ-ray (of energy $\geq$ 1.022 MeV). This γ-ray while moving in matter undergoes pair production in the Coulomb field of the electron or nucleus.
- In internal pair conversion, the excited nucleus does not emits a γ-ray, but directly emits an electron–positron pair in its own Coulomb field.

Differences between Internal Conversion and Internal Pair Conversion

- In internal conversion, the excited nucleus transfers its excess energy to one of the orbital electrons and this electron is ejected from the atom. As there is vacancy in inner shell, an electron from higher shell makes a transition to the inner shell and thereby emitting a characteristic X-ray.
- In internal pair conversion, the excited nucleus emits a electron positron pair only. No X-ray is emitted as there is no vacancy in any atomic shells.

Differences between Internal Conversion and Electron Capture

- In internal conversion, the excited nucleus transfers its excess energy to one of the orbital electrons and this electron is ejected from the atom. As there is vacancy in inner shell, an electron from higher shell makes a transition to inner shell and thereby emitting a characteristic X-ray. The nucleus undergoing internal conversion remains the same.
- In electron capture, the nucleus captures an orbital electron. Inside the nucleus, the captured electron combines with a proton to form a neutron. Therefore, Z of the nucleus decreases by 1.
- Changes in mass and atomic numbers for various decay modes are summarized in Table 3.3.

TABLE 3.3 Variation in atomic mass and atomic number in different decays

Parameter	*α-decay*	*β-decay*			*γ-decay*		
		β^--decay	β^+-decay	Electron capture	γ-decay	Internal conversion	Internal pair conversion
Atomic no.	Decreases by 2 units	Increases by 1 unit	Decreases by 1 unit	Decreases by 1 unit	Remains same	Remains same	Remains same
Mass no.	Decreases by 4 amu	Remains same	Remains same	Remains same	Remains same	Remains same	Remains same

3.8 ARTIFICIAL OR INDUCED RADIOACTIVITY

Stable isotopes have fixed number of protons and neutrons in the nucleus. For example, sodium has 11 protons and 12 neutrons in its nucleus. When we bombard such stable isotopes with neutrons or high-energy charged particles like p, d, α or heavy ions, these particles enter the nucleus and disturb the stable balance of protons and neutrons. In some cases, this new balance constitutes a stable nucleus. But in majority of cases, the new system formed is unstable and it begins to emit α-particles, β-particles or γ-rays like natural radioactive isotopes.

This process of inducing radioactivity in a given stable isotope by bombarding it with some suitable particles of proper kinetic energy is known as *artificial* or *induced activity*.

In the course of a study of the effect of α-particles from the naturally occurring radioelement Polonium, on the nuclei of certain light elements, B, Mg and Al in particular, I. Joliot Curie and her husband F. Joliot found that in addition to protons, there was an emission of neutrons and positrons. This result was not at all extraordinary since the nuclear process appeared to be essentially of the (α, p) type discovered by Rutherford, and a neutron and a positron could be together regarded as the equivalent of a proton. But earlier in 1934, the surprising fact was noted that although the formation of protons and neutrons ceased when the source of α-particles was removed, the positrons continued to be emitted. The foil of Al continued to emit positrons and the emission of positrons decays exponentially as in case of an ordinary or naturally occurring radio element. The same phenomenon was observed with B and Mg. The transmutation of B, Mg and Al by alphas has given birth to new radio elements emitting positrons.

According to Joliot, the reaction was (α, n) reaction as

$$^{27}_{13}\text{Al} + ^{4}_{2}\text{He} \rightarrow ^{30}_{15}\text{P} + ^{1}_{0}\text{n}$$

The $^{30}_{15}$P isotope of P obtained in this manner, which is not found in nature, would then be a radioactive species which decays by emission of positrons as under

$$^{30}_{15}\text{P} \rightarrow ^{30}_{14}\text{Si} + ^{0}_{+}e$$

The final product $^{30}_{14}$Si is stable naturally occurring isotope of Si. With this discovery, artificial radioactivity came into existence. In this experiment artificial radioactive nucleus formed is $^{30}_{15}$P.

$$^{4}_{2}\text{He} + ^{27}_{13}\text{Al} \rightarrow ^{30}_{15}\text{P} + n$$

$$^{30}_{15}\text{P} \rightarrow ^{30}_{14}\text{Si} + ^{0}_{+1}e + \nu$$

These artificial radioisotopes once produced decay like natural radioisotopes.

Though the first radioisotope produced was by bombarding α-particles from polonium, but nowadays, we have a large number of particle accelerators and nuclear reactors at our disposal, which are used for producing radioisotopes. Radioisotopes in a reactor are produced mostly by (n, γ) reaction. Some examples are:

$$^{23}_{11}\text{Na} + ^{1}_{0}n \rightarrow ^{24}_{11}\text{Na} + \gamma$$

$$^{59}_{27}\text{Co} + ^{1}_{0}n \rightarrow ^{60}_{27}\text{Co} + \gamma$$

$$^{109}_{47}\text{Ag} + ^{1}_{0}n \rightarrow ^{110}_{47}\text{Ag} + \gamma$$

Radioisotopes can also be produced by a variety of nuclear reactions in particle accelerators. Some such examples are:

$$^{58}_{28}\mathrm{Ni} + {}^{1}_{1}\mathrm{H} \rightarrow {}^{58}_{29}\mathrm{Cu} + {}^{1}_{0}n$$
$$^{114}_{48}\mathrm{Cd} + {}^{2}_{1}\mathrm{H} \rightarrow {}^{115}_{48}\mathrm{Cd} + {}^{1}_{1}\mathrm{H}$$
$$^{27}_{13}\mathrm{Al} + {}^{14}_{7}\mathrm{N} \rightarrow {}^{38}_{19}\mathrm{K} + {}^{1}_{1}\mathrm{H} + 2{}^{1}_{0}n$$

Today we have about 4000 artificial radioisotopes. These radioisotopes have found several uses in different branches of pure sciences, agriculture, medical sciences, etc. Some of these applications are discussed in the next section.

3.9 APPLICATIONS OF RADIOACTIVITY

Applications of radioactivity are varied and interesting. Before discussing these applications, we discuss some of the well-known cases where radioactivity has been utilized.

- Napoleon I Bonaparte, Emperor of the France died in 1821 at St. Helena. The exact cause of his death was not known at that time. It was only recently that by analyzing various elements present in his one scalp hair using radioactive technique, scientists were able to confirm that he died of repeatedly arsenic poisoning.
- In 1963 American President J.F. Kennedy was shot dead. It was not clear whether one or two assassins shot him because there were two bullet marks in his body. It was only by analyzing the bullet pieces recovered from his body using radioactive technique, it was confirmed that he was shot dead by a single assassin.
- Dinosaurs got extinct from the earth about 65.5 million years ago. The asteroid collision theory to account for the extinction of dinosaurs, which was first proposed by Walter Alvarez in the late 1970s, links the extinction event at the end of the Cretaceous period to a asteroid impact approximately 65.5 million years ago. Alvarez proposed that a sudden increase in iridium levels, recorded around the world in the period's rock stratum, was direct evidence of the impact. The bulk of the evidence now suggests that a 10 km-wide asteroid hit in the vicinity of the Yucatán Peninsula, creating the 170 km-wide Chicxulub Crater and triggering the mass extinction. The Yucatán Peninsula separates the Caribbean Sea from the Gulf of Mexico. The impact caused extinctions both directly (by heat from the meteorite impact) and also indirectly (via a worldwide cooling brought about when matter ejected from the impact crater reflected thermal radiation from the sun).
- We are familiar with un-manned space missions to moon and mars. In these missions one of the purpose is to know the elemental composition in the planets soil. For this purpose these flights are equipped with α-emitting radioactive source. The soil of the planet is exposed to α-particles. By measuring the energy of the back-scattered α-particles, scientists are able to get a clear picture of the elements present in a planet soil.

Having discussed some of the interesting applications of radioactivity, we discuss below some of the typical applications of radioactivity in various fields.

3.9.1 Medicine

Radioisotopes find wide applications in medicines. They are used both for diagnostic and therapeutic purposes.

For diagnostic purposes, X-rays are most common example, which locate any fractured bone inside the body. X-rays are also used to study the gastrointestinal tract. ^{131}I and most recently ^{132}I and ^{123}I are used to study malfunctioning thyroid gland. Kidney function is also studied using compound containing ^{131}I. Tumours in the brain are located by injecting intravenously ^{99m}Tc and then scanning the head with suitable scanners.

A more recent development is Positron Emission Tomography (PET), which is a more precise and accurate technique for locating tumours in the body. A positron emitting radionuclide (e.g. ^{18}F) is injected to the patient and it accumulates in the target tissue. As it emits positrons which promptly combine with nearby electrons, it results in the simultaneous emission of two γ-rays in opposite directions. These γ-rays are detected by a PET camera and give precise indication of their origin, i.e. depth also. This technique is also used in cardiac and brain imaging.

The most common therapeutic use of radioisotopes is in treating cancer. In this technique, cancerous cells are destroyed using γ-rays from a ^{60}Co radioactive source. Sometimes wires or sealed needles containing radioactive isotope such as ^{192}Ir or ^{125}I are directly placed into the cancerous tissue. The radiations from the radioisotope attack the tumour as long as the needle/wire is in place. When the treatment is complete, these are removed. This technique is frequently used to treat mouth, breast, lung and uterine cancer.

^{131}I is used to treat thyroid for cancers and other abnormal conditions of thyroid. ^{32}P is used to treat excess of red blood cells produced in the bone marrow.

3.9.2 Food and Agriculture

One of the more controversial uses of radiation today is food irradiation. High doses of γ-radiation do not make food radioactive. Gamma Irradiation kills bacteria, insects, and parasites, and retards spoilage in some foods like potatoes, onions, cooked food, etc. Irradiated foods are regularly eaten by astronauts on space missions, as well as by hospitalized patients with weak immune systems who need extra protection from microorganisms in food.

The irradiation process involves exposing food to intense controlled amounts of gamma rays from ^{60}Co or ^{137}Cs, X-rays, or electron beams from particle accelerators. The process has about the same effect on food as canning, cooking, or freezing. It kills pests and extends shelf life, but also reduces the food's nutritional value somewhat by destroying vitamins A, B1 (thiamin), C, and E. No radiation remains in the food after treatment.

3.9.3 Insect Control

About 10% of the world's crops are destroyed by insects. In efforts to control insect plagues, authorities often release sterile laboratory-raised insects into the wild. These insects are made sterile using ionizing radiation. They are irradiated with this radiation before they hatch. Female insects that mate with sterile male insects do not reproduce, and the population of the insect pests can be quickly curbed as a consequence. This technique of releasing sterile insects into the wild, called the *sterile insect technique* (SIT), is commonly used in protecting agricultural industries in many countries around the world.

3.9.4 Reactors

Atomic nuclei are storehouse of huge amounts of energy. This huge amount of energy can be explored when a heavy nucleus is broken into two or more lighter fragments or when two light nuclei are fused into a heavier one. Let us calculate energy released when we break a heavy nucleus into two lighter fragments.

If we bombard ^{235}U with neutron, the relevant nuclear reaction is:

$$^{235}\text{U} + {}^{1}n \rightarrow \text{fission products} + \text{neutrons} + \text{energy} \ (\sim 200 \text{ MeV})$$

The average number of neutrons emitted in this reaction is 2.43. Some of the neutrons in reaction can be used to induce fission in another ^{235}U nucleus, thus continuing a controlled, self-perpetuating nuclear chain reaction.

It has been estimated that 1 g of U or Pu on fission would produce about 1 MW of power for one day. The same energy production would require more than 3 tons of coal or about 2300 litres of fuel oil per day (see Section 4.7.2).

Other possible fission reactions are:

$$^{233}\text{U} + {}^{1}n \rightarrow \text{fission products} + \text{neutrons} + \text{energy} \ (\sim 200 \text{ MeV})$$
$$^{239}\text{Pu} + {}^{1}n \rightarrow \text{fission products} + \text{neutrons} + \text{energy} \ (\sim 200 \text{ MeV})$$

Similarly, if we fuse two lighter nuclei into a heavier one, then

$$^{2}\text{H} + {}^{3}\text{H} \rightarrow {}^{4}\text{He} + n + 17.6 \text{ MeV}$$

It has been estimated that if 0.5 g of ^{2}H and 0.5 g of ^{3}H is fused together it would produce about 5 MW of power for one day. The same energy production would require more than 15 tons of coal or about 12,000 litres of fuel oil for one day (see Section 4.8.1). This shows that huge amount of energy is stored in atomic nuclei. This energy is released in controlled manner in nuclear reactors to generate electricity and in uncontrolled manner in atomic and thermonuclear (hydrogen) bombs.

In India we have 19 commercial power reactors in operation and seven reactors under construction. They are shown in Tables 3.4 and 3.5.

TABLE 3.4 Commercial power producing reactors in India

S.No.	*Reactor*	*Location*	*Type**	*Capacity* (MWe)+	*Year of start*
1.	TAPS-1	Tarapur, Maharashtra	BWR	160	1969
2.	TAPS-2	Tarapur, Maharashtra	BWR	160	1969
3.	TAPS-3	Tarapur, Maharashtra	PHWR	540	2006
4.	TAPS-4	Tarapur, Maharashtra	PHWR	540	2005
5.	RAPS-1	Rawatbhata, Rajasthan	PHWR	100	1973
6.	RAPS-2	Rawatbhata, Rajasthan	PHWR	200	1981
7.	RAPS-3	Rawatbhata, Rajasthan	PHWR	220	2000
8.	RAPS-4	Rawatbhata, Rajasthan	PHWR	220	2000
9.	RAPS-5	Rawatbhata, Rajasthan	PHWR	220	2010

(*Contd.*)

TABLE 3.4 Commercial power producing reactors in India (*Contd.*)

S.No.	*Reactor*	*Location*	*Type**	*Capacity* (MWe)$^+$	*Year of start*
10.	RAPS-6	Rawatbhata, Rajasthan	PHWR	220	2010
11.	MAPS-1	Kalpakkam, Tamil Nadu	PHWR	220	1984
12.	MAPS-2	Kalpakkam, Tamil Nadu	PHWR	220	1986
13.	NAPS-1	Narora, Uttar Pradesh	PHWR	220	1991
14.	NAPS-2	Narora, Uttar Pradesh	PHWR	220	1992
15.	KAPS-1	Kakarpar, Gujarat	PHWR	220	1993
16.	KAPS-2	Kakarpar, Gujarat	PHWR	220	1995
17.	KAIGA-1	Kaiga, Karnataka	PHWR	220	2000
18.	KAIGA-2	Kaiga, Karnataka	PHWR	220	2000
19.	KAIGA-3	Kaiga, Karnataka	PHWR	220	2007
20.	KAIGA-4	Kaiga, Karnataka	PWHR	220	2012
21.	Kudankulam	Tamil Nadu	PWR	1000	2014

Total nuclear power plant capacity: 5780 MWe
*PHWR: Pressurized Heavy Water Reactor, BWR: Boiling Water Reactor.
$^+$MWe: megawatt electric.

TABLE 3.5 Commercial power producing reactors in India under construction

S.No.	*Reactor*	*Location*	*Type**	*Capacity* (MWe)	*Estimated year of start*
1.	KAPP-2	Kudankulam, Tamil Nadu	PWR	1000	2015
2.	RAPS-7	Rawatbhata, Rajasthan	PHWR	700	2016
3.	RAPS-8	Rawatbhata, Rajasthan	PHWR	700	2016
4.	KAPS-3	Kakarpar, Gujarat	PHWR	700	2015
5.	KAPS-4	Kakarpar, Gujarat	PHWR	700	2015
6.	PFBR	Kalpakkam	PWR	500	2015

Total estimated nuclear power plant capacity: 4300 MWe
*PWR: Pressurized Water Reactor
In addition to these reactors, 22 new reactors are planned which will add 21,300 MWe and will be operational by the year 2021. A total of 35 new reactors are proposed which will add another 40,000 MWe.

3.9.5 Research Reactors

Research reactors are essentially nothing more than a source of neutrons. They are much smaller in comparison to nuclear power reactors and, therefore, use less fuel. They pose less of a risk to the environment.

Neutrons produced by research reactors are used for experiments as well as the manufacture of artificial radioactive isotopes. These artificial radioisotopes can be used in various fields including medicine, industry and agriculture.

Hundreds of research reactors exist in the world. They are invaluable tools for research and development in science. In India we have nine research reactors and the various parameters of these research reactors are as shown in Table 3.6.

TABLE 3.6 Research reactors in India

Reactor	*Year of start*	*Power (MWe)*	*Fuel*	*Moderator*	*Coolant*	*Neutron flux*	*Location*
Apsara	1956	1	Enriched-U	H_2O	H_2O	10^{13}	BARC, Mumbai
Cirus	1960	40	Natural-U	D_2O	H_2O	6.7×10^{13}	BARC, Mumbai
Zerlina*	1961	100	Natural-U	D_2O	D_2O	0.5×10^{8}	BARC, Mumbai
Purnima-I	1972	–	PuO_2–Al	–	–	–	BARC, Mumbai
Purnima-II	1984	–	^{233}U(uranyl nitrate solution)	H_2O	–	–	BARC, Mumbai
Purnima-III	1990	–	^{233}U rods	H_2O	–	–	BARC, Mumbai
Dhruva	1985	100	Natural-U	D_2O	H_2O	1.8×10^{14}	BARC, Mumbai
FBTR	1985	40	U–Pu carbide	–	Liquid sodium	–	IGCAR, Kalpakkam
Kamini	1996	30	^{233}U	H_2O	H_2O	10^{12}	IGCAR, Kalpakkam

*Decommissioned in 1983.

3.9.6 Geology and Element Identification

Radioactivity is used to identify the location of deposits of uranium and other radioactive minerals. This is useful in mining exploration. The intensity of detected radiation also is an indication of the amount of radioactive element that may be located there.

3.9.7 Radioactive Dating

The principle of radioactive decay is applied in the technique of *radioactive dating*, a process widely used by scientists to determine the age of materials and artifacts.

Radioactive dating is defined as the method of determining the age of biological or geological samples by using the radioactive technique.

There are many radioisotopes by which we can determine the age of a given object, but the two most commonly used methods are:

- Radiocarbon dating.
- Uranium dating.

Radiocarbon Dating

Radioactive ^{14}C atoms exist naturally in very minute quantities. ^{14}C is everywhere around us: in our clothes, in the food we eat, even in the air we breathe. The ratio of radioactive ^{14}C atoms to stable ^{12}C atoms in the atmosphere has remained constant over thousands of years. Although ^{14}C naturally decays, it is also continually being formed. ^{14}C atoms are formed when neutrons from the cosmic radiation collide with ^{14}N atoms in the atmosphere. Thus, the decay of ^{14}C is reasonably balanced with its production, resulting in a constant ratio of ^{14}C to ^{12}C.

Carbon dioxide (CO_2) molecules in the air contain both isotopes of carbon. This CO_2 is continually used by plants to grow. Because the ratio of ^{14}C to ^{12}C in atmospheric CO_2 is constant, the intake of CO_2 by a plant results in a constant ratio of the two isotopes in the plant's body while it is alive. However, when the plant dies it no longer takes in CO_2. As a result, the ^{14}C decaying in the dead plant can not be replenished by a "fresh supply" of more CO_2, resulting in the decrease in activity of ^{14}C with time. Because animals eat plants and inhale air, the activity of ^{14}C decreases once they die, since the ^{14}C cannot be replenished. In order to determine the radioactivity of ^{14}C, a small portion of the sample is burnt, so that carbon present in it reacts with oxygen to form CO_2. The CO_2 that contains ^{14}C is radioactive, and the amount is measured using a radiation counter. Burning is done to facilitate measuring the level of ^{14}C. ^{14}C has a half-life of about 5730 years. This means that in a given sample of a carbon-containing substance, (without the ^{14}C being replenished) the activity of ^{14}C decreases by half every 5730 years. Suppose someone discovers an ancient manuscript and finds that the activity of ^{14}C in the paper is half of that found in living trees. This would mean that the manuscripts would be about 5730 years old.

The use of radioactive ^{14}C for dating was first done by William Libby, at the University of Chicago, USA, in 1947.

The age of a given sample can be determined using Eq. (3.7)

$$A = A_0 e^{-\lambda t}$$

In calculating the age of a sample, we make the following assumption:

The activity of ^{14}C in a living sample has remained same over thousands of years.

In the above equations A_0 is the activity of the sample, when it is living. A is the activity of the sample whose age is to be determined. For example, if we wish to determine the age of a given piece of wood, A_0 is the activity for a fresh wood piece taken from a presently living tree. A is the activity of the sample whose age is to be determined.

The relatively short half-life of ^{14}C (5730 years) means that the amount of ^{14}C remaining in materials and objects older than about 30,000 years is too small to be measured experimentally. Thus, carbon dating has following two limitations:

- It cannot measure the age of an object older than about 30,000 years.
- It cannot measure the age of non-living objects like rocks and minerals.

Uranium Dating

Uranium dating method is applicable to ores and rocks, etc. We have seen in Section 3.2 that ^{238}U decays into ^{206}Pb with a half-life of 4.47×10^9 years.

Let N_0 be the amount of ^{238}U in a given sample, when it was initially formed. N_U is the amount of ^{238}U still left in the sample after the lapse of time t, which represents the age of the sample. All the ^{238}U, which has decayed, has been converted into ^{206}Pb. Let the quantity of ^{206}Pb, be N_{Pb}. The quantity of ^{238}U, when the sample was formed, is

$$N_0 = N_U + N_{Pb}$$

Therefore, we have Eq. (3.6) as

$$N = N_0 e^{-\lambda t}$$

Here, $N_0 = N_U + N_{Pb}$ and $N = N_U$. Therefore,

$$N_U = (N_U + N_{Pb}) e^{-\lambda t}$$

Thus, knowing the amounts N_U (of ^{238}U) and N_{Pb} (of ^{206}Pb) and λ (decay constant of ^{238}U) in a given sample, the age t of the sample can be calculated.

The limitations of this method are:

- It is applicable to rocks, minerals, etc. containing ^{238}U only.
- This method is applicable to samples having age $\approx 10^9$ years only.

Geologists measure the amounts of other radioisotopes such as ^{84}Rb and ^{40}K, etc. in rocks to determine their age. Measurements show that the oldest rocks on earth are about 4.6 billion years old, which is a reasonably accurate estimate of the earth's age. Similarly, analysis of fossilized plants shows that they first occurred on earth about 3 billion years ago.

3.9.8 Radioactive Tracers

Radioisotopes can be used to help understand chemical and biological processes in plants and other living beings. This is true for the following two characteristic properties.

- Radioisotopes are chemically identical with other isotopes of the same element and are substituted in chemical reactions.
- Radioactive forms of the element can be easily detected by β and α radiations emitted by them.

Therefore, these properties of the radioisotope, namely its radioactivity, can act as a tag or label, which permits the fate of the element or its compound containing this element to be traced through a series of chemical or physical changes. The element which has been labelled or tagged is called tracer element. For example, if in $CaPO_4$, ordinary phosphorus is replaced by ^{32}P, which is radioactive, then phosphorus in $CaPO_4$ is known as tracer element and this technique is known as tracer technique. The tracer technique finds wide applications in various fields. Some of the applications of tracer technique are discussed below.

Agriculture

To see how plants utilize a given fertilizer, a solution of phosphate containing radioactive ^{32}P is injected into the root system of a plant. Since ^{32}P behaves identically to that of ^{31}P, a more common and non-radioactive form of the element, it is used by the plant in the same way. A GM counter is then used to detect the movement of the radioactive ^{32}P throughout the plant. This information helps scientists to understand the detailed mechanism of how plants utilize phosphorus to grow and reproduce.

Medicine

A **radioactive tracer**, or **radioactive label**, is a chemical compound in which one or more atoms have been replaced by a radioisotope. It finds wide applications in medicines. Radioisotopes of hydrogen, carbon, phosphorus, sulphur and iodine have been used extensively to trace the path of biochemical reactions. A radioactive tracer can also be used to track the distribution of a substance within a natural system such as in a cell or in a tissue. Radioactive tracers form the basis of a variety of imaging systems, such as positron emission tomography (PET) scans, single-photon emission computed tomography (SPECT) scans and technetium scans.

Since hydrogen atoms are present in all organic compounds, so tritium is frequently used as a tracer in biochemical studies. ^{14}C has been used extensively to trace the progress of organic molecules through metabolic pathways.

^{13}N, ^{15}O, ^{18}F etc. are used in PET scan. ^{33}P is used in DNA sequencing.

In metabolism research, tritium and ^{14}C-labelled glucose are commonly used in glucose clamps to measure rates of glucose uptake, fatty acid synthesis and other metabolic processes.

Industry

Radioisotopes are commonly used in industry for checking blocked water pipes, detecting leakage in oil pipes, etc. For example, small quantity of radioactive ^{24}Na is placed in a small enclosed ball and is allowed to move in pipe with water. The moving ball containing radioisotope is monitored with a detector. If the movement of ball stops, indicating the blocked pipe. Similarly, radioisotope ^{24}Na is mixed with oil flowing in an underground pipe. With radiation detector, the radioactivity over the pipe is monitored. If there is a leakage at a particular place, the radiation detector will show large activity at that particular place.

3.9.9 Neutron Activation Analysis (NAA)

It is one of the most sensitive and nondestructive techniques used for qualitative and quantitative elemental analysis of multiple major, minor and trace elements in samples from almost every conceivable field of scientific or technical interest. For many elements, NAA offers sensitivities that are superior to those possible by any other technique. In this technique, a small sample of the material to be analysed is exposed to a high flux ($\sim 10^{13}$ n/cm^2/s) of thermal neutrons from a reactor. Nuclei of elements present in the given sample undergo (n, γ) reaction and may become radioactive. For example, sodium present in the sample undergoes the following reaction:

$$^{23}_{11}\text{Na} + ^{1}_{0}n \rightarrow ^{24}_{11}\text{Na} + \gamma$$

$^{24}_{11}$Na so formed is radioactive. Many of such radioactive nuclei decay through β and subsequent by γ emission. These γ-rays are characteristic of the nuclei emitting γ-rays. By precise measurement of the γ-ray energies emitted by neutron exposed sample, helps us in determining the presence of elements in the sample. From the γ-ray intensities, it can be determined, exactly how much of the original neutron capturing isotope is present and consequently the amount of that element present in the sample can be determined. The applications for NAA are practically limitless for solid, liquid or gaseous samples. Most of the elements can be measured with an accuracy of 1 part per million, ~20 elements can be measured with two parts per billion levels and about 5–6 elements with 1–2 parts per trillion levels.

3.9.10 In Space Flights

Radioisotopes find wide applications in space flights. Suppose a spacecraft is somewhere near Neptune, this craft lacks the warm touch of sunrays. Thus, solar panels cannot provide power. Chemical batteries providing necessary power for the spacecraft would be too heavy to carry on board. The question is how to get sufficient power for the spacecraft. A radioisotope thermoelectric generator (RTG) is an answer to this question. An RTG generates electric power by converting heat released during the nuclear decay of radioactive isotope into electricity. This device has been in use since 1956. For example, an RTG fabricated in 1959 weighing about 2 kg and providing power of 2.5 watts was capable of delivering 11.60 kW-h of electric energy over a period of 280 days. This is the energy produced by nickel–Cadmium batteries weighing about 320 kg. Almost all Appolo missions to moon, Viking missions to Mars, the Pioneer, Voyager, etc. missions to outer solar system carried RTG as a source of power. The radioisotopes commonly used in RTG are either ^{210}Po or ^{238}Pu. Recent mission to Mars in which Curiosity landed on Mars is powered by ^{238}Pu-based RTG providing 110 W of electricity.

3.9.11 In Sea

Nuclear power is particularly suitable for vessels which need to be at sea for long periods without refueling. These vessels (ships/submarines) have a small nuclear reactor on board, which produces steam to run turbines. These turbines directly drive the propellers as well as electrical generators. Today, a large number of nuclear-powered ships and submarines are cruising the oceans.

NUMERICAL PROBLEMS

Section 3.2

Solved Problems

1. How many curies are there in 10^{10} Bq?

Solution: We know that 3.7×10^{10} Bq = 1 Ci

Therefore,

$$10^{10}\ \text{Bq} = \frac{1}{3.7 \times 10^{10}} \times 10^{10}\ \text{Ci}$$

$$= 0.27\ \text{Ci}$$

2. Calculate the activity of 10 g of ^{232}Th. Given: $\lambda_{^{232}\text{Th}} = 1.58 \times 10^{-18}\ \text{s}^{-1}$.

Solution:

$$1\ \text{g mole of}\ ^{232}\text{Th} = 232\ \text{g of}\ ^{232}\text{Th}$$

$$232\ \text{g of}\ ^{232}\text{Th contains} = 6.023 \times 10^{23}\ \text{atoms of}\ ^{232}\text{Th}$$

$$10\ \text{g of}\ ^{232}\text{Th contains} = \frac{6.023 \times 10^{23}}{232} \times 10 = 2.596 \times 10^{22}\ \text{atoms}$$

Therefore,

$$N = 2.596 \times 10^{22}$$

$$\text{Activity}\ N\lambda = 2.596 \times 10^{22} \times 1.58 \times 10^{-18}$$

$$= 4.104 \times 10^{4}\ \text{dps}$$

3. What will be the mass of 1 curie sample of a radioactive substance of atomic mass 214? Its half-life is 26.8 min.

Solution: Given:

$$\text{Activity } A = 1 \text{ Ci} = 3.7 \times 10^{10} \text{ dps}$$

$$\text{Half-life } t_{1/2} = 26.8 \text{ min} = 1608 \text{ s}$$

$$\text{Decay constant } \lambda = \frac{0.6931}{t_{1/2}}$$

$$= \frac{0.6931}{1608}$$

$$= 4.31 \times 10^{-4} \text{ s}^{-1}$$

Let m be the mass of sample having activity 1 Ci.

Therefore,

$$214 \text{ g of radioactive substance contains} = 6.023 \times 10^{23} \text{ atoms}$$

$$m \text{ g of radioactive substance contains } N = \frac{6.023 \times 10^{23} \times m}{214} \text{ atoms}$$

Now, we have

$$\text{Activity } A = \lambda \times N$$

Therefore,

$$3.7 \times 10^{10} = 4.31 \times 10^{-4} \times \frac{6.023 \times 10^{23} \times m}{214}$$

or

$$m = \frac{3.7 \times 10^{10} \times 214}{4.31 \times 10^{-4} \times 6.023 \times 10^{23}}$$

$$= 3.05 \times 10^{-8} \text{ g}$$

4. The half-life of $^{238}_{92}$U against α-decay is 4.5×10^9 years. Find the activity of 1 kg of $^{238}_{92}$U.

Solution: Half-life $t_{1/2} = 4.5 \times 10^9$ years $= 1.419 \times 10^{17}$ s

Therefore,

$$\text{Decay constant } \lambda = \frac{0.6931}{t_{1/2}}$$

$$= \frac{0.6931}{1.419 \times 10^{17}}$$

$$= 4.88 \times 10^{-18} \text{ s}^{-1}$$

$$238 \text{ g of } ^{238}_{92}\text{U contains} = 6.023 \times 10^{23} \text{ atoms}$$

$$1000 \text{ g of } ^{238}_{92}\text{U contains } N = \frac{6.023 \times 10^{23} \times 1000}{238} \text{ atoms}$$

Now, we have

$$\text{Activity } A = \lambda \times N$$

Therefore,

$$A = 4.88 \times 10^{-18} \times 2.53 \times 10^{24}$$
$$= 1.236 \times 10^{7} \text{ dps}$$
$$= 3.34 \times 10^{-4} \text{ Ci}$$

5. One milligram of a radioactive material with half-life of 1600 years is kept for 2000 years. Calculate the mass, which would have decayed by this time.

Solution: Given:

$$\text{Half-life } t_{1/2} = 1600 \text{ years}$$

Therefore,

$$\text{Decay constant } \lambda = \frac{0.6931}{t_{1/2}}$$
$$= \frac{0.6931}{1600} = 4.332 \times 10^{-4} \text{ years}^{-1}$$

Now, we have from Eq. (3.6)

$$N = N_0 e^{-\lambda t} \text{ or } \frac{N_0}{N} = e^{\lambda t}$$

Therefore,

$$\frac{N_0}{N} = e^{4.332 \times 10^{-4} \times 2000}$$
$$= e^{0.8663}$$
$$= 2.378$$

If N_0 is the initial number of nuclei at time $t = 0$, and the corresponding mass of the radioactive substance is m_0. Similarly, if N is the number of nuclei left at time t (= 2000 years), the corresponding mass be m, then we must have

$$\frac{N_0}{N} = \frac{m_0}{m}$$

or

$$\frac{N_0}{N} = \frac{m_0}{m} = 2.378$$

Given that $m_0 = 1$ mg
Therefore,

$$\frac{m_0}{m} = 2.378$$

or

$$m = \frac{1}{2.378} = 0.4205 \text{ mg}$$

Therefore, the amount of radioactive substance decayed is

$$= 1 - 0.4205 = 0.5795 \text{ mg}$$

6. The activity of a certain nuclide decreases to 15% of its original value in 10 days. Find its half-life.

Solution: Given:

$$t = 10 \text{ days}$$

$$\text{Let the initial activity} = A_0$$

$$\text{Activity left after 10 days} = \frac{15}{100} A_0$$

We have from Eq. (3.7)

$$A = A_0 e^{-\lambda \times t}$$

Therefore,

$$\frac{15}{100} A_0 = A_0 e^{-\lambda \times 10}$$

or

$$e^{-10 \times \lambda} = 0.15$$

or

$$e^{10 \times \lambda} = \frac{1}{0.15} = 6.6667$$

or

$$10 \times \lambda = \log_e 6.6667 = 1.8971$$

Therefore,

$$\lambda = 0.18971 \text{ day}^{-1}$$

$$t_{1/2} = \frac{0.6931}{\lambda}$$

$$= \frac{0.6931}{0.18931}$$

$$= 3.65 \text{ days}$$

7. Calculate the mass in kg of 1 curie and 1 rutherford in a sample of ^{226}Ra, whose half-life is 1620 years.

Solution: Given:

$$\text{Half-life } t_{1/2} = 1620 \text{ years}$$

Therefore,

$$\text{Decay constant } \lambda = \frac{0.6931}{t_{1/2}}$$

$$= \frac{0.6931}{1620 \times 365 \times 24 \times 3600}$$

$$= 1.3567 \times 10^{-11} \text{ s}^{-1}$$

$$\text{Activity } A = 1 \text{ Ci} = 3.7 \times 10^{10} = N\lambda$$

or

$$3.7 \times 10^{10} = N \times 1.3567 \times 10^{-11}$$

or

$$N = \frac{3.7 \times 10^{10}}{1.3567 \times 10^{-11}}$$

$$= 2.727 \times 10^{21} \text{ atoms}$$

6.023×10^{26} atoms have mass = 226 kg

$$2.272 \times 10^{21} \text{ atoms have mass} = \frac{226 \times 2.727 \times 10^{21}}{6.023 \times 10^{26}} = 1.023 \times 10^{-3} \text{ kg}$$

Therefore, the mass in kg of 1 curie in a sample of $^{226}Ra = 1.023 \times 10^{-3}$ kg

$$\text{Activity } A = 1R = 10^6 = N\lambda$$

or

$$10^6 = N \times 1.3567 \times 10^{-11}$$

or

$$N = \frac{10^6}{1.3567 \times 10^{-11}} = 7.37 \times 10^{16} \text{ atoms}$$

6.023×10^{26} atoms have mass = 226 kg

$$7.37 \times 10^{16} \text{ atoms have mass} = \frac{226 \times 7.37 \times 10^{16}}{6.023 \times 10^{26}} = 2.77 \times 10^{-8} \text{ kg}$$

Therefore, the mass in kg of 1 rutherford in a sample of $^{226}Ra = 2.77 \times 10^{-8}$ kg.

8. If a radioactive material initially contains 3.0 mg of ^{234}U, how much it will contain after 1,50,000 years? What will be its activity at the end of this time? Given: $t_{1/2} = 2.5 \times 10^5$ years and $\lambda = 8.8 \times 10^{-14}\ s^{-1}$

Solution: 234 g of ^{234}U contains = 6.023×10^{23} atoms

$$3 \times 10^{-3} \text{ g of } ^{234}U \text{ contains } N = \frac{6.023 \times 10^{23} \times 3 \times 10^{-3}}{234} = 7.721 \times 10^{18} \text{ atoms}$$

Number of atoms left after 1,50,000 years is given by

$$N = N_0 e^{-\lambda \times t}$$

Now,

$$\lambda = \frac{0.6931}{t_{1/2}}$$

$$= \frac{0.6931}{2,50,000}$$

$$= 2.7724 \times 10^{-6} \text{ years}^{-1}$$

Therefore,

$$N = 7.721 \times 10^{18} e^{-2.7724 \times 10^{-6} \times 150,000}$$

$$= 7.721 \times 10^{18} e^{-0.41586}$$

$$= 7.721 \times 10^{18} \times 0.65977$$

$$= 5.094 \times 10^{18} \text{ atoms}$$

6.023×10^{23} atoms have mass = 234 g

$$5.094 \times 10^{18} \text{ atoms have mass} = \frac{234 \times 5.094 \times 10^{18}}{6.023 \times 10^{23}} = 1.979 \text{ mg}$$

Therefore, the weight of the sample left after 1,50,000 years will be 1.979 mg.

$$\begin{aligned}\text{Activity } N\lambda &= 5.094 \times 10^{18} \times 8.8 \times 10^{-14}\\ &= 448{,}272 \text{ dps}\\ &= 1.212 \times 10^{-5} \text{ Ci}\\ &= 0.01212 \text{ mCi}\\ &= 12.12 \text{ μCi}\end{aligned}$$

9. Calculate the activity of ^{40}K in a human body weighing 100 kg. Assume that 0.35% of the body weight is potassium. In natural potassium, the abundance of ^{40}K is 0.012%. Half-life of $^{40}K = 1.31 \times 10^9$ years.

Solution: Total weight of potassium in 100 kg human body = $100 \times 0.35 \times 10^{-2} = 0.35$ kg
Weight of ^{40}K with abundance 0.012% = $0.35 \times 0.012 \times 10^{-2} = 4.2 \times 10^{-5}$ kg
From Avogadro's hypothesis, number of atoms in 4.2×10^{-5} kg of ^{40}K

$$\begin{aligned}N &= \frac{6.023 \times 10^{26}}{40} \times 4.2 \times 10^{-5}\\ &= 6.324 \times 10^{20} \text{ atoms}\end{aligned}$$

$$\begin{aligned}\text{Therefore, activity of } ^{40}K &= \lambda \times N = \frac{0.693 \times N}{\lambda}\\ &= \frac{0.693 \times 6.324 \times 10^{20}}{1.31 \times 10^9 \times 365 \times 24 \times 60 \times 60}\\ &= 1.061 \times 10^4 \text{ dps}\end{aligned}$$

$$\begin{aligned}\text{Activity in human body} &= 1.061 \times 10^4 \text{ dps}\\ &= 0.287 \text{ μCi}\end{aligned}$$

10. Radionuclides are useful source of small amounts of energy in space vehicles, remote communication systems etc. One such radionuclide is ^{210}Po. Calculate the power available in watts from a 10 mg of ^{210}Po. It is an α-emitter with an energy of 5.3 MeV. Given $t_{1/2} = 138$ days and $m_{Po} = 209.982848$ amu.

Solution: 210 g of ^{210}Po contains = 6.023×10^{23} nuclei.

$$10 \times 10^{-3} \text{g of } ^{210}Po \text{ contains } \frac{6.023 \times 10^{26}}{210} \times 10 \times 10^{-3} \text{ nuclei.}$$

$$N = 2.87 \times 10^{19} \text{ nuclei}$$

$$\text{Decay constant } \lambda = \frac{\ln 2}{t_{1/2}} = \frac{0.6931}{138 \times 24 \times 3600} = 5.81 \times 10^{-8}\,\text{s}^{-1}$$

Therefore,

$$\text{Activity} = N \times \lambda = 2.87 \times 10^{19} \times 5.81 \times 10^{-8} = 1.667 \times 10^{12}\ \text{dps.}$$

1 disintegration/s produces power = 5.3 MeV/s.

1.667×10^{12} dps will produce power = $5.3 \times 1.667 \times 10^{12}$ MeV/s

$$= \frac{5.3 \times 1.667 \times 10^{12} \times 1.6 \times 10^{-6}}{10^{7}} = 1.4\ \text{W}$$

Unsolved Problems

1. How many Bq are there in 2700 disintegrations per second? [**Ans.** 2700 Bq]

2. How many Bq are there in 95 mCi? [**Ans.** 3.515×10^8 Bq]

3. What is the activity of 3.5×10^{10} atoms of ^{60}Co? λ for ^{60}Co = 0.132 year^{-1}.
[**Ans.** 146 dps]

4. How many atoms are there in 1.2×10^{10} dps of ^{54}Co? Half-life of ^{54}Co = 1.46 min.
[**Ans.** 1.51×10^{12} atoms]

5. A ^{131}I source has an activity of 30 mCi today. What was the activity of the source one month ago? Given: $\lambda = 0.086\ \text{d}^{-1}$. [**Ans.** 0.395 Ci]

6. A ^{88}Rb source has an activity of 1200 Bq. What is the activity of the source 30 minutes back? Given: $\lambda = 0.039\ \text{min}^{-1}$. [**Ans.** 3866 dps]

7. The decay constant for ^{133}Xe is 0.13 d^{-1}. What is its half-life and average life?
[**Ans.** 5.38 days, 7.69 days]

8. A radioisotope decays in such a manner that after 1 hour, only 25% of the initial amount remains. Calculate the decay constant and half-life of the isotope.
[**Ans.** 1.386 hour^{-1}, 30 min]

9. How much time will it take 100 mCi of ^{3}H to decay to 25 mCi. Given: $\lambda = 0.056$ year^{-1}.
[**Ans.** 24.75 years]

10. The half-life of α-emitter ^{210}Po is 138 days. What mass of ^{210}Po is needed for a 10 mCi source? [**Ans.** 2.22×10^{-9} kg]

11. A beam of thermal neutrons having energy 0.025 eV passes through an apparatus of 50 cm long at the rate of $10^9\ \text{s}^{-1}$. $t_{1/2}$ for neutrons is 12 min. Calculate the rate at which neutrons decay are expected in the apparatus. [**Ans.** ~220 neutrons/s]

12. Find the number of α-particles emitted per second by 3 mg of ^{226}Ra. Also find the activity in microcuries. $t_{1/2}$ for ^{226}Ra = 1620 years.
[**Ans.** 1.1×10^8, 3000 μCi]

13. Estimate the number of α-particles emitted per second by 2 cc of ^{222}Rn at NTP. Also estimate the activity in millicuries. $t_{1/2}$ for ^{222}Rn = 3.82 days.
[**Ans.** 1.13×10^{14}, 3×10^6 mCi]

14. Calculate the amount of ^{210}Po to give us an α-particle source of 5 mCi.
[**Ans.** 1.1 μg]

15. Find the number of β-particles emitted per minute by natural KCl of 0.2 kg. Given that 90% of ^{40}K disintegrations results in β-emission and the abundance of ^{40}K in natural potassium is 0.0117%. [**Ans.** 9.12×10^4]

16. If one starts with 1 kg of ^3H, how long will it be until only 1 g is left? Half-life of ^3H is 12.26 years. [**Ans.** 122.13 years]

17. The activity of ^{32}P today is 50 μCi. What will be the activity in 6 weeks from now? Given: λ for $^{32}P = 0.0495\ d^{-1}$. [**Ans.** 6.25 μCi]

18. What is the fraction of atoms left after 10 half-lives of a substance?

[**Ans.** 1/1024]

19. Suppose you begin with 1.0×10^{-2} g of a pure radioactive substance and 4 h later determine that 0.25×10^{-2} g remain. What is the half-life of the substance?

[**Ans.** 2h]

20. The half-life of ^{32}P is 14.3 days. What interval of time is required for 100 mCi of ^{32}P to decay to 25 mCi? What time is required for decay of 7/8 of the ^{32}P atoms? [**Ans.** 42.9 days]

21. What is the mass in grams of 1000 MBq of pure ^{32}P? How many ^{32}P atoms constitute 1000 MBq? What is the mass in grams of 1000 MBq of Na_3PO_4 if all the phosphorus in the compound is radioactive? [**Ans.** 9.47×10^{-8}g, 1.78×10^{15} atoms, 4.88×10^{-7}g]

22. How many MBq of ^{132}I ($t_{1/2}$ = 2.3 h) should be ordered so that the sample activity will be 500 MBq when it arrives 24 h later? [**Ans.** 6,91,733 MBq]

Section 3.3

Solved Problems

1. Complete the following reactions by putting appropriate particle on the arrow:

$$^{238}_{92}U \rightarrow {}^{234}_{90}Th \rightarrow {}^{234}_{91}Pa \rightarrow {}^{234}_{92}U$$

$$^{211}_{82}Pb \rightarrow {}^{211}_{83}Bi \rightarrow {}^{207}_{81}Tl \rightarrow {}^{207}_{82}Pb$$

$$\downarrow \rightarrow {}^{211}_{84}Pa \rightarrow \uparrow$$

Solution:

$$^{238}_{92}U \xrightarrow{\alpha} {}^{234}_{90}Th \xrightarrow{\beta^-} {}^{234}_{91}Pa \xrightarrow{\beta^-} {}^{234}_{92}U$$

$$^{211}_{82}Pb \xrightarrow{\beta^-} {}^{211}_{83}Bi \xrightarrow{\alpha} {}^{207}_{81}Tl \xrightarrow{\beta^-} {}^{207}_{82}Pb$$

$$\downarrow \xrightarrow{\beta^-} {}^{211}_{84}Pa \xrightarrow{\alpha} \uparrow$$

2. $^{238}_{92}U$ decays to Pb through the successive emission of 6 electrons and 8 α-particles. What is the mass number of Pb isotope? What is the total energy evolved in the decay? Given:

$$M(^{238}U) = 238.050786 \text{ amu}$$

$$M(Pb) = 205.9744550 \text{ amu}$$

$$M(\alpha) = 4.002603 \text{ amu}$$
$$M(e) = 5.486 \times 10^{-4} \text{ amu}$$

Solution: Number of nucleons in 8 α-particles = $8 \times 4 = 32$

Therefore, the mass of Pb = $238 - 32 = 206$

or

The daughter nucleus is $^{206}_{82}\text{Pb}$

$$\text{Total mass of products} = M(^{206}\text{Pb}) + 8M(\alpha) + 6M(e)$$
$$= 205.974455 + 8 \times 4.002603 + 6 \times 5.486 \times 10^{-4}$$
$$= 237.9985706 \text{ amu}$$
$$\text{Therefore, decrease in mass} = 238.050786 - 237.9985706$$
$$= 0.0522154 \text{ amu}$$

or

$$\text{Energy evolved} = 0.0522154 \times 931.47$$
$$= 48.6 \text{ MeV}$$

Unsolved Problems

1. Complete the following reactions by putting appropriate particle on the arrow

$$^{232}_{90}\text{Th} \rightarrow {}^{228}_{88}\text{Ra} \rightarrow {}^{228}_{89}\text{Ac} \rightarrow {}^{228}_{90}\text{Th}$$
$$^{214}_{82}\text{Pb} \rightarrow {}^{214}_{83}\text{Bi} \rightarrow {}^{213}_{84}\text{Po} \rightarrow {}^{210}_{82}\text{Pb} \rightarrow {}^{210}_{83}\text{Bi} \rightarrow {}^{210}_{84}\text{Po} \rightarrow {}^{206}_{82}\text{Pb}$$
$$\downarrow\rightarrow {}^{210}_{81}\text{Tl} \rightarrow\downarrow$$

2. $^{237}_{93}\text{Np}$ decays to a nucleus after 7 α-emissions and 4 β^--emissions. What is the resultant nucleus and how much energy is evolved in this process? Given:

$$M(^{237}\text{Np}) = 237.048169 \text{ amu}$$
$$M(\text{final nucleus}) = 208.980388 \text{ amu}$$
$$M(\alpha) = 4.002603 \text{ amu}$$
$$M(e) = 5.486 \times 10^{-4} \text{ amu}$$

[Ans. ^{209}Bi, 44.12 MeV]

Section 3.4

Solved Problems

1. In the following radioactive series, write down the atomic number and mass number of B, C, D, E.

$$^{238}_{90}\text{A} \xrightarrow{\alpha} \text{B} \xrightarrow{\beta} \text{C} \xrightarrow{\alpha} \text{D} \xrightarrow{\beta} \text{E}$$

Solution:

$$^{238}_{90}\text{A} \xrightarrow{\alpha} {}^{234}_{88}B \xrightarrow{\beta} {}^{234}_{89}\text{C} \xrightarrow{\alpha} {}^{230}_{87}\text{D} \xrightarrow{\beta} {}^{230}_{88}\text{E}$$

2. ^{226}Ra decays to ^{222}Rn with a half-life of 1620 years. ^{222}Rn decays to ^{218}Po with a shorter half-life of 3.82 days. Starting with an initially pure sample of ^{226}Ra, find the number of ^{222}Rn half-lives that have elapsed when the ^{222}Rn reaches 99% of its equilibrium concentration.

Solution: We have from Eq. (3.24)

$$N_b(t) = N_a \frac{\lambda_a}{\lambda_b}[1 - e^{-\lambda_b t}]$$

and from Eq. (3.25)

$$N_b(\infty) = N_a \frac{\lambda_a}{\lambda_b}$$

Therefore,

$$\frac{N_b(t)}{N_b(\infty)} = 1 - e^{-\lambda_b t} = \frac{99}{100}$$

Therefore,

$$e^{-\lambda_b t} = \frac{1}{100}$$

or

$$t = \frac{1}{\lambda_b}\log_e(100)$$

$$= \frac{T_{1/2}}{\log_e(2)}\log_e(100)$$

$$\frac{t}{T_{1/2}} = \frac{\log_e(100)}{\log_e(2)} = 6.64$$

Therefore, number of half-lives elapsed is 6.64.

3. ^{212}Bi decays to ^{212}Ra by β^--emission in 66% of the disintegrations and to ^{208}Th by α-emission in 34% of the cases. The total half-value period is 60.5 minutes. Find the decay constants for β^-, α and total emission.

Solution: Suppose λ_β and λ_α be the decay constants for β- and α-decay respectively. The total decay constant $\lambda_T = \lambda_\beta + \lambda_\alpha$.

In this case, $\lambda_\beta = \frac{66}{100}\lambda_T$ and $\lambda_\alpha = \frac{34}{100}\lambda_T$

Total half-value period = 60.5 min = 60.5 × 60 = 3630 s

Therefore,

$$\lambda_T = \frac{0.693}{3630} = 1.909 \times 10^{-4}\ \text{s}^{-1}.$$

$$\lambda_\beta = \frac{66}{100} \times 1.909 \times 10^{-4} = 1.24 \times 10^{-4}\ \text{s}^{-1}.$$

$$\lambda_\alpha = \frac{34}{100} \times 1.909 \times 10^{-4} = 6.49 \times 10^{-5}\ \text{s}^{-1}.$$

4. The atomic ratio between ${}^{238}_{92}U$ and ${}^{234}_{92}U$ in a mineral sample is found to be 1.8×10^4. The half-life of ${}^{238}_{92}U$ is 2.5×10^5 years. Find the half-life of ${}^{234}_{92}U$.

Solution: Let number of atoms of ${}^{238}_{92}U$ is N_{238} and the number of atoms of ${}^{234}_{92}U$ is N_{234}. Let half-life of ${}^{238}_{92}U$ be T_{238} and that of ${}^{234}_{92}U$ be T_{234}. Using Eq. (3.27), we have

$$\frac{N_{238}}{N_{234}} = \frac{T_{238}}{T_{234}}$$

Given:

$$\frac{N_{238}}{N_{234}} = 1.8 \times 10^4$$

and

$$T_{234} = 2.5 \times 10^4 \text{ years}$$

Therefore,

$$\frac{T_{238}}{2.5 \times 10^5} = 1.8 \times 10^4$$

or

$$T_{238} = 4.5 \times 10^9 \text{ years}$$

Unsolved Problems

1. The half-life of a radionuclide A is 10 min and that of its daughter B is 5 min. The daughter of B, i.e. C is stable. Calculate the number of atoms of A, B and C at an interval of 5 min up to 50 min. Given, initially at $t = 0$, number of atoms of A equal to 10^{10}; B and C equal to zero.
2. Repeat Problem 1 taking initial activity (10 mCi) instead of number of atoms.
3. A radioactive series is as under

$$^{140}\text{Ba} \xrightarrow{12.8\text{d}} {}^{140}\text{La} \xrightarrow{40.2\text{h}}$$

Initially given 10,000 disintegrations per second (dps) of ^{140}Ba, no dps for ^{140}La. Calculate (i) dps of ^{140}Ba after 30 days and (ii) dps of ^{140}La after 12 hours and after 30 days.

[Ans. (i) 1970, (ii) 241, 296]

4. Secular equilibrium exists between ^{226}Ra and ^{222}Rn. The half-lives of the two are respectively 1.62×10^3 years and 3.82 days. Calculate the weight of ^{222}Rn, which is in secular equilibrium with 10 g of ^{226}Ra.

[Ans. $\approx$ 63 μg]

Section 3.5

Solved Problems

1. Are the following decays possible?

(i) $^{232}\text{U} \rightarrow {}^{231}\text{U} + {}^{1}n$ (ii) $^{232}\text{U} \rightarrow {}^{231}\text{Pa} + p$

(iii) $^{239}Pu \rightarrow {}^{235}U + {}^{4}He$

Given:

$$M(^{239}Pu) = 239.052158 \text{ amu}$$
$$M(^{235}U) = 235.043925 \text{ amu}$$
$$M(^{231}U) = 231.036270 \text{ amu}$$
$$M(^{231}Pa) = 231.035880 \text{ amu}$$
$$M(^{4}He) = 4.002603 \text{ amu}$$
$$M(^{1}n) = 1.008665 \text{ amu}$$
$$M(p) = 1.007825 \text{ amu}$$

Solution:

(i) $^{232}U \rightarrow {}^{231}U + {}^{1}n$

This decay is possible if the total mass of the products is smaller than that of ^{232}U. Total mass of the products = 231.036270 + 1.008665 = 232.044935 amu.

So, the total mass of the products is greater than that of the mass of ^{232}U. Hence, this decay is not possible.

(ii) $^{232}U \rightarrow {}^{231}Pa + p$

In this case, total mass of the products = 231.035880 + 1.007825 = 232.043705 amu.

So, the total mass of the products is greater than that of the mass of ^{232}U. Hence, this decay is not possible.

(iii) $^{239}Pu \rightarrow {}^{235}U + {}^{4}He$

In this case, total mass of the products = 235.043925 + 4.002603 = 239.046528 amu.

So, the total mass of the products is smaller than that of the mass of ^{239}Pu. Hence, this decay is possible.

2. Calculate the kinetic energy of α-particles in the following decay.

$$^{239}Pu \rightarrow {}^{235}U + {}^{4}He$$

Given:

$$M(^{239}Pu) = 239.052158 \text{ amu}$$
$$M(^{235}U) = 235.043925 \text{ amu}$$
$$M(^{4}He) = 4.002603 \text{ amu}$$

Solution:

$$^{239}Pu \rightarrow {}^{235}U + {}^{4}He$$

The total disintegration energy Q is the difference in masses of the initial nucleus and the final products, or

$$\begin{aligned} Q &= M(^{239}Pu) - M(^{235}U) - M(^{4}He) \\ &= 239.052158 - 235.043925 - 4.002603 \\ &= 0.00563 \text{ amu} \\ &= 0.00563 \times 931.47 = 5.244 \text{ MeV} \end{aligned}$$

Now, from Eq. (3.32)

$$K_\alpha = Q\frac{A-4}{A}$$

$$= 5.244 \times \frac{237}{241}$$

$$= 5.16 \text{ MeV}$$

Therefore, the kinetic energy of α-particle = 5.16 MeV

3. Calculate the height of potential barrier faced by an α-particle inside the ${}^{226}_{88}$Ra nucleus.

Solution: The height of Coulomb barrier is given by the relation

$$B = \frac{2(Z-2)e^2}{4\pi\varepsilon_0 R}$$

$$= \frac{2(Z-2)e^2}{4\pi\varepsilon_0 R_0 A^{1/3}}$$

$$= \frac{8.98 \times 10^9 \times 2 \times 86 \times (1.6 \times 10^{-19})^2}{1.3 \times 10^{-15} \times 226^{1/3}}$$

$$= 4.9934 \times 10^{-12} \text{ J}$$

$$= 31.2 \text{ MeV}$$

4. Rutherford bombarded 7.7 MeV α-particles from ^{214}Po on ^{14}N to initiate the nuclear reaction ${}^{14}N(\alpha, p){}^{17}O$. Find the height of Coulomb barrier faced by α-particles.

Solution: Height of Coulomb barrier is given by the relation

$$V_c = \frac{Z_1 Z_2 e^2}{4\pi\varepsilon_0 (R_1 + R_2)}$$

where Z_1 is the charge of the incident projectile, Z_2 is the charge of the target. R_1 and R_2 are the radii of the projectile and target nuclei respectively. We have $R = R_0 A^{1/3}$, therefore,

$$V_c = \frac{Z_1 Z_2 e^2}{4\pi\varepsilon_0 R_0 (A_1^{1/3} + A_2^{1/3})}$$

Substituting various values, we get

$$V_c = \frac{8.98 \times 10^9 \times 2 \times 7 \times (1.6 \times 10^{-19})^2}{1.5 \times 10^{-15} \times (4^{1/3} + 14^{1/3})}$$

$$= 5.367 \times 10^{-13} \text{ J}$$

$$= 3.35 \text{ MeV}$$

5. Find the kinetic energy required by a proton to penetrate Coulomb barrier of a hydrogen nucleus.

Solution: When a proton is bombarded on hydrogen nucleus, the Coulomb barrier E_b is given by

$$E_b = \frac{Z_1 Z_2 e^2}{r}$$

Here $Z_1 = Z_2 = 1$ and $r = 1.2$ fm

$$E_b = \frac{e^2}{r} = \frac{\hbar c}{r} \times \frac{e^2}{\hbar c}$$

or

$$E_b = \frac{197}{1.2} \times \frac{1}{137}$$

$$= 1.2 \text{ MeV}$$

Unsolved Problems

1. Find the speeds of α-particles with energies 1 MeV and 1000 MeV.

   [Ans. 6.95×10^6 m/s, 2.2×10^8 m/s]

2. The total energy librated in the α-decay of $^{226}_{88}$Ra is 4.87 MeV.
   (i) Identify the daughter nucleus.
   (ii) Find the kinetic energy of α-particle.
   (iii) Find the recoil energy of the nucleus.

   [Ans. (i) $^{222}_{86}$Rn (ii) 4.78 MeV (iii) 0.09 MeV)]

3. $^{210}_{84}$Po decays by emitting a 5.30 MeV α-particle. Find the disintegration energy for this decay. Identify the daughter nucleus and find its atomic mass. Given:

$$M(^{210}\text{Po}) = 209.982876 \text{ amu}$$

$$M(^{4}\text{He}) = 4.002603 \text{ amu}$$

   [Ans. 5.40 MeV, $^{206}_{82}$Pb, 205.974476 amu]

4. Determine the kinetic energy of α-particle emitted in an α-decay in terms of Q-value of the reaction

$$^{A}_{Z}\text{P} \rightarrow {}^{A-4}_{Z-2}\text{D} + {}^{4}_{2}\text{He}$$

5. Find the height of Coulomb barrier, when $^{208}_{82}$Pb ions are bombarded on $^{208}_{82}$Pb target.

   [Ans. 596.1 MeV]

6. Calculate the barrier height for an α-particle inside the $^{238}_{92}$U nucleus.

   [Ans. 32.2 MeV]

7. Calculate the height of potential barrier faced by an α-particle inside the $^{252}_{98}$Cf nucleus.

   [Ans. 33.6 MeV]

8. Find the energy released in the α-decay of ^{238}U

$$^{238}_{92}\text{U} \rightarrow {}^{234}_{90}\text{Th} + {}^{4}_{2}\text{He}$$

   Given $m_U = 238.050786$ amu. $m_{TH} = 234.043583$ amu and $m_{He} = 4.002603$ amu.

   [Ans. 4.28 MeV]

9. The straggling of heavy ions at low energy is mostly a consequence of

(a) finite momentum
(b) fluctuating state of ionization
(c) multiple scattering

[**Ans.** (c) Multiple scattering changes the ions' direction of motion. This makes them straggle. Thus, straggling is due to multiple scattering.]

10. ^{212}Po decays by α-particle emission. Energy of these α-particles is 8.787 MeV. Calculate the velocity of these α-particles. [**Ans.** 2.06×10^7 m/s]

Section 3.6

Solved Problems

1. ^{14}C decays by β^--emission. The end point energy of β^--particles in this decay is 0.156 MeV. Given the mass of ^{14}C = 14.007685 amu, find the mass of daughter nucleus.

Solution: The given decay is

$$^{14}C \rightarrow {}^{14}N + \beta^- + \overline{\nu}_e$$

For β^--emission with energy equal to end point energy, the energy carried by $\overline{\nu}_e = 0$ MeV and mass of $\overline{\nu}_e \approx 0$ MeV.

Now,

$$M(^{14}N) = M(^{14}C) - \text{K.E. of } \beta^- \text{ (in amu)}$$

$$= 14.007685 - \frac{0.156}{931.47}$$

$$= 14.007517 \text{ amu}$$

Therefore, the mass of ^{14}N = 14.007517 amu

2. Find whether ^{7}Be is stable or not. If not, find the decay mode. Given:

$$M(^7Be) = 7.016928 \text{ amu}$$
$$M(^7Li) = 7.0169003 \text{ amu}$$
$$M(^7B) = 7.029920 \text{ amu}$$

Solution: For β^--decay, we have the following equation:

$$^7Be \rightarrow {}^7B + \beta^- + \overline{\nu}_e$$

From the given masses, it is clear that the mass of daughter nucleus is larger than the parent nucleus, so β^--decay of ^{7}Be is not possible.

For β^+-decay, we have the following equation:

$$^7Be \rightarrow {}^7Li + \beta^+ + \nu_e$$

The condition for β^+-decay is that the mass of the parent nucleus should be greater than the daughter mass by $2m_ec^2$.

Mass of the daughter nucleus = 7.016003 amu

$$\text{Mass } 2m_ec^2 \text{ and parent nucleus} = \frac{1.022}{931.47} + 7.016003 = 7.017100 \text{ amu}$$

This mass 7.017100 amu is greater than the parent mass of ^{7}Be, hence this decay mode is also not possible.

For electron capture, we have the following equation:

$${}^{7}_{4}\text{Be} + {}^{0}_{-1}e \rightarrow {}^{7}_{3}\text{Li}$$

This decay mode is possible as mass of ^{7}Be is greater than the mass of ^{7}Li.

3. A reservoir contains 20,000 metric tons (1 metric ton = 1000 kg) of water. If the proton mean life is 10^{32} years, how many decays you expect to observe in one year? Assume that your detector is 50% efficient and free protons and those bound in nuclei have same decay probability.

Solution: Weight of water in the reservoir = 2×10^3 metric tons = 2×10^7 kg. Each H_2O molecule contains 10 protons and its molecular weight is 18.

Therefore, 18 kg of water contains $10 \times 6.023 \times 10^{26}$ protons.

$$2 \times 10^7 \text{ kg of water have } \frac{10}{18} \times 6.023 \times 10^{26} \times 2 \times 10^7 \text{ protons}$$
$$= 6.70 \times 10^{33} \text{ protons}$$

Therefore, the expected number of proton decays observed per year

$$= \frac{6.70 \times 10^{33}}{\tau_p} = \frac{6.70 \times 10^{33}}{10^{32}}$$
$$= 67.0 \text{ decays/year}$$

Since the efficiency of the detector is 50%, therefore the number of proton decays observed per year = 67.0 × 0.5 = 33.5 decays/year

Unsolved Problems

1. Find whether ${}^{64}_{29}$Cu is stable or not. If not, find the decay mode. Given:

$$M(^{64}\text{Cu}) = 63.929766 \text{ amu}$$
$$M(^{64}\text{Zn}) = 63.929145 \text{ amu}$$
$$M(^{64}\text{Ni}) = 63.927968 \text{ amu}$$

2. Calculate the maximum energy of electrons emitted in β^--decay of ^{12}B. Given:

$$M(^{12}\text{B}) = 12.014353 \text{ amu}$$
$$M(^{12}\text{C}) = 12.0000 \text{ amu}$$

[**Ans.** 12.9 MeV]

3. Sixty milligrams of naturally occurring potassium shows β^--activity of 1.9 disintegrations per second. Isotope ^{40}K present in natural potassium is responsible for this activity. ^{40}K is 0.012% of the natural potassium. Find the half-life of ^{40}K.

[**Ans.** 1.25×10^9 years]

4. One student claims that he has observed ^{16}N to decay by β^--particle emission. Other student claims that ^{16}N decays by α-emission. Which student is correct? Given:

$$M(^{16}N) = 16.006100 \text{ amu}, M(^{16}O) = 15.9945915 \text{ amu}$$
$$M(^{12}B) = 12.014353 \text{ amu}, M(^{4}\mu_e) = 4.002603 \text{ amu}$$
$$M(\beta^-) = 0.0005486 \text{ amu}.$$

[**Ans.** first student is correct]

5. A radioisotope emits β-particles. The counts due to β-activity of a freshly prepared sample of this element is measured with a GM counter, the following data are obtained.

Time (minute)	*Counts*
0	1080
20	732
40	509
70	300
100	175
130	101
160	59
190	37

(i) Plot the logarithm of the counts against time.
(ii) Determine the disintegration constant from the slope of the curve.
(iii) Find the half-life of the element.

[**Ans.** (ii) 2.94×10^{-4} s^{-1}, (iii) 39.2 min]

6. ^{64}Cu nucleus decays by β^-, β^+ and electron capture with a half-life of ~13 hrs. Write down the nuclear reactions for these three processes and identify the daughter nucleus in each case. [**Ans.** ^{64}Zn, ^{64}Ni, ^{64}Ni]

7. ^{126}I nuclei decay by negatron emission, positron emission and electron capture. Write the decay equation for each mode of decay and identify the daughter nuclide.

[**Ans.** ${}^{126}_{53}I \rightarrow {}^{126}_{54}Xe + \beta^- + \overline{\nu}$, ${}^{126}_{53}I \rightarrow {}^{126}_{52}Te + \beta^+ + \nu$, ${}^{126}_{53}I + e^- \rightarrow {}^{126}_{52}Te + \nu$]

8. Out of the two nuclei ^{64}Ni and ^{64}Cu, which can undergo β^--decay? Given the masses of $^{64}Ni = 63.927966$ amu, $^{64}Cu = 63.929764$ amu, $^{64}Zn = 63.929142$ amu and $m_e = 5.4858 \times 10^{-4}$ amu. [**Ans.** ^{64}Ni decay not possible, ^{64}Cu decay possible]

Section 3.7

Solved Problems

1. Refer to Figure 3.24, the first excited state of ^{60}Ni is at 1332 keV and the second is at 2505 keV. Second excited state decays to first state, which in turn decay to ground state. Find the energies of the two γ-rays.

Solution: When nucleus makes a transition from 2505 keV state to 1332 keV state, energy of the γ-rays emitted = 2505 – 1332 = 1173 keV

When nucleus makes a transition from first excited state to ground state, energy of the γ-rays emitted = 1332 – 0 = 1332 keV.

2. ^{137}Cs decays by β^- emission as shown in Figure 3.23. Find the energies of K and L shell conversion electrons. The binding energies of K and L shell electrons in ^{137}Ba is 37.4 and 6.0 keV respectively.

Solution: Energy of the conversion electron $E_c = E_\gamma - I_{be}$.

When conversion takes place from K shell

$$E_{cK} = E_\gamma - I_{beK} = 662 - 37.4 = 624.6 \text{ keV}$$

When conversion takes place from K shell

$$0E_{cL} = E_\gamma - I_{beL} = 662 - 6.0 = 656 \text{ keV}$$

3. ^{137}Cs decays by emitting two groups of β^--particles of energies 0.51 and 1.17 MeV. The daughter nucleus ^{137}Ba also emits a γ-ray of energy 662 keV. Use this information to sketch the decay scheme. You may assume that the most energetic β^--particle leaves the daughter nucleus in its ground state.

Solution: While constructing this decay scheme, we allow a possible error of about 1 keV. As the most energetic β^--particle leaves the daughter nucleus in the ground state, therefore the 1.17 MeVβ-particle populates the ground state of the daughter nucleus. The energy of the ground state is assumed to be 0. β-particle of energy 0.51 MeV leaves the nucleus in the first excited state. The energy of the first excited state will be 1.17 – 0.51 = 0.66 MeV or 660 keV.

The decay scheme constructed is as shown.

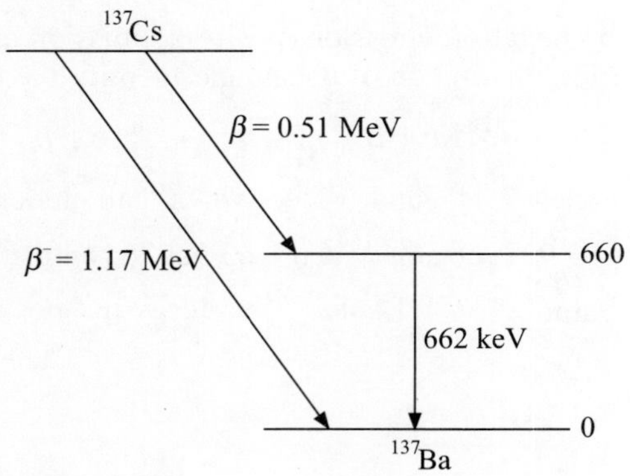

4. The daughter nucleus of a given α emitter has several excited states and therefore the kinetic energy of an emitted particle can have one of several possible values. For a particular heavy nucleus, the energies (in MeV) of α-particles are:

5.545, 5.513, 5.486, 5.469, 5.443, 5.417, 5.389.

The daughter nucleus formed after α decay emits several γ-rays with the following energies (in keV): 26, 33, 43, 56, 60, 99, 103, 125.

Use this data to draw a decay scheme indicating the energy levels and marking the γ-ray transitions. You may assume that the most energetic α-particle leaves the daughter nucleus in its ground state.

Solution: While constructing this decay scheme, we allow a possible error of about 1 keV. As the most energetic α-particle leaves the daughter nucleus in the ground state, therefore the 5.545 MeV α-particle populates the ground state of the daughter nucleus. The energy of the ground state is assumed to be 0. α-particle of energy 5.513 MeV leaves the nucleus in the first excited state. The energy of the first excited state will be 5.545 – 5.513 = 0.032 MeV or 32 keV. Similarly, 5.486 MeV α-particle leaves the daughter nucleus in the second excited state, whose energy is 5.545 – 5.486 = 0.059 MeV or 59 keV. In this way, we construct the level scheme of the daughter nucleus.

After constructing the level scheme, the energies of γ-rays emitted by the daughter nucleus can be estimated by measuring the energy difference between various levels by hit and trial. If the energy difference between any two levels matches with the γ-ray energy in the given list, then most probably that γ-ray is due to transition between those two levels.

The decay scheme constructed and the position of various γ-rays is as shown.

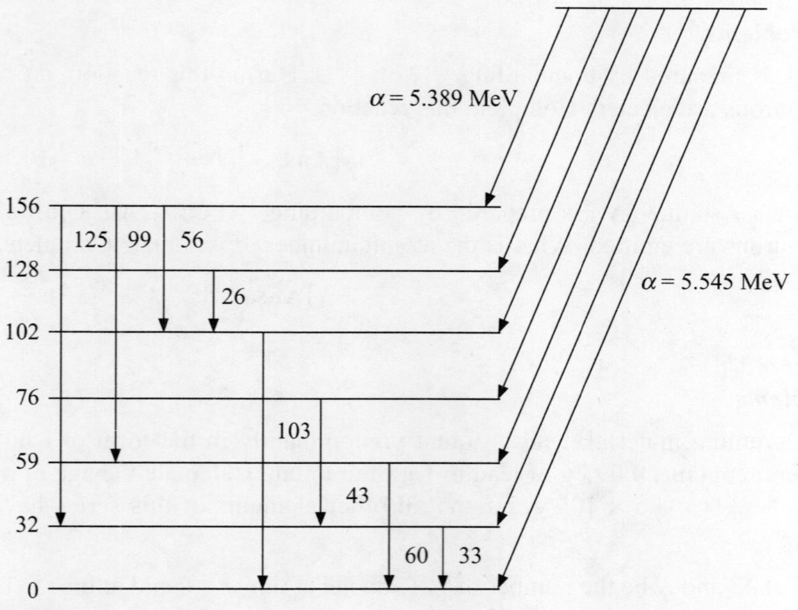

Unsolved Problems

1. The measured width of the γ-ray of ^{55}Fe is 3.4×10^{-8} eV. Determine the lifetime of this excited state. The relation between the width (γ) and the lifetime (τ) is $\gamma \times \tau \cong h$, where h is Planck constant. [**Ans.** 19.36 ns]

2. The lifetime of the excited state of ^{203}Hg is 8×10^{-13} s. Determine the width of the γ-ray emitted when the nucleus goes to the ground state. The relation between the width (γ) and the lifetime (τ) is $\gamma \times \tau \cong h$, where h is Planck constant.

   [Ans. 5.17×10^{-3} eV]

3. ^{226}Ra decays by emitting two groups of α-particles of energies 4.785 and 4.602 MeV. The daughter nucleus ^{222}Rn also emits a γ-ray of energy 186 keV. Use this information to sketch the decay scheme. You may assume that the most energetic α-particle leaves the daughter nucleus in its ground state.

4. For a nuclide X with the decay scheme how many γ-rays are emitted per 100 disintegrations of X if the coefficient for internal conversion is 0.25?

Section 3.8

Solved Problem

1. An isotope of element 105 was prepared by bombarding a $^{249}_{98}$Cf target with $^{15}_{7}$N, $^{249}_{98}$Cf emits four neutrons. Write the equation for the nuclear reaction.

Solution: The reaction is

$$^{249}_{98}\text{Cf} + ^{15}_{7}\text{N} \rightarrow ^{260}_{105}\text{X} + 4\,^{1}_{0}n$$

Unsolved Problems

1. $^{243}_{97}$Bk is prepared by bombarding ^{14}N on ^{238}U. During this reaction, an α-particle and 5 neutrons are ejected. Complete this reaction.

   [Ans. $^{14}_{7}\text{N} + ^{238}_{92}\text{U} \rightarrow ^{243}_{97}\text{Bk} + ^{4}_{2}\text{He} + 5\,^{1}_{0}n$]

2. Element Astatine (At) is prepared by bombarding ^{12}C on $^{197}_{79}$Au. During this process, 4 neutrons are emitted. What is the atomic number of Astatine? Complete the reaction.

   [Ans. 85; $^{12}_{6}\text{C} + ^{197}_{79}\text{Au} \rightarrow ^{205}_{85}\text{At} + 4\,^{1}_{0}n$]

Section 3.9

Solved Problems

1. In a uranium material, lead is found predominantly in the form of isotope 206. The mineral contains 0.093 g of lead in 1 g of uranium. Calculate the age of mineral. Half-life of $^{238}_{92}$U $= 4.5 \times 10^9$ years and all other elements of this series have very short-lives.

Solution: Let N_0 and N be the number of ^{238}U nuclei at time $t = 0$ and at time t. The difference $N_0 - N$ is the number of nuclei of ^{206}Pb at time t.

Now, $N = N_0 e^{-\lambda t}$, where λ is decay constant of ^{238}U

Now, $t_{1/2} = 4.5 \times 10^9$ years

So, $$\lambda = \frac{0.6931}{4.5 \times 10^9}\ \text{years}^{-1}$$

$$= 1.54 \times 10^{10} \text{ years}^{-1}$$

Also,

$$N_0 - N = N_1 = N_0(1 - e^{-\lambda t})$$

$$\frac{\text{Number of } ^{206}\text{Pb atoms at time } t}{\text{Number of } ^{238}\text{U atoms at time } t} = \frac{N_1}{N}$$

$$= \frac{1 - e^{\lambda t}}{e^{-\lambda t}}$$

$$= e^{\lambda t} - 1$$

So,

$$e^{\lambda t} = 1 + \frac{N_1}{N}$$

or

$$\lambda t = \log_e \left(1 + \frac{N_1}{N}\right)$$

So,

$$t = \frac{1}{\lambda} \log_e \left(1 + \frac{N_1}{N}\right)$$

Number of nuclei in 238 g of ^{238}U = 6.023×10^{23}

Number of nuclei in 1 g of ^{238}U (N) = $6.023 \times 10^{23} \times \frac{1}{238}$

Number of nuclei in 206 g of ^{206}Pb = 6.023×10^{23}

Number of nuclei in 0.093 g of ^{206}Pb (N_1) = $6.023 \times 10^{23} \times \frac{0.093}{206}$

Now,

$$\frac{N_1}{N} = 6.023 \times 10^{23} \times \frac{0.093}{206} \times \frac{238}{6.023 \times 10^{23}}$$

$$= 0.1074$$

Therefore,

$$t = \frac{1}{\lambda} \log_e \left(1 + \frac{N_1}{N}\right)$$

$$= \frac{1}{1.54 \times 10^{-10}} \log_e(1 + 0.1074)$$

or

Age of the mineral $t = 0.6626 \times 10^9$ years

2. A sample of carbon from an ancient wooden boat piece gives 5 count/min/g of carbon due to ^{14}C present in it. If freshly cut wooden piece gives 16 count/min, what is the age of the boat? Half-life of ^{14}C = 5760 years.

Solution: We have

$$N = N_0 e^{-\lambda t}$$

or

$$e^{\lambda t} = \frac{N_0}{N} \Rightarrow \lambda t = \log_e \frac{N_0}{N}$$

or

$$t = \frac{1}{\lambda} \log_e \frac{N_0}{N}$$

Since

$$\lambda = \frac{0.693}{T_{1/2}}$$

Therefore,

$$t = \frac{T_{1/2}}{0.693} \log_e \frac{N_0}{N}$$

Substituting various values, we get

$$t = \frac{5760}{0.693} \log_e \frac{16}{5}$$

or

$$t = 9668 \text{ years}$$

3. (a) What type of reaction is taking place in a nuclear reactor?
(b) What is the average energy released per reaction in a reactor?
(c) Why are the reaction products radioactive?
(d) What is the necessity of a moderator in a nuclear reactor? Are light or heavy elements preferred for moderators and why?

Solution:

(a) In nuclear reactors, fission of heavy nuclei like ^{233}U, ^{235}U, ^{239}Pu, take place.
(b) In fission of heavy nuclei, about 200 MeV of energy is released per fission.
(c) In fission, some of the fission fragments are left with high excitation energy.
(d) The fission is caused by thermal neutrons (having average energy of about 0.025 eV). However, in fission, fast neutrons are produced. To slow down these fast neutrons, moderator is required. Light nuclei (like H_2, D_2, D_2O, Carbon, etc.) are good moderators as energy transferred in a single collision with light moderator is quite large.

4. A by-product of fast breeder reactor is ^{239}Pu. It can be used as fuel in nuclear reactors. It is an α-emitter with a half-life of 24,000 years. Consider 10 g of ^{239}Pu at $t = 0$.

(a) What is the number of nuclei at $t = 0$?
(b) What is the initial activity?
(c) For how long you need to store ^{239}Pu until it has decayed to a safe activity level of 0.1 Bq?

Solution:

(a) 239 kg of ^{239}Pu contains 6.023×10^{26} nuclei.

0.01 kg of ^{239}Pu will contain $\dfrac{6.023 \times 10^{26}}{239} \times 0.01$ nuclei

$$N = 2.52 \times 10^{22} \text{ nuclei}$$

(b) Half-life of ^{239}Pu = 24,000 years

Therefore, decay constant

$$\lambda = \frac{\ln 2}{t_{1/2}} = \frac{0.6931}{24,000 \times 365 \times 2400 \times 3600}$$

$$= 9.158 \times 10^{-13} s^{-1}$$

Therefore,

$$\text{Initial activity} = \lambda \times N$$

$$= 9.158 \times 10^{-13}\ 2.52 \times 10^{22}$$

$$= 2.31 \times 10^{10} \text{ disintegrations per second}$$

(c) We have

$$A = A_0 e^{-\lambda t}$$

which gives

$$t = \frac{1}{\lambda} \ln \frac{A_0}{A}$$

Substituting various values, we get

$$t = \frac{1}{9.158 \times 10^{-13}} \ln \frac{2.31 \times 10^{10}}{0.1}$$

or

$$t = 2.86 \times 10^{13} \text{ s} = 0.91 \times 10^{6} \text{ years}$$

5. Radioactive dating of once living objects like trees, etc. is done using the isotope
(a) ^{40}K (b) ^{12}C (c) ^{14}C

Solution: ^{14}C maintains a small but fixed proportion in the CO_2 present in the atmosphere. A living species also maintains the same isotopic proportion of ^{14}C. After it dies, this proportion starts changing. This provides a means of dating the time of death. ^{12}C is stable and cannot be used for dating purposes, ^{40}K has a half-life of 1.25×10^9 years and is also not present in atmosphere and hence cannot be used for dating purposes for once living species.

Unsolved Problems

1. A typical smoke detector used in homes has approximately 0.5 μCi of ^{241}Am in it. How many atoms of ^{241}Am does this represent? Given: $\lambda = 0.0015$ year^{-1}.
[**Ans.** 1.05×10^4 atoms]

2. A sample of carbon from an ancient tree gives 3.2 count/min/g of carbon due to ^{14}C present in it. What is the approximate age of the tree? Half-life of ^{14}C = 5760 years and a freshly cut wooden piece gives 16 count/min? [**Ans.** 13,380 years]

3. Some of the carbon atoms in trees and lumber are ^{14}C. There are none among the carbon atoms in the petroleum products. Why?

4. A piece of charcoal recovered from a fire pit in an ancient camp site has average β-activity due to ^{14}C content of charcoal as 9.2 count/min/gm. What is the age of the charcoal sample? Given the absolute activity of the ^{14}C in the wood from a living tree is independent of the species of the tree and its average value is 16.0 disintegrations per minute per gram of sample. **[Ans.** 4600 years]

5. ^{206}Pb is found in certain U ore. This indicates that the lead is of radioactive origin. What is the age of the U ore if it now contains 0.80 g of ^{206}Pb for each gram of ^{238}U? **[Ans.** 4.25×10^9 years]

6. At present the abundance of the isotopes in natural uranium are 0.72% of $^{235}_{92}U$ and 99.28% of ^{238}U. What were the values at the time of formation of the earth, i.e. 4.55×10^9 years ago? The half-lives of $^{235}_{92}U$ and ^{238}U are 7.13×10^8 years and 4.51×10^9 years respectively. **[Ans.** 23.1%, 76.9%]

7. ^{232}Th have a half-life of 1.41×10^{10} years. Its successive radioactive decays lead to the stable isotope ^{208}Pb. A piece of rock contains 3.65 g of ^{232}Th and 1.25 g of ^{208}Pb. Assume that all the ^{208}Pb has come from the decay of ^{232}Th, what is the age of the rock as determined from the Th/Pb ratio? **[Ans.** 6.58×10^9 years]

8. The ^{14}C content decreases after the death of a living system with a half-life of 5600 years. If the ^{14}C content of an old piece of wood is found to be 6.5% of that of an equivalent present-day sample, how old is the piece of wood? **[Ans.** 22,088 years]

9. ^{14}C decays with a half-life of 5600 years.
   (a) What is the nature of this decay and what are the final products?
   (b) Find the age of a dead tree whose radioactivity is 1/4 of that of a comparable but living young tree. **[Ans.** $^{14}C \rightarrow {}^{14}N + e^- + \bar{\nu}_e$, 11,200 year]

10. A human body weighing 70 kg contains about 23% carbon, out of which about 10^{-10}% are ^{14}C atoms. Determine the total number of ^{14}C atoms in the body and also determine its activity. Given half-life of ^{14}C is 5715 years. **[Ans.** 8.08×10^{14}, 84nCi]

11. A normal human body contains about 150 g of potassium. Determine the total number of atoms of ^{40}K present in the body. Also determine its activity. Given abundance of ^{40}K in natural potassium is 0.0117% and its half-life is 1.27×10^9 years. **[Ans.** 2.71×10^{20}, 127nCi]

REVIEW QUESTIONS

Short Answer Type

1. Why do the number of neutrons tend to exceed the number of protons in stable nuclei?
2. Explain the difference between X-rays and γ-rays.
3. Explain the nuclear potential barrier for α-decay with suitable diagram and hence show how this decay is not possible classically.
4. In case of successive radioactive disintegration, what is the meaning of permanent equilibrium?
5. Differentiate between a nuclear decay process and a nuclear reaction.
6. A certain radioactive element disintegrates for an interval of time equal to its mean life. What fraction of element has disintegrated?

7. Discuss the factors which support the neutrino hypothesis of β-decay.
8. What are radioisotopes? Discuss their three uses.
9. Describe the law of absorption of γ-rays in matter.
10. Explain Geiger–Nuttal law.
11. α-rays and γ-rays can have the same energy. What is the difference between them?
12. What are internal conversion and electron capture processes?
13. What is neutrino? How does its emission explain β-spectrum?
14. What is radioactive dating?
15. Why is β-spectrum continuous?
16. What does the term nuclear winter mean?
17. How are atomic number and mass number changes during α-, β- and γ-decays?
18. What is a radioactive tracer? Give its two applications.
19. What are the important features of β-ray spectra? What is end point energy?
20. What do you understand by internal conversion?
21. How is the age of earth calculated from the radioactive decay?
22. Define two units to measure the intensity of radioactivity.
23. Define decay constant.
24. What is the principle of radioactive dating?
25. Although nucleus is positively charged, how will you explain β-rays coming out of it?
26. What do you mean by end point energy of a β-particle?
27. List some similarities and differences between the properties of photons and neutrinos.
28. Explain the concept of parity violation in β-decay.
29. Name three types of radiations which are emitted by radioactive substances.
30. Why is energy of α-particles discrete in α-decay?
31. Explain the phenomenon of α-decay on the basis of uncertainty principle.
32. What happens to the atomic number and mass number of a nucleus when it
    (i) emits electron,
    (ii) emits positron,
    (iii) captures electron.
33. What do you mean by radioactivity?
34. Explain the different modes of radioactive disintegration.
35. What is the S.I. unit of radioactivity? Define it.
36. Define average life of a radioactive material.
37. Why is neptunium series not observed in nature?
38. Is α-spectrum a line spectrum or continuous spectrum? Explain.
39. Write few important properties of α-particles.
40. Define the range of α-particles.
41. What is internal conversion process?
42. Define internal conversion coefficient.
43. What is induced radioactivity? Give one example.

Long Answer Type

1. What is β-decay? Explain its spectrum. What led Pauli to predict the existence of neutrino?
2. Give the α-decay scheme and discuss the Geiger–Nuttal law completely.
3. What is β-decay? Describe neutrino hypothesis of β-decay. What is the evidence for existence of neutrino?
4. Discuss the theory of successive decay of radioactive substances and obtain the conditions for transient and secular equilibrium.
5. What is neutrino hypothesis? What is the need and significance of this hypothesis?
6. Explain how quantum mechanics helps to explain the α-decay process. Also show mathematically that α-particle takes up most of the Q-value as its kinetic energy.
7. Compare the positron emission and electron capture.
8. Explain the Geiger–Nuttal law in terms of the range of α-particles and also in terms of the energy of α-particles. Also compare the graphs obtained for different radioactive series in accordance with this law.
9. Describe internal conversion process.
10. What do you understand by successive disintegration and what is the condition for permanent equilibrium?
11. Describe the important features of β-ray spectrum emitted by radioactive nuclides. Discuss the arguments which led Pauli to postulate the existence of neutrino. Outline the neutrino theory of β-decay. How does it explain the continuous energy spectrum in β-decay?
12. What is the range of α-particles? Discuss the Geiger–Nuttal law, giving its importance.
13. What is radioactivity? Deduce statistical law of radioactivity. What is decay constant, half-life and average life of radioactive substances?
14. Give the theory of α-decay.
15. Briefly explain the phenomena:
    (i) Electron capture.
    (ii) Internal conversion.
16. What do you understand by half-life and mean life of a radioactive substance? Derive expressions for them.
17. Explain the Geiger–Nuttal law and discuss its importance.
18. Describe neutrino hypothesis of β-decay. What is the evidence for the existence of neutrino? Give the qualitative account of the theory of β-decay.
19. What do you mean by radioactive series? Explain thorium series. Why is neptunium series generally not considered?
20. Explain the theory of β-decay. How did it help Pauli in predicting neutrino?
21. Write short notes on:
    (i) Geiger–Nuttal law.
    (ii) Internal conversion.

22. Explain the term radioactive dating, radioactive series, radioactive tracers, and radioactive decay constant.
23. Give the theory of β-decay.
24. What is neutrino? What were the factors which led Pauli to give his hypothesis? Show that this hypothesis has been able to explain them.
25. What is α-decay? Discuss the energetics of α-decay.
26. Explain neutrino hypothesis.
27. Explain the processes of β-decay, positron emission and electron capture.
28. What is Geiger–Nuttal law?
29. What are the different modes of β-radioactivity?
30. Explain the energetics of β-decay.
31. Explain how neutrino hypothesis explains the process of β-decay.
32. Describe the salient features of β-decay process.
33. Describe the salient features of β-decay of nuclei. What is the evidence for the existence of neutrino?
34. Discuss the phenomenon of α-decay qualitatively.
35. State and explain the laws of radioactive disintegration. Define disintegration constant and half-life.
36. Explain qualitatively the phenomenon of γ-decay in nuclei.
37. Explain the method of radioactive dating.
38. Discuss parity violation in β-decay.
39. Explain the theory of α-decay and derive the Geiger–Nuttal law.
40. Determine the kinetic energy of α-particles emitted in an α-decay in terms of Q-value reaction.
41. What are the laws of radioactive disintegration? Derive the relation $N = N_0 e^{-\lambda t}$.
42. Show that radioactive decay is exponential in nature.
43. Prove that it will take infinite time to disintegrate whole of the radioactive material.
44. Define activity. Derive the relation $A = A_0 e^{-\lambda t}$.
45. Define half-life of a radioactive material. Find the relation between half-life and disintegration constant.
46. What is average life of a radioactive material? Prove that the average life of a radioactive material is given by the relation $\tau = \dfrac{t_{1/2}}{0.6931}$.
47. Explain secular and transient equilibrium.
48. Define α-decay. Find the expression for the disintegration energy in the α-decay process.
49. Define the range of α-particles. On what factors does it depend?
50. Explain the conditions for spontaneous emission of β-particles.
51. Write short notes on detection of neutrino (ν) and antineutrino ($\overline{\nu}$).
52. Discuss briefly the applications of radioactivity.

Chapter 4

Nuclear Reactions

4.1 INTRODUCTION

In Chapter 1, we have discussed how Rutherford discovered the existence of the nucleus by bombarding fast moving α-particles on thin gold foil. Similarly, Chadwick in 1932 bombarded fast moving α-particles on ^{9}Be nucleus, which led to the discovery of neutrons.

The question arises, what happens when we bombard fast moving, charged/neutral particles (also known as projectiles) on a stable nucleus (also called target nucleus)? Experimentally, it has been observed that these particles can interact with the nucleus or can be captured by the nucleus. The nucleus so formed can be different from the initial nucleus and is invariably unstable. It decays by emitting γ-ray/a particle/a group of particles. The daughter nucleus so produced could be stable or unstable. In case it is unstable, it follows the laws of radioactive decay.

Nuclear Reaction

This process of bombarding a target nucleus by fast moving projectiles and the subsequent interaction between the two, which alters the composition, energy, etc. of the target nucleus is known as *nuclear reaction.* This is also known as transmutation of one element into another.

Suppose a target nucleus X is bombarded by particle a. During this process a new nucleus Y is formed and a particle (or γ-ray) b is emitted, then this nuclear reaction is written as:

$$a + X \rightarrow Y + b$$

This reaction is also denoted as $X(a, b)Y$. For example, consider the following reaction:

$$^4_2\text{He} + {}^9_4\text{Be} \rightarrow {}^{12}_6\text{C} + {}^1_0n$$

It can be written as:

$$^9_4\text{Be}(^4_2\text{He}, {}^1_0n)\,^{12}_6\text{C}$$

or

$$^9_4\text{Be}(\alpha, n)\,^{12}_6\text{C}$$

If more than one particle say b_1, b_2 and b_3 are emitted, then the reaction is written as:

$$a + X \rightarrow Y + b_1 + b_2 + b_3$$

and it is denoted as $X(a, b_1\ b_2\ b_3)Y$. Consider the following reaction:

$$^{238}_{92}\text{U} + {}^{14}_{7}\text{N} \rightarrow {}^{243}_{97}\text{Bk} + {}^{4}_{2}\text{He} + {}^{1}_{0}n + {}^{1}_{0}n + {}^{1}_{0}n + {}^{1}_{0}n + {}^{1}_{0}n$$

Alternatively, we can write it as:

$$^{238}_{92}\text{U} + {}^{14}_{7}\text{N} \rightarrow {}^{243}_{97}\text{Bk} + {}^{4}_{2}\text{He} + 5\,{}^{1}_{0}n$$

The concise version is:

$$^{238}_{92}\text{U}({}^{14}_{7}\text{N}, \alpha\ 5n)\,{}^{243}_{97}\text{Bk}$$

In most of the nuclear reactions, mass and atomic number of both target and resulting nuclei are also written. Further, if the incident particle, emitted particles are composite particles (like heavy ions), their mass and charge numbers are also displayed. If there is no confusion, then in some cases, charge number is dropped.

The first nuclear reaction was performed by Rutherford in 1919. He bombarded nitrogen gas with α-particles obtained from natural radioactive source. The reaction is:

$$^{4}_{2}\text{He} + {}^{14}_{7}\text{N} \rightarrow {}^{17}_{8}\text{O} + {}^{1}_{1}\text{H}$$

or

$$^{14}_{7}\text{N}(\alpha, p)\,{}^{17}_{8}\text{O}$$

or

$$^{14}_{7}\text{N}({}^{4}_{2}\text{He}, {}^{1}_{1}\text{H})\,{}^{17}_{8}\text{O}$$

Similarly, the reaction used by Chadwick to discover neutron can be written as:

$$^{4}_{2}\text{He} + {}^{9}_{4}\text{Be} \rightarrow {}^{12}_{6}\text{C} + {}^{1}_{0}n$$

or

$$^{9}_{4}\text{Be}({}^{4}_{2}\text{He}, {}^{1}_{0}n)\,{}^{12}_{6}\text{C}$$

It is not essential that nuclear reaction gets initiated only by particles emitted by radioactive materials. Nuclear reactions can be initiated by high energy charged particles. The charged particles can be imparted high energy by devices called *particle accelerators* like Cyclotron, Linear Accelerator, etc. (to be discussed in Chapter 6). In these devices charged particles are accelerated to very high energy and then they are bombarded on a target. These days accelerators are available that can accelerate particles starting from protons to uranium nuclei. The energy of α-particles emitted during radioactivity or by radioactive elements is limited up to 10 MeV. But with these devices, changed particles can be accelerated to very, very high energies, greater than few thousands of MeV.

The first nuclear reaction by accelerating protons was performed by Cockcroft and Walton in 1932.

$${}^{7}_{3}\text{Li} + {}^{1}_{1}\text{H} \rightarrow {}^{4}_{2}\text{He} + {}^{4}_{2}\text{He}$$

The study of nuclear reactions is extremely important. Almost all the properties of the nucleus such as mass, size, charge distribution, etc. have been obtained from such a study. Further, this study provides information about the nature of nuclear forces and reaction mechanisms.

Below we discuss some important types of nuclear reactions.

4.2 TYPES OF NUCLEAR REACTIONS

There are many ways of classifying nuclear reactions. Two commonly used ways to classify nuclear reactions are:

1. Reactions based on the reaction mechanism.
2. Reactions based on the mass of projectile.

4.2.1 Reactions Based on the Reaction Mechanism

Elastic Scattering

In elastic scattering incident particle strikes the target nucleus and leaves the nucleus without any energy loss or in elastic scattering there is no loss or gain of energy of the incident projectile. The direction of the incident particle generally changes. Some examples of elestic scattering are:

$${}^{4}_{2}\text{He} + {}^{208}_{82}\text{Pb} \rightarrow {}^{208}_{82}\text{Pb} + {}^{4}_{2}\text{He}$$

$${}^{1}_{0}n + {}^{208}_{92}\text{U} \rightarrow {}^{208}_{92}\text{U} + {}^{1}_{0}n$$

Inelastic Scattering

In inelastic scattering the incident particle loses some of its kinetic energy to the target nucleus. This raises the internal energy of the nucleus. The target nucleus is raised to the higher excited quantum state. Generally, this excited nucleus decays by γ-emission. Some examples of inelastic scattering are:

$${}^{1}_{1}\text{H} + {}^{27}_{13}\text{Al} \rightarrow {}^{27}_{13}\text{Al}^{*} + {}^{1}_{1}\text{H}$$

$${}^{4}_{2}\text{He} + {}^{56}_{26}\text{Fe} \rightarrow {}^{56}_{26}\text{Fe}^{*} + {}^{4}_{2}\text{He}$$

where * on the residual or product nucleus indicates that the nucleus is in excited state.

Radiative Capture

In such reactions incident particle is captured by the nucleus with the emission of γ-rays. Some examples of radiative capture reaction are as under:

$${}^{1}_{1}\text{H} + {}^{27}_{13}\text{Al} \rightarrow \left[{}^{28}_{14}\text{Si}\right]^{*} \rightarrow {}^{28}_{14}\text{Si} + \gamma$$

$${}^{4}_{2}\text{He} + {}^{12}_{6}\text{C} \rightarrow \left[{}^{16}_{8}\text{O}\right]^{*} \rightarrow {}^{16}_{8}\text{O} + \gamma$$

here * indicates that the nucleus is in excited state.

Compound Nuclear Reactions

The incident particle is absorbed by the target nucleus and a compound system is formed. This compound system lives for a long time $\sim 10^{-14}$ to 10^{-15} seconds (compared to the time $\sim 10^{-21}$ to 10^{-22} seconds taken by a low energy particle ~ 5 MeV to cross a distance equivalent to nuclear diameter) and then a particle or group of particles is ejected. Some examples of compound nuclear reactions are:

$$^{4}_{2}\text{He} + ^{27}_{13}\text{Al} \rightarrow \left[^{31}_{15}\text{P}\right] \rightarrow ^{30}_{15}\text{P} + ^{1}_{0}n$$

$$^{4}_{2}\text{He} + ^{60}_{28}\text{Ni} \rightarrow \left[^{64}_{30}\text{Zn}\right] \rightarrow ^{62}_{29}\text{Cu} + ^{1}_{1}\text{H} + ^{1}_{0}n$$

Photodisintegration

High energy γ-ray photons are absorbed by the nucleus and the nucleus ejects a particle or a group of particles. For example,

$$^{2}_{1}\text{H} + \gamma \rightarrow ^{1}_{1}\text{H} + ^{1}_{0}n$$

$$^{16}_{8}\text{O} + \gamma \rightarrow ^{11}_{6}\text{C} + 2^{1}_{1}\text{H} + 3^{1}_{0}n$$

Direct Reactions

In some cases fast moving incident particle while crossing through the nucleus or very close to the nucleus either picks up one or a few nucleons from the nucleus and keeps on moving in the forward direction or it loses one or a few nucleons to the nucleus and keep on moving in the forward direction.

First type of reactions are known as *pickup reactions* while the second type of reactions are known as *stripping reactions*. Both these reactions together are known as *direct reactions*. The typical reaction time of direct reactions is of the order of 10^{-21} to 10^{-22} seconds (compared with reaction time of compound reactions which is $\sim 10^{-14}$ to 10^{-15} seconds). Some of the pickup reactions are:

$$^{59}_{27}\text{Co} + ^{2}_{1}\text{H} \rightarrow ^{57}_{26}\text{Fe} + ^{4}_{2}\text{He}$$

$$^{45}_{21}\text{Sc} + ^{3}_{2}\text{He} \rightarrow ^{44}_{21}\text{Sc} + ^{4}_{2}\text{He}$$

And some of the stripping reactions are:

$$^{63}_{29}\text{Cu} + ^{2}_{1}\text{H} \rightarrow ^{64}_{29}\text{Cu} + ^{1}_{1}\text{H}$$

$$^{197}_{79}\text{Au} + ^{2}_{1}\text{H} \rightarrow ^{198}_{80}\text{Hg} + ^{1}_{0}n$$

Heavy Ion Reactions

The reaction in which an incident particle is a fast moving heavy ion is known as *heavy ion reaction*. For example, when $^{207}_{82}\text{Pb}$ is accelerated to high energy, i.e. $^{207}_{82}\text{Pb}$ is a projectile and is bombarded on say $^{107}_{47}\text{Ag}$ target

$$^{207}_{82}\text{Pb} + ^{107}_{47}\text{Ag} \rightarrow ^{106}_{47}\text{Ag} + ^{208}_{82}\text{Pb}$$

Another such reaction is

$$^{27}_{13}\text{Al} + ^{14}_{7}\text{N} \rightarrow ^{16}_{8}\text{O} + ^{25}_{12}\text{Mg}$$

Radioactive Decay

Typical examples of radioactive decay are α-, β- and γ-decay. These have been discussed in detail in Chapter 3.

Fission Reactions

When heavy nuclei capture incident particle, they break into generally two fragments with the emission of few particles. The most common example is:

$$^{235}_{92}\text{U} + ^{1}_{0}n \rightarrow ^{141}_{56}\text{Ba} + ^{92}_{36}\text{Kr} + 3^{1}_{0}n$$

Fusion Reactions

When two light nuclei combine to form a heavier nucleus, the reaction is known as *fusion reaction*. For example,

$$^{2}_{1}\text{H} + ^{2}_{1}\text{H} \rightarrow ^{3}_{2}\text{He} + ^{1}_{0}n$$

$$^{3}_{2}\text{He} + ^{3}_{2}\text{He} \rightarrow ^{4}_{2}\text{He} + ^{1}_{1}\text{H} + ^{1}_{1}\text{H}$$

Elementary Particle Production Reactions

When very high energy (E > 300 MeV) incident particles like protons, etc. interact with protons or neutrons, a variety of elementary particles like mesons, pions, etc. are produced. We shall discuss these reactions in detail in Chapter 8.

4.2.2 Reactions Based on the Mass of Projectile

Reactions Induced by Particles of Mass 1

Mass 1 particles are protons and neutrons. We discuss both the cases as follows:

(i) **Proton-induced reactions:**

(a) *(p, γ)-reactions:* Radiative capture reactions have high probability for lighter elements and at low energy protons ($E_p < 1$ MeV). General form of this reaction is:

$$\boxed{^{A}_{Z}X + ^{1}_{1}\text{H} \rightarrow ^{A+1}_{Z+1}Y + \gamma}$$

Some specific cases are:

$$^{27}_{13}\text{Al} + ^{1}_{1}\text{H} \rightarrow ^{28}_{14}\text{Si} + \gamma$$

$$^{59}_{27}\text{Co} + ^{1}_{1}\text{H} \rightarrow ^{60}_{28}\text{Ni} + \gamma$$

(b) *(p, α)-reactions:* For majority of low Z nuclei, Q-value for (p, α)-reactions is +ve, i.e. these reactions are exothermic. These reactions predominate, if proton energy is between 2 MeV and 5 MeV. General form of this reaction is:

$$\boxed{^{A}_{Z}X + ^{1}_{1}\text{H} \rightarrow ^{A-3}_{Z-1}Y + ^{4}_{2}\text{He}}$$

Some examples are:

$$ {}^{7}_{3}\mathrm{Li} + {}^{1}_{1}\mathrm{H} \rightarrow {}^{4}_{2}\mathrm{He} + {}^{4}_{2}\mathrm{He} $$

$$ {}^{14}_{7}\mathrm{N} + {}^{1}_{1}\mathrm{H} \rightarrow {}^{11}_{6}\mathrm{C} + {}^{4}_{2}\mathrm{He} $$

(c) *(p, n)-reactions:* The probability of (p, n)-reactions increases with increase in energy of protons. General form of this reaction is:

$$ \boxed{{}^{A}_{Z}X + {}^{1}_{1}\mathrm{H} \rightarrow {}_{Z+1}{}^{A}Y + {}^{1}_{0}n} $$

Some typical cases are:

$$ {}^{23}_{11}\mathrm{Na} + {}^{1}_{1}\mathrm{H} \rightarrow {}^{23}_{12}\mathrm{Mg} + {}^{1}_{0}n $$

$$ {}^{65}_{29}\mathrm{Cu} + {}^{1}_{1}\mathrm{H} \rightarrow {}^{65}_{30}\mathrm{Zn} + {}^{1}_{0}n $$

At higher energies $(p, 2n)$, $(p, 3n)$, etc. reactions occur.

(ii) **Neutrons-induced reactions:** Neutrons have no electric charge and, therefore, can penetrate atomic nuclei without experiencing any Coulomb repulsion. So these reactions can occur at very low energies (~0.025 eV). At these low energies reactions with charged particles are not possible. Some of the neutron induced reactions are:

(a) *(n, γ)-reactions:* These reactions have very high probability at energies ~0.025 eV (called thermal energy). In most of (n, γ)-reactions, product nucleus formed is radioactive. The general form of these reactions is:

$$ \boxed{{}^{A}_{Z}X + {}^{1}_{0}n \rightarrow {}^{A+1}{}_{Z}X + \gamma} $$

Some typical examples are:

$$ {}^{23}_{11}\mathrm{Na} + {}^{1}_{0}n \rightarrow {}^{24}_{11}\mathrm{Na} + \gamma \qquad (t_{1/2} \text{ of } {}^{24}_{11}\mathrm{Na} = 15 \text{ hours}) $$

$$ {}^{59}_{27}\mathrm{Co} + {}^{1}_{0}n \rightarrow {}^{60}_{27}\mathrm{Co} + \gamma \qquad (t_{1/2} \text{ of } {}^{60}_{27}\mathrm{Co} = 5 \text{ years}) $$

$$ {}^{24}_{12}\mathrm{Mg} + {}^{1}_{0}n \rightarrow {}^{25}_{12}\mathrm{Mg} + \gamma \qquad ({}^{25}_{12}\mathrm{Mg} \text{ stable}) $$

(b) *(n, p)-reactions:* Since these reactions are endothermic (Q-value is –ve) so they occur at relatively high energies. These reactions are represented by the relation

$$ \boxed{{}^{A}_{Z}X + {}^{1}_{0}n \rightarrow {}_{Z-1}{}^{A}Y + {}^{1}_{1}\mathrm{H}} $$

Some examples are:

$$ {}^{27}_{13}\mathrm{Al} + {}^{1}_{0}n \rightarrow {}^{27}_{12}\mathrm{Mg} + {}^{1}_{1}\mathrm{H} $$

$$ {}^{64}_{30}\mathrm{Zn} + {}^{1}_{0}n \rightarrow {}^{64}_{29}\mathrm{Cu} + {}^{1}_{1}\mathrm{H} $$

(c) *(n, α)-reactions:* General form of these reactions is:

$$ \boxed{{}^{A}_{Z}X + {}^{1}_{0}n \rightarrow {}^{A-3}_{Z-2}Y + {}^{4}_{2}\mathrm{He}} $$

Some typical examples are:

$$^{27}_{13}\text{Al} + ^{1}_{0}n \rightarrow ^{24}_{11}\text{Na} + ^{4}_{2}\text{He}$$

$$^{64}_{30}\text{Zn} + ^{1}_{0}n \rightarrow ^{61}_{28}\text{Ni} + ^{4}_{2}\text{He}$$

(d) *(n, 2n)-reactions:* These reactions have –ve *Q*-value, so high energy neutrons are needed for these reactions. These reactions are represented by

$$\boxed{^{A}_{Z}X + ^{1}_{0}n \rightarrow ^{A-1}_{Z}X + ^{1}_{0}n + ^{1}_{0}n}$$

Typical examples are:

$$^{27}_{13}\text{Al} + ^{1}_{0}n \rightarrow ^{26}_{13}\text{Al} + ^{1}_{0}n + ^{1}_{0}n$$

$$^{64}_{30}\text{Zn} + ^{1}_{0}n \rightarrow ^{63}_{30}\text{Zn} + ^{1}_{0}n + ^{1}_{0}n$$

Reactions Induced by Particles of Mass 2

Only projectile of mass 2 is deuteron.

(i) **Deuteron-induced reactions:** Deuteron is very loosely bound system (binding energy of deuteron ~ 2.2 MeV), so a large number of reactions are possible with deuterons and the most predominant being stripping and pickup reactions.

(a) *(d, p)-reactions:* The general (*d*, *p*)-reaction is:

$$\boxed{^{A}_{Z}X + ^{2}_{1}\text{H} \rightarrow ^{A+1}_{Z}X + ^{1}_{1}\text{H}}$$

Examples are:

$$^{27}_{13}\text{Al} + ^{2}_{1}\text{H} \rightarrow ^{28}_{13}\text{Al} + ^{1}_{1}\text{H}$$

$$^{59}_{27}\text{Co} + ^{2}_{1}\text{H} \rightarrow ^{60}_{27}\text{Co} + ^{1}_{1}\text{H}$$

These reactions are also known as stripping reactions.

(b) *(d, n)-reactions:* The general form of this reaction is:

$$\boxed{^{A}_{Z}X + ^{2}_{1}\text{H} \rightarrow ^{A+1}_{Z+1}Y + ^{1}_{0}n}$$

Examples are:

$$^{27}_{13}\text{Al} + ^{2}_{1}\text{H} \rightarrow ^{28}_{14}\text{Si} + ^{1}_{0}n$$

$$^{59}_{27}\text{Co} + ^{2}_{1}\text{H} \rightarrow ^{60}_{28}\text{Ni} + ^{1}_{0}n$$

These are other examples of stripping reactions.

(c) *(d, t)-reactions:* These reactions are represented as:

$$\boxed{^{A}_{Z}X + ^{2}_{1}\text{H} \rightarrow ^{A-1}_{Z}X + ^{3}_{1}\text{H}}$$

Typical examples are:

$$^{27}_{13}\text{Al} + ^{2}_{1}\text{H} \rightarrow ^{26}_{13}\text{Al} + ^{3}_{1}\text{H}$$

$$^{59}_{27}\text{Co} + ^{2}_{1}\text{H} \rightarrow ^{58}_{27}\text{Co} + ^{3}_{1}\text{H}$$

These reactions are also known as pick-up reactions. At high incident energies reactions such as $(d, 2n)$, $(d, 3n)$, $(d, p\alpha)$, etc. also become possible. However, the probability of (d, γ)-reaction is very small.

Reactions Induced by Particles of Mass 3

In mass 3 category, we have 2 projectiles ${}^3_1\text{H}$ and ${}^3_2\text{He}$. Some of the reactions induced by ${}^3_1\text{H}$ are:

$${}^{59}_{27}\text{Co} + {}^3_1\text{H} \rightarrow {}^{60}_{27}\text{Co} + {}^2_1\text{H} \qquad (t, d)\text{-reaction}$$

$${}^{59}_{27}\text{Co} + {}^3_1\text{H} \rightarrow {}^{61}_{28}\text{Ni} + {}^1_0n \qquad (t, n)\text{-reaction}$$

$${}^{59}_{27}\text{Co} + {}^3_1\text{H} \rightarrow {}^{59}_{26}\text{Fe} + {}^3_2\text{He} \qquad (t, {}^3_2\text{He})\text{-reaction}$$

$${}^{59}_{27}\text{Co} + {}^3_1\text{H} \rightarrow {}^{58}_{26}\text{Fe} + {}^4_2\text{He} \qquad (t, {}^4_2\text{He})\text{-reaction}$$

The reactions induced by ${}^3_2\text{He}$ are:

$${}^{59}_{27}\text{Co} + {}^3_2\text{He} \rightarrow {}^{61}_{28}\text{Ni} + {}^1_1\text{H} \qquad ({}^3_2\text{He}, p)\text{-reaction}$$

$${}^{59}_{27}\text{Co} + {}^3_2\text{He} \rightarrow {}^{61}_{29}\text{Cu} + {}^1_0n \qquad ({}^3_2\text{He}, n)\text{-reaction}$$

$${}^{59}_{27}\text{Co} + {}^3_2\text{He} \rightarrow {}^{59}_{28}\text{Ni} + {}^3_1\text{H} \qquad ({}^3_2\text{He}, t)\text{-reaction}$$

$${}^{59}_{27}\text{Co} + {}^3_2\text{He} \rightarrow {}^{58}_{27}\text{Co} + {}^4_2\text{He} \qquad ({}^3_2\text{He}, {}^4_2\text{He})\text{-reaction}$$

$${}^{59}_{27}\text{Co} + {}^3_2\text{He} \rightarrow {}^{60}_{27}\text{Co} + {}^1_1\text{H} + {}^1_1\text{H} \qquad ({}^3_2\text{He}, 2p)\text{-reaction}$$

Reactions Induced by Particles of Mass 4

The only projectile with mass 4 is α-particle. The various reactions are as under.

(a) *(α, p)-reactions:* The general form of the reaction is:

$$\boxed{{}^A_Z X + {}^4_2\text{He} \rightarrow {}^{A+3}_{Z+1}Y + {}^1_1\text{H}}$$

Examples are:

$${}^{27}_{13}\text{Al} + {}^4_2\text{He} \rightarrow {}^{30}_{14}\text{Si} + {}^1_1\text{H}$$

$${}^{59}_{27}\text{Co} + {}^4_2\text{He} \rightarrow {}^{62}_{28}\text{Ni} + {}^1_1\text{H}$$

(b) *(α, n)-reactions:* The general form of (α, n)-reaction is:

$$\boxed{{}^A_Z X + {}^4_2\text{He} \rightarrow {}^{A+3}_{Z+2}Y + {}^1_0n}$$

Examples are:

$${}^{27}_{13}\text{Al} + {}^4_2\text{He} \rightarrow {}^{30}_{15}\text{P} + {}^1_0n$$

$${}^{59}_{27}\text{Co} + {}^4_2\text{He} \rightarrow {}^{62}_{29}\text{Cu} + {}^1_0n$$

(c) *$(\alpha, 2n)$-reactions:* The general form of this reaction is:

$$\boxed{{}^A_Z X + {}^4_2\text{He} \rightarrow {}^{A+2}_{Z+2}Y + 2{}^1_0n}$$

Typical examples are:

$$^{27}_{13}\mathrm{Al} + ^{4}_{2}\mathrm{He} \rightarrow ^{29}_{15}\mathrm{P} + 2\,^{1}_{0}n$$

$$^{59}_{27}\mathrm{Co} + ^{4}_{2}\mathrm{He} \rightarrow ^{61}_{29}\mathrm{Cu} + 2\,^{1}_{0}n$$

At higher energies reactions such as (α, 3n), (α, 4n), (α, 5n), etc. have been observed.

Reactions Induced by Heavy Ions

Heavy ions are generally projectiles with mass greater than α-particles. Depending upon the energy of heavy ions large variety of reactions are possible. Some of the heavy ion reactions are:

$$^{27}_{13}\mathrm{Al}(^{14}_{7}\mathrm{N},\ ^{16}_{7}\mathrm{N})\,^{25}_{13}\mathrm{Al}$$

$$^{27}_{13}\mathrm{Al}(^{14}_{7}\mathrm{N},\ ^{14}_{8}\mathrm{O})\,^{27}_{12}\mathrm{Mg}$$

$$^{27}_{13}\mathrm{Al}(^{12}_{6}\mathrm{C},\ \alpha\ 2n)\,^{33}_{17}\mathrm{Cl}$$

$$^{197}_{79}\mathrm{Au}(^{12}_{6}\mathrm{C},\ 4n)\,^{205}_{85}\mathrm{At}$$

$$^{238}_{92}\mathrm{U}(^{14}_{7}\mathrm{N},\ \alpha\ 5n)\,^{243}_{97}\mathrm{Bk}$$

Generally, heavy ion reactions have been used to synthesize elements with atomic number > 92.

Reactions Induced by γ-rays

The first reactions induced by γ-rays are:

$$^{2}_{1}\mathrm{H} + \gamma \rightarrow ^{1}_{1}\mathrm{H} + ^{1}_{0}n$$

$$^{9}_{4}\mathrm{Be} + \gamma \rightarrow ^{8}_{4}\mathrm{Be} + ^{1}_{0}n$$

Gamma rays used for these reactions were 2.614 MeV from ^{208}Tl. When γ-rays of very high energy are available, number of reactions induced by γ-rays are observed. Some of these reactions are:

$$^{31}_{15}\mathrm{P} + \gamma \rightarrow ^{30}_{15}\mathrm{P} + ^{1}_{0}n$$

$$^{10}_{5}\mathrm{B} + \gamma \rightarrow ^{8}_{4}\mathrm{Be} + ^{2}_{1}\mathrm{H}$$

$$^{11}_{5}\mathrm{B} + \gamma \rightarrow ^{8}_{4}\mathrm{Be} + ^{3}_{1}\mathrm{H}$$

These reactions induced by γ-rays are also known as *photodisintegrations* or *photonuclear reactions*.

Nuclear Decay and Nuclear Reactions

Dominant differences between nuclear decay (Chapter 3) and nuclear reactions are shown in Table 4.1

TABLE 4.1 Difference between nuclear decay and nuclear reactions

Nuclear decay	*Nuclear reaction*
It is a spontaneous process (i.e. no external aid is needed)	It is not a spontaneous process (i.e. external aid is needed to start a reaction)
Only α-, β- and γ-emissions are possible	In nuclear reactions all possible particles like p, n, ^2H, ^3He, ^4He, ^6Li, etc. can be emitted
Mass of the daughter nucleus either remains same or decreases	Mass of the product nucleus can also increase
Energy of the emitted particles or γ-rays is generally low	Energy of the emitted particles or γ-rays can be very high

4.3 NUCLEAR REACTION CROSS-SECTION

A question often arises that when some projectiles are bombarded on a target, an interaction or reaction between these two takes place, then how many nuclei are taking part in this reaction? Is this number small or large? Depending upon the numbers, the probability of a particular reaction becomes small or large. There must be a quantitative measure of this probability. This quantity should be easily measurable and also theoretically calculable, so that we can compare the experimental results with the theoretical calculations.

The quantity, which defines the probability of a given reaction, is called the *nuclear reaction cross-section* and it is denoted as σ.

The reaction cross-section can be visualized as an effective area around the nucleus such that if the incident particles cross that area, a nuclear reaction takes place, otherwise not. This area is also known as effective area. It is not exactly same as the geometrical area (πr^2), where r is the radius of the nucleus. This effective area could be larger, smaller or equal to the geometrical area of the nucleus. Thus, nuclear cross-section is a measure of the probability that bombarding particles at a particular energy would interact with the target.

4.3.1 Measurement of Cross-Section

Consider a beam of particles of flux I (number of particles passing a unit area per unit time) incident on a thin sheet of a material having thickness dx and face area A as shown in Figure 4.1. Further assume that thickness is so small that none of the nuclei of the thin sheet overlap each other. Thus, each nucleus has an equal probability to cause a nuclear reaction with the incident beam. Let σ be the effective area of the target nucleus. As stated earlier, this area is such that when the incident particle crosses this area the reaction always takes place. Let n be the number of target nuclei per cc in this thin sheet.

Number of nuclei in the sheet per unit face area = $n \times dx$

Total number of nuclei in the sheet = $A \times n \times dx$

Since each nucleus has an effective area σ, so the total effective area available for the reaction = $A \times n \times \sigma \times dx$

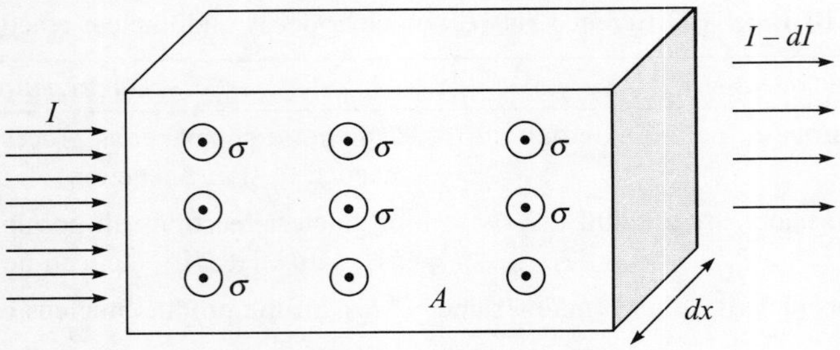

Figure 4.1 A beam of particles incident on a thin foil of face area A and thickness dx.

$$\text{Fractional effective area } f = \frac{\text{Total effective area}}{\text{Total face area}} = \frac{\sigma \times A \times n \times dx}{A}$$

$$= \sigma \times n \times dx$$

When a projectile takes part in a nuclear reaction, it is absorbed by the target nucleus, so the flux of incident particles decreases. If the fractional effective area (f) is large, then the corresponding decrease in the flux (I) is large and vice versa.

Thus, the change in the intensity dI is given by

$$dI = -fI$$

substituting for f

$$dI = -I\sigma n dx$$

or

$$\frac{dI}{I} = -\sigma n dx$$

−ve sign has been introduced because as dx increases the flux I decreases.

Assuming that at $x = 0$, $I = I_0$, integrating equation with respect to x, we get

$$\boxed{I = I_0 e^{-n\sigma x}}$$

Thus, the flux of the incident beam exponentially decreases as the thickness of the sheet increases.

4.3.2 Units of Cross-Section

Units of reaction cross-sections are the units of area. These cross-sections are normally very small ~10^{-27} to 10^{-28} m^2. Since this value is small so it is convenient to use a unit called *barn*.

$$\boxed{1 \text{ barn (b)} = 10^{-28} \text{ m}^2}$$

Smaller units of cross-sections are millibarn (mb), microbarn (μb), etc.

$$\boxed{1 \text{ mb} = 10^{-3} \text{ barn} = 10^{-31} \text{ m}^2}$$

$$\boxed{1\ \mu\text{b} = 10^{-6} \text{ barn} = 10^{-34} \text{ m}^2}$$

4.3.3 Different Types of Cross-Sections

In nuclear reactions we come across different types of reaction cross-sections. They are:

1. Partial reaction cross-section.
2. Total reaction cross-section.
3. Differential reaction cross-section.

Partial Reaction Cross-Section

For a given energy and target projectile combination, several nuclear reactions such as elastic scattering, inelastic scattering, capture reaction, disintegration reactions, etc. are possible simultaneously. There is a definite reaction cross-section for each such reaction denoted by σ_1, σ_2, σ_3 ,..., etc. These cross-sections are known as *partial reaction cross-sections.*

Total Reaction Cross-Section

As stated earlier, for a given energy and target projectile combination, many partial cross-sections may exist simultaneously. Sum of all such partial cross-sections is known as *total reaction cross-section* σ.

$$\sigma = \sigma_1 + \sigma_2 + \sigma_3 + \cdots$$

Differential Reaction Cross-Section

When the incoming particles interact with the target nuclei nuclear reaction or scattering takes place. Very often the outgoing particles have an anisotropic distribution. This means that the number of outgoing particles at different angles will be different. Also in most of the cases, the energy of the outgoing particles may be different at different angles. We measure the number of outgoing particles per second into a solid angle $d\Omega$ making an angle θ with the incident direction. For measuring this quantity, another type of cross-section is introduced. This cross-section depends upon the angle and is called *differential cross-section.* This cross-section is defined as the cross-section per unit solid angle and is denoted by $\sigma(\theta, \phi)$.

$$\cdot\, \sigma(\theta, \phi) = \frac{d\sigma}{d\Omega}$$

And the total cross-section is given by

$$\sigma_T = \int_\Omega \frac{d\sigma}{d\Omega} d\Omega$$

4.4 CONSERVATION LAWS IN NUCLEAR REACTIONS

In nuclear reactions certain physical quantities never change during and after the reaction. We say that these quantities are always conserved in nuclear reactions. Some of the conservation laws are as under.

4.4.1 Conservation of Mass-Energy

In a nuclear reaction total energy of reactants (rest mass plus kinetic energy for a non-relativistic case) is always equal to the total energy of the products (rest mass plus kinetic energy). Consider a reaction

$$a + X \rightarrow Y + b$$

If K_a, K_b and K_Y are kinetic energies of a, b and Y respectively, then

$$\boxed{K_a + m_a c^2 + m_X c^2 = K_Y + m_Y c^2 + K_b + m_b c^2}$$

assuming target nucleus X is at rest initially. However, for relativistic case, we have

$$E_a + m_X c^2 = E_Y + E_b$$

where

$$E_i = \sqrt{p_i^2 c^2 + m_i^2 c^4}$$

where $i = a, y$ or b

4.4.2 Conservation of Linear Momentum

In a nuclear reaction total initial momentum must be equal to the total final momentum. Assuming the nucleus of the target is at rest, the momentum brought into the nucleus by the projectile is distributed among the products.

4.4.3 Conservation of Charge/Atomic Number

In a nuclear reaction total charge on the reactants must be equal to the total charge on products, i.e. nuclear charge cannot be created or destroyed.

$$^{27}_{13}\text{Al} + {}^{1}_{1}\text{H} \rightarrow {}^{24}_{11}\text{Na} + 3{}^{1}_{1}\text{H} + {}^{1}_{0}n$$

Here, total charge on reactants is 13 + 1 = 14. Total charge on products is 11 + 3 × 1 = 14. Similarly,

$$^{1}_{0}n \rightarrow {}^{1}_{1}\text{H} + {}^{0}_{-1}e + \overline{\nu}$$

Total charge on the reactants is 0. Total charge on products is 1 + (–1) = 0.

4.4.4 Conservation of Nucleons/Mass Number

In a nuclear reaction total number of nucleons before and after the reaction remains the same, i.e. nucleons cannot be created nor destroyed in a nuclear reaction. For example,

$$^{235}_{92}\text{U} + {}^{1}_{0}n \rightarrow {}^{141}_{56}\text{Ba} + {}^{92}_{36}\text{Kr} + 3{}^{1}_{0}n$$

There are 235 + 1 = 236 nucleons before the reaction and 141 + 92 + 3 × 1 = 236 nucleons after the reaction.

4.4.5 Conservation of Angular Momentum

The total angular momentum I which is the vector sum of spin angular momentum s and relative orbital angular momentum l of the two sides must be equal.

4.4.6 Conservation of Spin

This law states that if the vectorial sum of the spins of the reactants is an odd integer multiple of $\hbar/2$, then the vectorial sum of the spin of products must also be an odd integer multiple of $\hbar/2$. Consider the decay of neutron

$$ {}_0^1n \rightarrow {}_1^1\text{H} + {}_{-1}^{\ 0}e + \overline{\nu} $$

Since the spin of all the particles involved in the reaction is 1/2, so the left-hand side has spin 1/2 and the right-hand side can have spin 1/2 or 3/2. Thus, this satisfies the law of conservation of spin.

Similarly, if the vectorial sum of the spins of the reactants is an integer multiple of $\hbar$, then the vectorial sum of the spin of products must also be an integer multiple of $\hbar$.

4.4.7 Conservation of Statistics

Spin ½ particles follow Fermi–Dirac statistics, while integer spin particles follow Bose–Einstein statistics. If the number of particles involved as reactants is odd, then the number of product particles must be odd, so the reaction follows the Fermi–Dirac statistics.

Similarly, if the number of particles involved as reactants is even, then the number of product particles must be even, so the reaction follows the Bose–Einstein statistics. This behaviour is consistent with conservation of nucleon number as discussed before. For example, consider the reaction

$$ {}_{13}^{27}\text{Al} + {}_2^4\text{He} \rightarrow {}_{15}^{30}\text{P} + {}_0^1n $$

Total mass number (A) on right-hand side and left-hand side is 31. So, both sides follow the Fermi–Dirac statistics.

4.4.8 Conservation of Parity

Total parity in a nuclear reaction must be conserved, i.e. the total parity of the reactants must be equal to the total parity of the products. No violation of parity conservation has been reported in nuclear reactions based on strong interactions.

4.4.9 Conservation of Lepton Number

Lepton number of the reactants must be equal to that of products. Leptons like e^-, ν, μ^- etc. are assigned total lepton number as +1, while their antiparticles have been assigned total lepton number –1. For the reaction

$$ {}_1^1\text{H} \rightarrow {}_0^1n + e^+ + \nu $$

For left-hand side of the equation, there are no leptons, so the lepton number is 0. On right-hand side, positron has lepton number –1, while for neutrino, lepton number is +1. So the total lepton number on right-hand side is $-1 + 1 = 0$ (see also Section 8.7.2).

4.5 KINEMATICS OF NUCLEAR REACTIONS

In any nuclear reaction, laws of conservation of energy and momentum always hold good. These conservation laws impose certain conditions on nuclear reactions like that the energy of the outgoing particles must be real, etc. These restrictions are called kinematical restrictions and the mathematical relations derived by imposing these restrictions are known as *kinematics*.

Consider the following nuclear reaction in which a light projectile of mass M_1 of kinetic energy E_1 interacts with a nucleus of mass M_2 (assumed stationary), the outgoing light particle is of mass M_3 of energy E_3 and the product heavy nucleus is of mass M_4 of energy E_4.

$$M_1 + M_2 \rightarrow M_3 + M_4$$

Applying the law of conservation of mass-energy assuming that the nucleus of mass M_2 initially is at rest, then

$$M_1c^2 + E_1 + M_2c^2 = E_3 + M_3c^2 + E_4 + M_4c^2$$

$$E_3 + E_4 - E_1 = [(M_1 + M_2) - (M_3 + M_4)]c^2 \tag{4.1}$$

Define a quantity Q as the difference between the kinetic energies of products and that of the incident particle in a reaction. Therefore,

$$Q = E_3 + E_4 - E_1$$

Using this value of Q, Eq. (4.1) becomes

$$Q = (M_1 + M_2)c^2 - (M_3 + M_4)c^2 \tag{4.2}$$

Thus, Q-value of the reaction is the difference between masses of the interacting particles to that of the products.

If Q is positive, i.e. total mass of the reactants is more than the total mass of the products, then such reactions are known as *Exoergic* or *Exothermic reactions.* The second possibility is that Q is negative, i.e. mass of the reactants is less than that of products, such reactions are known as *Endoergic* or *Endothermic reactions.*

Schematically a nuclear reaction is represented in Figure 4.2, where a particle of mass M_1, kinetic energy E_1 and momentum $\sqrt{2M_1E_1}$ strikes a stationary target of mass M_2 resulting in the formation of a compound nucleus of mass M_c. This compound nucleus decays into lighter particle of mass M_3, kinetic energy E_3 and momentum $\sqrt{2M_3E_3}$ and the heavy nucleus of mass M_4, kinetic energy E_4 and momentum $\sqrt{2M_4E_4}$.

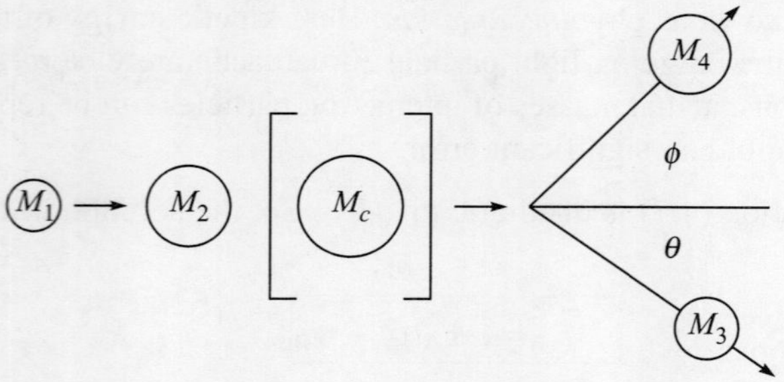

Figure 4.2 Dynamics of two-body nuclear reaction in laboratory frame.

Applying law of conservation of energy to this reaction, we get

$$E_1 + Q = E_3 + E_4 \tag{4.3}$$

Using law of conservation of linear momentum in a direction of incident particle and perpendicular to it, we get the following equations:

$$\sqrt{2M_1E_1} = \sqrt{2M_3E_3}\cos\theta + \sqrt{2M_4E_4}\cos\phi \tag{4.4}$$

$$0 = \sqrt{2M_3E_3}\sin\theta - \sqrt{2M_4E_4}\sin\phi \tag{4.5}$$

Equations (4.2) to (4.5) are valid only if the system of masses M_1, M_2, M_3 and M_4 is an isolated system. (By isolated system we mean that none of the particles M_1, M_2, M_3 and M_4 are interacting with any other particle present in the system.) The most common practice is to perform experiments on ejected light particle of mass M_3. Therefore, using Eqs. (4.3) to (4.5), we eliminate E_4 and angle ϕ.

Equations (4.4) and (4.5) can be written as:

$$\sqrt{2M_4E_4}\cos(\phi) = \sqrt{2M_1E_1} - \sqrt{2M_3E_3}\cos\theta$$

$$\sqrt{2M_4E_4}\sin(\phi) = \sqrt{2M_3E_3}\sin\theta$$

Squaring these two equations and adding, we get

$$2M_4E_4 = 2M_1E_1 + 2M_3E_3 - 4\sqrt{M_1M_3E_1E_3}\cos\theta$$

or

$$E_4 = \frac{M_1}{M_4}E_1 + \frac{M_3}{M_4}E_3 - \frac{2}{M_4}\sqrt{M_1M_3E_1E_3}\cos\theta$$

Substituting E_4 in Eq. (4.3), we get

$$E_1 + Q = E_3 + \frac{M_1}{M_4}E_1 + \frac{M_3}{M_4}E_3 - \frac{2}{M_4}\sqrt{M_1M_3E_1E_3}\cos\theta$$

or

$$Q = \frac{M_1}{M_4}E_1 - E_1 + \frac{M_3}{M_4}E_3 + E_3 - \frac{2}{M_4}\sqrt{M_1M_3E_1E_3}\cos\theta \tag{4.6}$$

$$\boxed{Q = \frac{M_1 - M_4}{M_4}E_1 + \frac{M_3 + M_4}{M_4}E_3 - \frac{2}{M_4}\sqrt{M_1M_3E_1E_3}\cos\theta} \tag{4.7}$$

This equation is known as *Q-value equation*. Here kinetic energy of the incident particle E_1 and kinetic energy of the outgoing light particle E_3 can accurately be measured experimentally.

In many applications, actual masses of interacting particles can be replaced by mass number A_1, A_2, A_3 and A_4 without any significant error.

The Q-value equation (4.7) is quadratic in $\sqrt{E_3}$. So, the solution of this quadratic equation is:

$$\boxed{\sqrt{E_3} = v \pm \sqrt{v^2 + w}} \tag{4.8}$$

where

$$\boxed{v = \frac{\sqrt{M_1 M_3 E_1}}{M_3 + M_4} \cos\theta} \tag{4.9}$$

$$\boxed{w = \frac{M_4 Q + E_1(M_4 - M_1)}{M_3 + M_4}} \tag{4.10}$$

According to this equation only those reactions are energetically possible for which $\sqrt{E_3}$ is real and positive.

4.5.1 Exoergic or Exothermic Reactions

In such reactions the total mass of the reactants is more than that of the products. This excess mass appears as the kinetic energy of the products. Let us study E_3 and direction θ of the particle M_3 as a function of incident particle energy E_1, which is gradually increased.

Zero Bombarding Energy ($E_1 \cong 0$)

If $E_1 \to 0$, then from Eq. (4.9)

$$v \to 0$$

and from Eq. (4.10)

$$w = \frac{QM_4}{M_3 + M_4}$$

From Eq. (4.8)

$$\boxed{E_3 = \frac{QM_4}{M_3 + M_4}}$$

Therefore, the kinetic energy E_3 of the outgoing particle is independent of angle θ.

Finite Bombarding Energy ($E_1 > 0$)

Generally, the projectile of mass M_1 is lighter than the unobserved nucleus of mass M_4. Therefore, the term $M_4 - M_1$ is always positive or from Eq. (4.10), w is always positive for all values of E_1. Then from Eq. (4.8) only one value of E_3 is positive. We call this situation as E_3 is single-valued for all bombarding energies and is given by Eq. (4.8). Here E_3 depends upon angle θ and would be smallest at $\theta = 180°$.

4.5.2 Endoergic or Endothermic Reactions

When $E_1 \to 0$ or zero bombarding energy and Q is negative, then from Eqs. (4.9) and (4.10), $v \to 0$ and w is negative. Therefore, $v^2 + w$ is always negative. Hence, $\sqrt{E_3}$ is imaginary. This means no endothermic reaction is possible.

4.5.3 Threshold Energy

The smallest bombarding energy at which nuclear reaction just starts is known as threshold energy, $(E_1)_{\text{th}}$. According to Eq. (4.8) reaction takes place only if E_1 is large enough to make $v^2 + w = 0$. From Eqs. (4.9) and (4.10)

$$\frac{M_1 E_1 M_3}{(M_3 + M_4)^2} \cos^2\theta + \frac{M_4 Q + E_1(M_4 - M_1)}{M_3 + M_4} = 0$$

Since $M_3 + M_4 \neq 0$, so multiplying by $M_3 + M_4$ on both sides, we have

$$\frac{M_1 E_1 M_3}{(M_3 + M_4)} \cos^2\theta + M_4 Q + E_1(M_4 - M_1) = 0$$

Substituting for $\cos^2(\theta) = 1 - \sin^2\theta$, we get

$$\frac{M_1 E_1 M_3}{(M_3 + M_4)} [1 - \sin^2\theta] + E_1(M_4 - M_1) = -M_4 Q$$

Taking E_1 as common and rearranging, we have

$$E_1\left[\frac{M_1 M_3}{M_3 + M_4} + M_4 - M_1 - \frac{M_1 M_3}{M_3 + M_4}\sin^2\theta\right] = -M_4 Q$$

Multiplying by $M_3 + M_4$ on both sides

$$E_1\left[M_3 M_4 + M_4^2 - M_1 M_4 - M_1 M_3 \sin^2\theta\right] = -M_4 Q(M_3 + M_4)$$

or

$$E_1 = \frac{-M_4 Q(M_3 + M_4)}{M_3 M_4 + M_4^2 - M_1 M_4 - M_1 M_3 \sin^2\theta}$$

Dividing the numerator and denominator by M_4, E_1 at an angle θ is given by

$$(E_1)_\theta = -Q \frac{M_3 + M_4}{M_3 + M_4 - M_1 - \dfrac{M_1 M_3}{M_4}\sin^2\theta} \tag{4.11}$$

If light outgoing particle M_3 is observed at an angle $\theta = 0°$, then $(E_1)_\theta$ has its minimum possible value, known as *threshold energy* and is given by

$$(E_1)_{\text{th}} = -Q \frac{M_3 + M_4}{M_3 + M_4 - M_1} \tag{4.12}$$

Equation (4.2) can be written as:

$$M_1 + M_2 = M_3 + M_4 + \frac{Q}{c^2}$$

Generally,

$$M_2 >> \frac{Q}{c^2}$$

Therefore, neglecting $\frac{Q}{c^2}$, we have

$$M_1 + M_2 = M_3 + M_4$$

Using this result, Eq. (4.12) becomes

$$\boxed{(E_1)_{\text{th}} = -Q\frac{M_1 + M_2}{M_2}}$$

At the threshold energy, outgoing particle with smaller mass (M_3) appears at $\theta = 0°$. When the bombarding energy is increased M_3 appears at larger angles i.e. $\theta > 0°$ also.

4.6 COMPOUND NUCLEUS

In 1936 Bohr proposed the compound nucleus hypothesis. According to this hypothesis, nuclear reaction takes place in two stages:

1. Formation of compound nucleus.
2. Decay of compound nucleus.

Formation of Compound Nucleus

The incident particle strongly interacts with the nucleons present in the nucleus, giving away its energy to all the nucleons, thereby losing its identity and becomes a part of the nucleus. This new nucleus formed is in a highly excited state and is called compound nucleus.

Decay of Compound Nucleus

There is a finite probability that one or a group of nucleons residing on the surface or near the surface of the nucleus and having enough energy can escape the compound nucleus, thereby leading to the decay of compound nucleus. This reaction can be written as

$$M_1 + M_2 \rightarrow [M_c] \rightarrow M_3 + M_4$$

This hypothesis also implies that the decay of compound nucleus is independent of the way it was formed. This independence of mode of formation and decay is true only if the decay time of the compound nucleus is much longer than the nuclear time. Nuclear time is defined as the time taken by an incident particle to cross the diameter of the nucleus. For example, 5 MeV proton has velocity ~3×10^7 m/s. Nuclear diameter of a nucleus with mass 200 is about ~7×10^{-15} m. Thus, nuclear time is ~ $7 \times 10^{-15}/3 \times 10^7 = 2 \times 10^{-22}$ s. Decay time of the compound nucleus has been experimentally measured and is about 10^{-14} to 10^{-15} s. This is about 10^7 times larger than the nuclear time.

The compound nucleus forgets the mode of formation. However, decay of the compound nucleus depends upon the energy, angular momentum and parity brought in by the incident particle.

Evidence to the compound nucleus hypothesis was provided by the series of experiments of Ghoshal. In his experiments compound nucleus $\left[{}^{64}_{30}\text{Zn}\right]^*$ was formed in two different modes as under:

$$ {}^{1}_{1}\text{H} + {}^{63}_{29}\text{Cu} \rightarrow \left[{}^{64}_{30}\text{Zn}\right]^* $$

and

$$ {}^{4}_{2}\text{He} + {}^{60}_{28}\text{Ni} \rightarrow \left[{}^{64}_{30}\text{Zn}\right]^* $$

This compound nucleus decays by one of the following modes:

$$ \left[{}^{64}_{30}\text{Zn}\right]^* \rightarrow {}^{63}_{30}\text{Zn} + {}^{1}_{0}n $$

$$ \rightarrow {}^{62}_{30}\text{Zn} + 2{}^{1}_{0}n $$

$$ \rightarrow {}^{62}_{29}\text{Cu} + {}^{1}_{0}n + {}^{1}_{1}\text{H} $$

Ghoshal found that the decay probability of the compound nucleus was independent of the mode of its formation.

4.7 NUCLEAR FISSION

Nuclear fission is a process of splitting a nucleus, generally heavy atom (target) into two or more lighter atoms (fission fragments) when it is bombarded by neutrons or charged particles. A few radioactive nuclides can also spontaneously fission such as ^{254}Cf and ^{256}Fm. Fission releases large amount of energy along with one or more neutrons. The large amount of energy is released as sum of masses of the fission products is less than the original mass of the heavy atom and bombarding particle. This missing mass ($\approx$0.1% of the heavy atom mass) is converted into energy according to Einstein's equation $E = mc^2$.

4.7.1 Neutron-Induced Fission

Fermi and his associates bombarded natural uranium with slow neutrons and found at least four β^--activities having different half-lives. They assumed that these activities could be the product of an (n, γ) reaction leading to the formation of elements with atomic numbers 93, 94, etc. through successive β^--decay as shown under.

$$ {}^{238}_{92}\text{U} + n \rightarrow {}^{239}_{92}\text{U} $$

$$ {}^{239}_{92}\text{U} \rightarrow {}^{239}_{93}\text{Np} + \beta^- $$

$$ {}^{239}_{93}\text{Np} \rightarrow {}^{239}_{94}\text{Pu} + \beta^- $$

However, after extremely painstaking experiments, Hahn and Strassmann in 1938 established that isotopes of Ba, La, Sr, Y, Rb, Cs, etc. were among the sources of β^--activity. They concluded that these radioactive nuclides were formed as a consequence of the uranium nucleus splitting into two (not necessarily equal) parts following the bombardment of slow neutrons on uranium nucleus. This is known as *neutron-induced fission*. (Another type of fission is spontaneous fission, in which a heavy nucleus with excess of neutrons fissions into two or more smaller fragments. This process is similar to neutron-induced fission, with the exception that no neutrons are needed to initiate the fission. Examples of such fissionable nuclei are ^{254}Cf, ^{256}Fm, etc.)

Meitner and Frisch theoretically predicted that during this splitting process, ≈ 200 MeV energy is released and they named this process as *Nuclear Fission*. This prediction was experimentally verified by Frisch. In 1939 Bohr and Wheeler explained the phenomenon of fission on the basis of liquid drop model of the nucleus. They also established that with slow neutrons only $^{235}_{92}$U undergo fission. The fission of $^{235}_{92}$U with slow neutrons is diagrammatically shown in Figure 4.3. Later on it was established that fission can be induced in almost all the nuclei through interaction of neutrons, protons, α-particles or even heavy nuclei of suitable kinetic energy.

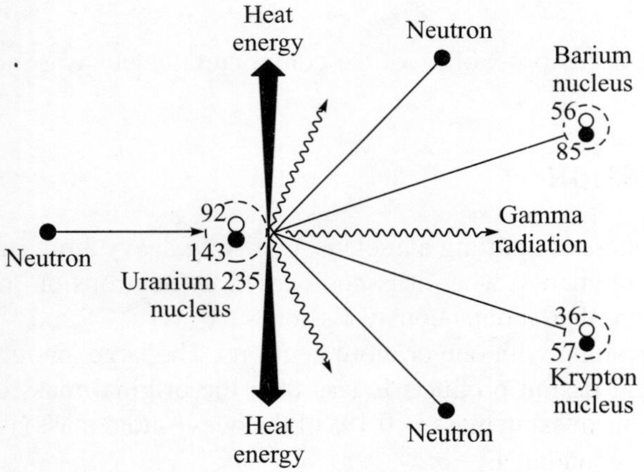

Figure 4.3 Schematic diagram of fission of ^{235}U by neutrons.

Some of the fission reactions are

$$^{235}_{92}\text{U} + {}^{1}_{0}n \rightarrow {}^{141}_{56}\text{Ba} + {}^{92}_{36}\text{Kr} + 3\,{}^{1}_{0}n \quad Q = +175 \text{ MeV}$$

$$^{63}_{29}\text{Cu} + p \rightarrow {}^{24}_{11}\text{Na} + {}^{39}_{19}\text{K} + {}^{1}_{0}n \quad Q = -24 \text{ MeV}$$

$$^{2}_{1}\text{H} + \gamma \rightarrow {}^{1}_{1}\text{H} + {}^{1}_{0}n \quad Q = -2.2 \text{ MeV}$$

Out of all the possible fission reactions, fission induced by slow or thermal neutrons (having energy ≈ 0.025 eV) is most commonly studied fission reaction. This reaction is utilized in nuclear reactors to generate electricity and to produce radioactive isotopes. It has been found that along with $^{235}_{92}$U, other heavy nuclei which can undergo fission with thermal neutrons are $^{233}_{92}$U and $^{239}_{94}$Pu.

When nuclei undergo fission, the fission products are not determined uniquely. There is a distribution of masses of two fission products as shown in Figure 4.4. The distribution is symmetric. About 97% of the total fission products fall within the narrow range from mass 85 to 104 (lighter fragment group) and the other range from mass 130 to 149 (heavier group). For heavy fragment, there is always a corresponding light fragment. The maximum yield of fragments is at about $A_1 = 95$ and $A_2 = 140$. In all more than 200 fission fragments have been observed.

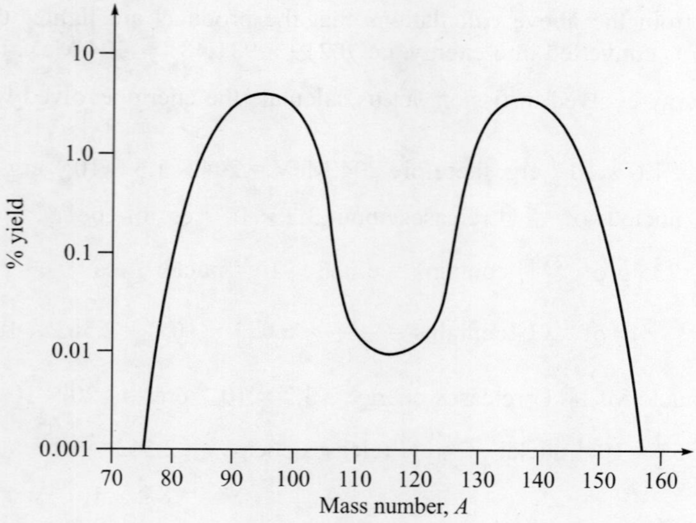

Figure 4.4 Per cent yield of different fission fragments in fission of ^{235}U.

4.7.2 Energy Released in Fission

When heavy nuclei undergo fission, large amount of energy is released. It can be shown by considering the fission of $^{235}_{92}U$ by thermal neutrons. The fission fragments formed have mass numbers about 95 and 139. These fission fragments are radioactive and undergo several stages of β^--decay. Ultimately after emission of seven β^--particles, stable nuclides ^{95}Mo and ^{139}La are formed. This event is finally represented by the following reaction:

$$^{235}_{92}U + ^{1}_{0}n \rightarrow ^{95}_{42}Mo + ^{139}_{57}La + 7\,^{0}_{-1}e + 2\,^{1}_{0}n$$

Let us consider the masses of reactants and products in this reaction.

Masses of reactants:

$$\text{Mass of } ^{235}_{92}U = 235.0439 \text{ amu}$$

$$\text{Mass of } ^{1}_{0}n = 1.0087 \text{ amu}$$

$$\text{Total mass} = 236.0526 \text{ amu}$$

Masses of products:

$$\text{Mass of } ^{139}_{57}La = 138.9061 \text{ amu}$$

$$\text{Mass of } ^{95}_{42}Mo = 94.9058 \text{ amu}$$

$$\text{Mass of } 7\,{}_{-1}^{\;0}e = 7 \times 0.00055 = 0.00385 \text{ amu}$$

$$\text{Mass of } 2\,{}_{0}^{1}n = 2 \times 1.0087 = 2.0174 \text{ amu}$$

$$\text{Total mass} = 235.8332 \text{ amu}$$

It is evident from the above calculations that the products are lighter than reactants by 0.219 amu, which is converted into energy, i.e. $0.219 \times 931.481 \approx 204$ MeV. In order to get an idea about the energy evolved in fission, let us calculate the energy evolved when 1 g of ${}_{92}^{235}$U undergoes fission.

Since 1 MeV = 1.6×10^{-6} erg, therefore 204 MeV = $204 \times 1.6 \times 10^{-6}$ erg = 3.2×10^{-4} erg. Thus, fission of 1 nucleus of ${}_{92}^{235}$U releases about 3.2×10^{-4} erg of energy.

$$235 \text{ g of } {}_{92}^{235}\text{U contains} = 6.023 \times 10^{23} \text{ nuclei}$$

$$1 \text{ g of } {}_{92}^{235}\text{U contains} = \frac{1}{235} \times 6.023 \times 10^{23} = 2.563 \times 10^{21} \text{ nuclei}$$

$$\text{Fission of 1 nuclei of } {}_{92}^{235}\text{U releases energy} = 3.2 \times 10^{-4} \text{ erg} = 3.20 \times 10^{-11} \text{ W s}$$

$$\begin{aligned}\text{Fission of } 2.563 \times 10^{21} \text{ nuclei of } {}_{92}^{235}\text{U releases energy} &= 2.563 \times 10^{21} \times 3.2 \times 10^{-11} \text{ W s}\\ &= 8.2 \times 10^{10} \text{ W s}\\ &= 8.2 \times 10^{7} \text{ kWs}\\ &= \frac{8.2 \times 10^{7}}{3600} = 2.3 \times 10^{4} \text{ kWh}\\ &= \frac{2.3 \times 10^{4}}{24} = 0.96 \times 10^{3} \text{ kWd}\end{aligned}$$

Therefore, 1 g of U or Pu on fission would produce about 1 MW of power for one day. The same energy production requires more than 3 tons of coal or about 2300 litres of fuel oil per day.

4.8 NUCLEAR FUSION

Figure 1.4 of Chapter 1 shows that the binding energy per nucleon in the lightest nuclei is (like that in heaviest nuclei) less than that for nuclei of intermediate mass number. It implies that the sum of the masses of individual light nuclei is more than would be the mass of the nuclei formed by their fusion. Therefore, the combination of two of the lightest nuclei by a process of fusion is thus energetically advantageous.

However, fusing two lighter nuclei into one heavier nucleus is not so easy. Before light nuclei can combine, their mutual Coulomb repulsion must be overcome. In contrast fission induced by thermal neutrons has no such Coulomb barrier and thus very low energy neutrons can induce fission in heavy nuclei.

Suppose we wish to fuse two ${}_{10}^{20}$Ne nuclei to form ${}_{20}^{40}$Ca nucleus. The Coulomb barrier for two ${}_{10}^{20}$Ne nuclei is about 21 MeV. Therefore, we have to accelerate one ${}_{10}^{20}$Ne nucleus to at least

21 MeV and bombard it on a stationary $^{20}_{10}$Ne nucleus. Only then the two fuse together. The Q-value for this reaction is about 20.7 MeV. Thus, the total energy released is 21 MeV kinetic energy plus 20.7 MeV Q-value or about 42 MeV. This means 21 MeV is supplied and 42 MeV comes out. This is not a promising situation. Accelerators accelerating $^{20}_{10}$Ne provide a beam current of about 10^{-6} A at about 20 MeV. The total power produced is 20×10^6 V $\times 10^{-6}$ A = 2 W. This small power is of no practical use.

Another approach is to heat a container of neon gas to very high temperature, so that thermal energy becomes larger than the Coulomb repulsion between the two interacting nuclei. This process requires a temperature of the order of 10^{11} K. If instead of $^{20}_{10}$Ne, we take hydrogen, the temperature required are of the order of 10^6–10^7 K.

Despite of the drawbacks, energy released in fusion is at present a subject of intensive research to perfect techniques of heating fusible nuclei so that large amount of power could be produced. It is fusion which powers sun and stars and is therefore ultimately responsible for the life on earth.

4.8.1 Energy Released in Fusion

Unlike fission, fusion is not a natural process on earth. Like we have spontaneous fission, spontaneous fusion does not exist. This is because of the limitation imposed due to the Coulomb barrier. However, once we overcome Coulomb barrier, fusion becomes a dominant mode as two fusing nuclei quickly reach a state of minimum energy.

The most elementary fusion reaction

$$p + p \rightarrow {}^{2}_{2}\text{He}$$

is not possible, as $^{2}_{2}$He is not a bound system. Once it is formed, it will immediately break up into the constituents.

Other possible fusion reaction is:

$${}^{2}_{1}\text{H} + {}^{2}_{1}\text{H} \rightarrow {}^{4}_{2}\text{He} + \gamma$$

The energy released (Q-value) of this reaction is 23.8 MeV. Other similar reactions are:

$${}^{2}_{1}\text{H} + {}^{2}_{1}\text{H} \rightarrow {}^{3}_{2}\text{He} + n \qquad (Q = 3.3\text{ MeV})$$

$${}^{2}_{1}\text{H} + {}^{2}_{1}\text{H} \rightarrow {}^{3}_{1}\text{H} + p \qquad (Q = 4.0\text{ MeV})$$

These reactions are generally called *deuterium–deuterium reactions or D–D reactions.*

Another fusion reaction is

$${}^{2}_{1}\text{H} + {}^{3}_{1}\text{H} \rightarrow {}^{4}_{2}\text{He} + n \qquad (Q = 17.6\text{ MeV})$$

This reaction is called *deuterium–tritium reaction* or *D–T reactions*, which provides mono-energetic neutrons of 14.1 MeV. The Q-value of this reaction is 17.6 MeV or 2.82×10^{-5} ergs or 2.82×10^{-12} W s.

Let us calculate the energy evolved in nuclear fusion, when 1 g of ^{2}H and ^{3}H undergoes fusion.

Approximately, 1 g of a mixture of ^{2}H and ^{3}H will contain about 0.5 g of ^{2}H and 0.5 g of ^{3}H.

Therefore, 0.5 g of ^{2}H or ^{3}H will contain nuclei $= \dfrac{0.5}{2} \times 6.023 \times 10^{23} = 1.5 \times 10^{23}$

$$1 \text{ fusion produces power} = 2.82 \times 10^{-12} \text{ W s}$$

$$1.5 \times 10^{23} \text{ fusions produce power} = 1.5 \times 10^{23} \times 2.82 \times 10^{-12} = 4.23 \times 10^{11} \text{ W s}$$

$$= 4.23 \times 10^{8} \text{ kW s}$$

$$= \frac{4.23 \times 10^8}{3600 \times 24} = 4.89 \times 10^3 \text{ kW d}$$

$$\sim 5 \text{ MW d}$$

which is much higher compared to power produced in case of fission.

4.8.2 Hydrogen Burning and Solar Energy

Sun and most other stars consist of largely hydrogen. The first fusion reaction occurring in sun and other stars is

$$^1_1\text{H} + {}^1_1\text{H} \rightarrow {}^2_1\text{H} + e^+ + \nu \qquad (Q = 1.44 \text{ MeV})$$

This reaction occurs via weak interaction discussed in Chapter 8, the cross-section for which is very small ($\approx 10^{-33}$ b) at proton mean energy of 1 keV which corresponds to central temperature of sun as 10^7 K. This is the slowest reaction occurring in the sun. It has been calculated that for 10^{18} protons, only 1 deuterium nucleus is formed per second. Since the number of deuterons is very small compared to protons (1 deuteron for 10^{18} protons); the next most likely reaction is

$$^2_1\text{H} + {}^1_1\text{H} \rightarrow {}^3_2\text{He} + \gamma \qquad (Q = 5.49 \text{ MeV})$$

The reaction of $^3_2$He with protons

$$^3_2\text{He} + {}^1_1\text{H} \rightarrow {}^4_3\text{Li}$$

is not possible as $^4_3$Li is not a bound system, so it immediately breaks up into constituents. Further, $^3_2$He is unlikely to interact with $^2_1$H as density of $^2_1$H is small and as soon as $^2_1$H is formed, it interact with $^1_1$H to form $^3_2$He. $^3_2$He thus formed keeps on wandering till it finds another $^3_2$He and interacts with it as under.

$$^3_2\text{He} + {}^3_2\text{He} \rightarrow {}^4_2\text{He} + 2\,{}^1_1\text{H} + \gamma \qquad (Q = 12.86 \text{ MeV})$$

In a nutshell, four protons are converted to helium

$$4\,{}^1_1\text{H} \rightarrow {}^4_2\text{He} + 2e^+ + 2\nu$$

with a total Q-value of 26.7 MeV. This cycle of fusion reactions is known as *proton–proton cycle.*

Heavier elements like ${}^{12}_{6}\mathrm{C}$ are also present in the interior of the sun. In the presence of heavy elements like ${}^{12}\mathrm{C}$ following types of fusion reactions occur:

$${}^{12}_{6}\mathrm{C} + {}^{1}_{1}\mathrm{H} \rightarrow {}^{13}_{7}\mathrm{N} + \gamma$$

$${}^{13}_{7}\mathrm{N} \rightarrow {}^{13}_{6}\mathrm{C} + e^{+} + \nu$$

$${}^{13}_{6}\mathrm{C} + {}^{1}_{1}\mathrm{H} \rightarrow {}^{14}_{7}\mathrm{N} + \gamma$$

$${}^{14}_{7}\mathrm{N} + {}^{1}_{1}\mathrm{H} \rightarrow {}^{15}_{8}\mathrm{O} + \gamma$$

$${}^{15}_{8}\mathrm{O} \rightarrow {}^{15}_{7}\mathrm{N} + e^{+} + \nu$$

$${}^{15}_{7}\mathrm{N} + {}^{1}_{1}\mathrm{H} \rightarrow {}^{12}_{6}\mathrm{C} + {}^{4}_{2}\mathrm{He}$$

The net result of these reactions is:

$$4{}^{1}_{1}\mathrm{H} \rightarrow {}^{4}_{2}\mathrm{He} + 2e^{+} + 2\nu \qquad (Q = 24.7\ \mathrm{MeV})$$

This cycle of fusion reaction is known as *carbon cycle* or *CNO cycle*. During this cycle ${}^{12}_{6}\mathrm{C}$ is neither created nor destroyed. ${}^{12}_{6}\mathrm{C}$ acts as a catalyst to aid fusion process. The two positrons in the final reaction on annihilation give another 2 MeV energy. Thus, total energy released in CNO cycle is 26.7 MeV.

One may ask the question, why four hydrogen atoms do not directly combine to give one He atom rather than going through CNO cycle. The reason is that even at high temperature available inside the sun, protons do not have sufficient kinetic energy to overcome the Coulomb repulsion.

It is not possible to get the CNO cycle on the earth as temperature needed is very high. Experiments show that there is enough hydrogen in the sun to maintain this CNO cycle, and the temperature of the sun will remain practically unchanged for 10^9 years.

4.8.3 Helium Burning in Stars

As the age of the star advances, hydrogen at the core of the star continues to get depleted as it is converted into helium. Eventually, a helium core develops and continues to grow at the centre of the star, while energy due to hydrogen gets restricted to the outer shell surrounding the helium core. As the helium core grows, the star continues to contract releasing gravitational energy and raising the core temperature. Due to these conditions, the condition of the onset of helium burning at the core is attained. For this to occur, mean temperature of the helium core should be of the order of 10^8 K. At this temperature again a state of equilibrium in the star is reached and this burning can continue for 10^8 years. This helium burning reaction, also known as triple alpha reaction takes place in two steps:

$${}^{4}_{2}\mathrm{He} + {}^{4}_{2}\mathrm{He} \rightarrow {}^{8}_{4}\mathrm{Be} + \gamma \qquad (Q = -0.095\ \mathrm{MeV})$$

$${}^{8}_{4}\mathrm{Be} + {}^{4}_{2}\mathrm{He} \rightarrow {}^{12}_{6}\mathrm{C} + \gamma \qquad (Q = 7.4\ \mathrm{MeV})$$

First reaction is endothermic and second reaction is exothermic. Overall energy released in helium burning is 7.3 MeV, which is much less than that released in hydrogen burning.

NUMERICAL PROBLEMS

Section 4.2

Solved Problems

1. Complete the following reactions:

(i) ${}^{59}_{27}\text{Co} + {}^{1}_{0}n \rightarrow ? + \gamma$ (ii) ${}^{14}_{7}\text{N} + {}^{1}_{1}\text{H} \rightarrow {}^{11}_{6}\text{C} + ?$

(iii) ${}^{64}_{30}\text{Zn} + {}^{1}_{0}n \rightarrow ? + {}^{1}_{0}n + {}^{1}_{0}n$ (iv) ${}^{238}_{92}\text{U} + {}^{14}_{7}\text{N} \rightarrow {}^{243}_{97}\text{Bk} + 5{}^{1}_{0}n + ?$

Solution:

(i) Here total number of protons of the reactants is 27 and total number of nucleons is 60. Therefore, the resultant nucleus will have 27 protons and 60 nucleons. So, the missing nucleus is ${}^{60}_{27}\text{Co}$ and the reaction is

$${}^{59}_{27}\text{Co} + {}^{1}_{0}n \rightarrow {}^{60}_{27}\text{Co} + \gamma$$

Similarly, for other reactions.

(ii) The missing particle is ${}^{4}_{2}\text{He}$ and the complete reaction is

$${}^{14}_{7}\text{N} + {}^{1}_{1}\text{H} \rightarrow {}^{11}_{6}\text{C} + {}^{4}_{2}\text{He}$$

(iii) The missing nucleus is ${}^{62}_{30}\text{Zn}$ and the complete reaction is

$${}^{64}_{30}\text{Zn} + {}^{1}_{0}n \rightarrow {}^{63}_{30}\text{Zn} + {}^{1}_{0}n + {}^{1}_{0}n$$

(iv) The missing particle is ${}^{4}_{2}\text{He}$ and the complete reaction is

$${}^{238}_{92}\text{U} + {}^{14}_{7}\text{N} \rightarrow {}^{243}_{97}\text{Bk} + 5{}^{1}_{0}n + {}^{4}_{2}\text{He}$$

2. The ratio of activities of $\frac{{}^{14}\text{C}}{{}^{32}\text{P}}$ today at $t = 0$ is 1. What will be the ratio after 14 days and after 70 days? Given half-life of ${}^{14}\text{C}$ = 5700 years and that of ${}^{32}\text{P}$ = 14 days

Solution: Let $A_{0\text{C}}$, and $A_{0\text{P}}$ be the initial activities of ${}^{14}\text{C}$ and ${}^{32}\text{P}$, respectively.

Therefore,

$$\frac{A_{0\text{C}}}{A_{0\text{P}}} = 1$$

We have ${}^{14}\text{C}$

$$A_\text{C} = A_{0\text{C}}e^{-\lambda_\text{C}t}$$

and for ${}^{32}P$

$$A_\text{P} = A_{0\text{P}}e^{-\lambda_\text{P}t}$$

Taking the ratio, we get

$$\frac{A_\text{C}}{A_\text{P}} = \frac{A_{0\text{C}}e^{-\lambda_\text{C}t}}{A_{0\text{P}}e^{-\lambda_\text{P}t}}$$

Given $\dfrac{A_{0C}}{A_{0P}} = 1$

and
$$\lambda_C = \frac{0.6931}{5700 \times 365 \times 24 \times 3600} = 3.86 \times 10^{-12}\,s^{-1}$$

$$\lambda_P = \frac{0.6931}{14 \times 24 \times 3600} = 0.57 \times 10^{-6}\,s^{-1}$$

Therefore, after 14 days

$$\frac{A_C}{A_P} = \frac{e^{-3.86 \times 10^{-12} \times 14 \times 24 \times 3600}}{e^{-0.57 \times 10^{-6} \times 14 \times 24 \times 3600}}$$

$$\frac{A_C}{A_P} \approx 2$$

and after 70 days

$$\frac{A_C}{A_P} = \frac{e^{-3.86 \times 10^{-12} \times 70 \times 24 \times 3600}}{e^{-0.57 \times 10^{-6} \times 70 \times 24 \times 3600}}$$

$$\frac{A_C}{A_P} \approx 31$$

Unsolved Problems

1. Find the missing nucleus/particle in the following reactions:

(i) ${}^{31}_{15}P + \gamma \rightarrow\ ? + {}^{1}_{0}n$

(ii) ${}^{27}_{13}Al + {}^{4}_{2}He \rightarrow {}^{30}_{15}P\ +\ ?$

(iii) ${}^{59}_{27}Co + {}^{4}_{2}He \rightarrow\ ? + 2{}^{1}_{0}n$

(iv) ${}^{59}_{27}Co + {}^{3}_{2}He \rightarrow\ ? + {}^{1}_{1}H + {}^{1}_{1}H$

(v) ${}^{27}_{13}Al + {}^{3}_{1}H \rightarrow\ ? + {}^{2}_{1}H$

(vi) ${}^{27}_{13}Al\ +\ ? \rightarrow {}^{26}_{13}Al + {}^{3}_{1}H$

(vii) $? + {}^{1}_{0}n \rightarrow {}^{63}_{30}Zn + {}^{1}_{0}n + {}^{1}_{0}n$

(viii) ${}^{24}_{12}Mg + {}^{1}_{0}n \rightarrow {}^{25}_{12}Mg\ +\ ?$

(ix) ${}^{197}_{79}Au + {}^{12}_{6}C \rightarrow {}^{205}_{85}At\ +\ ?$

(x) ${}^{27}_{13}Al + {}^{14}_{7}N \rightarrow {}^{25}_{13}Al\ +\ ?$

[**Ans.** (i) ${}^{30}_{15}P$, (ii) ${}^{1}_{0}n$, (iii) ${}^{61}_{29}Cu$, (iv) ${}^{60}_{27}Co$, (v) ${}^{28}_{13}Al$, (vi) ${}^{2}_{1}H$, (vii) ${}^{64}_{30}Zn$, (viii) γ, (ix) $4{}^{1}_{0}n$, (x) ${}^{16}_{7}N$]

2. ${}^{27}Al$ is bombarded with neutrons. Write the following reactions:

$$(n, p), (n, \alpha), (n, d), (n, \gamma)$$

3. Complete the following nuclear reactions:

(i) ${}^{27}Al(\alpha, p)$

(ii) ${}^{27}Al(\alpha, d)$

(iii) ${}^{27}Al(\alpha, \alpha)$

(iv) ${}^{27}Al(\alpha, n)$

(v) ${}^{27}Al(\alpha, \gamma)$

(vi) ${}^{27}Al(\alpha, {}^{3}He)$

(vii) ${}^{27}Al(\alpha, {}^{3}H)$

4. A neutron and a proton can undergo radioactive capture at rest:

$$p + n \rightarrow d + \gamma$$

Find the energy of the photon emitted in this capture
m_p = 1.00783 amu, m_n = 1.00867 amu, m_d = 2.01410 amu **[Ans.** 2.235 MeV]

5. Deuterons undergo photodisintegration reaction with γ-rays of energy 2.225 MeV. Given m_d = 2.0141022 amu, m_p = 1.0078252 amu and 1 amu = 931.5 MeV, find the mass of neutron. **[Ans.** 1.0086656 amu]

Section 4.3

Solved Problems

1. A 0.01 mm thick ${}^{7}_{3}$Li target is bombarded with a beam of flux of 10^{13} particles/cm^2-s. As a result 10^8 neutrons/s are produced. Calculate the cross-section for this reaction. Given: density of lithium = 500 kg/m^3.

Solution:

$$\text{Thickness of } {}^{7}_{3}\text{Li} = t = 0.01 \text{ mm} = 10^{-5} \text{ m}$$

$$\text{Density} = 500 \text{ kg/m}^3$$

$$\text{Number of } {}^{7}_{3}\text{Li nuclei/volume } n = \frac{\rho N}{M}$$

$$= \frac{500 \times 6.023 \times 10^{26}}{7}$$

$$= 4.302 \times 10^{28}/\text{m}^3$$

$$\text{Number of } {}^{7}_{3}\text{Li nuclei/area} = 4.302 \times 10^{28} \times t$$

$$= 4.302 \times 10^{28} \times 10^{-5}$$

$$= 4.302 \times 10^{23}$$

Number of nuclei undergoing interactions/s = Number of neutrons produced/s
Number of neutrons produced/s = 10^8
Number of incident particles striking/unit area of target $N_0 = 10^{13}$/unit area of target.

$$\text{Cross-section } \sigma = \frac{\text{Number of neutrons produced/s}}{N_0 \times n}$$

$$= \frac{10^8}{10^{13} \times 4.302 \times 10^{23}}$$

$$= 2.32 \times 10^{-29} \text{ m}^2$$

$$= 0.232 \text{ b } (1 \text{ b} = 10^{-28} \text{ m}^2)$$

2. Neutron capture cross-section of thermal neutrons for ^{113}Cd is 20,000 b. The mean atomic mass of natural cadmium is 112 amu and its density is 8.64×10^3 kg/m^3 What fraction of an incident beam of thermal neutrons is absorbed by a cadmium sheet of thickness 0.2 mm? Given abundance of ^{113}Cd in natural cadmium is 12%.

Solution: Number of natural cadmium atoms/volume = $\dfrac{\rho N}{M}$

Thus, number of ^{113}Cd atoms present in natural cadmium

$$n = 0.12\,\frac{6.023\times10^{26}\times8.64\times10^{3}}{112}$$

or

$$n = 5.58 \times 10^{27} \text{ atoms/m}^3.$$

Now,

$$\sigma = 20{,}000 = 20{,}000 \times 10^{-28} = 2 \times 10^{-24} \text{ m}^2$$

and

$$n\sigma = 5.58 \times 10^{27} \times 2 \times 10^{-24} = 1.16 \times 10^4 \text{ m}^{-1}$$

Now, we have $N = N_0 e^{-n\sigma x}$

Therefore, the fraction of incident neutrons absorbed is

$$\frac{N_0 - N}{N_0} = \frac{N_0 - N_0 e^{-n\sigma x}}{N_0} = 1 - e^{-n\sigma x}$$

$$x = 0.2 \text{ mm} = 2 \times 10^{-4} \text{ m}$$

$$\frac{N_0 - N}{N_0} = 1 - e^{-1.16\times10^4\times2\times10^{-4}} = 0.894$$

So, 89.4% of incident neutrons is absorbed in the sheet of natural cadmium.

3. A beam of thermal neutrons while traversing a 2 mm thick foil of ^{197}Au, some 70% of the neutrons are removed. What is the total thermal neutron cross-section? Comment on result obtained on cross–section, given that the radius of a gold nucleus is 6.5×10^{-15} m. (Density of gold: 19,300 kg m^{-3}.)

Solution: To solve this problem, we use the equation

$$N(x) = N_0 e^{-\sigma n x}$$

where σ is the cross-section, n is the number of absorbing or scattering nuclei per unit volume and x is the thickness of the absorber. The number of nuclei per unit volume is given by

$$n = \frac{N_{av}\rho}{A}$$

where N_{av} is Avogadros number, ρ is the density and A is the atomic mass number for gold. Putting the numbers give σ = 102 barns. Assuming this cross-section as area, or area = πr^2, which gives $r = \sqrt{\dfrac{\text{area}}{\pi}}$.

Interpreting this as an area yields a radius of 5.7×10^{-14} m which is almost nine times larger than the actual radius given. This indicates that resonance capture is occurring.

Unsolved Problems

1. A thin sheet of ^{59}Co, 0.05 cm thick is irradiated with a neutron beam of flux 10^{12} n/cm^2/s for 10 hours. If the cross-section of the neutron capture by ^{59}Co is 37 b, how many nuclei of the isotope ^{60}Co would be produced at the end of irradiation/cm^3 and what is the initial β^--activity of the sample? $t_{1/2}$ for ^{60}Co = 5.2 years and density of cobalt = 8.9 g/cc.
[**Ans.** 6.5×10^{15}, 690 μCi]
2. The capture cross-section of ^{59}Co for thermal neutrons (energy = 0.025 eV) is 37 b. What percentage of a beam of thermal neutrons will penetrate a 2.0 mm sheet of ^{59}Co? Density of cobalt = 8.9×10^3 kg/m^3. [**Ans.** 51%]
3. The mean free path of neutrons in bismuth is about 6.5 cm. Find the total neutron cross-section of bismuth. (Atomic number ~ 209, density ~9.8 g/cm^3.)
[**Ans.** 5.46 b]
4. Find the mean free path of neutrons in bismuth. Given the total neutron cross-section of bismuth 5.46 barns. (Atomic number ~209, density ~9.8 g/cm^3)
[**Ans.** 6.5 cm]

Section 4.4

Solved Problem

1. Are the following reactions/decays possible? If not, write the correct relation.
(i) ${}^{27}_{13}\text{Al} + {}^{4}_{2}\text{He} \rightarrow {}^{30}_{14}\text{Si} + {}^{1}_{0}n$ (ii) ${}^{235}_{92}\text{U} + {}^{1}_{0}n \rightarrow {}^{143}_{56}\text{Ba} + {}^{90}_{36}\text{Kr} + 2{}^{1}_{0}n$
(iii) ${}^{32}_{15}\text{P} \rightarrow {}^{32}_{16}\text{S} + {}^{0}_{-1}e + \nu_e$

Solution:

(i)
$${}^{27}_{13}\text{Al} + {}^{4}_{2}\text{He} \rightarrow {}^{30}_{14}\text{Si} + {}^{1}_{0}n$$

In the reaction, the number of protons in the reactants is 13 + 2 = 15. In the products, the number of protons is 14 + 0 = 14. Hence, number of nucleons is not conserved. In the products, ${}^{1}_{0}n$ should be replaced by ${}^{1}_{1}\text{H}$. The correct equation is

$${}^{27}_{13}\text{Al} + {}^{4}_{2}\text{He} \rightarrow {}^{30}_{14}\text{Si} + {}^{1}_{1}\text{H}$$

(ii)
$${}^{235}_{92}\text{U} + {}^{1}_{0}n \rightarrow {}^{143}_{56}\text{Ba} + {}^{90}_{36}\text{Kr} + 2{}^{1}_{0}n$$

In the reaction, the number of nucleons in the reactants is 235 + 1 = 236. In the products, the number of nucleons is 143 + 90 + 2 = 235. Hence, number of nucleons is not conserved. In the products, $2{}^{1}_{0}n$ should be replaced by $3{}^{1}_{0}n$. The correct equation is

$${}^{235}_{92}\text{U} + {}^{1}_{0}n \rightarrow {}^{143}_{56}\text{Ba} + {}^{90}_{36}\text{Kr} + 3{}^{1}_{0}n$$

(iii)
$${}^{32}_{15}\text{P} \rightarrow {}^{32}_{16}\text{S} + {}^{0}_{-1}e + \nu_e$$

In the reaction, the number of leptons in the reactants is 0 + 0 = 0. In the products, the number of leptons is 1 + 1 = 2. Hence, lepton number is not conserved. In the products, ν_e should be replaced by $\overline{\nu}_e$. The correct equation is

$${}^{32}_{15}\text{P} \rightarrow {}^{32}_{16}\text{S} + {}^{0}_{-1}e + \overline{\nu}_e$$

Unsolved Problem

1. Are the following reactions/decays possible? If not, write the correct relation.

(i) $^{137}_{55}Cs \rightarrow ^{136}_{56}Ba + ^{0}_{-1}e + \overline{\nu}_e$ (ii) $^{60}_{27}Co \rightarrow ^{60}_{28}Ni + ^{0}_{-1}e$

(iii) $^{136}_{55}Ba^{*} \rightarrow ^{136}_{56}Ba + \gamma$ (iv) $^{22}_{11}Na \rightarrow ^{22}_{10}Ne + ^{0}_{-1}e$

(v) $^{24}_{12}Mg + ^{2}_{1}H \rightarrow ^{25}_{10}Ne + ^{1}_{1}H$ (vi) $^{4}_{2}He + ^{9}_{4}Be \rightarrow ^{12}_{6}C + ^{1}_{0}n$

(vii) $^{115}_{49}In + ^{6}_{3}Li \rightarrow ^{118}_{50}Sn + ^{3}_{1}H$ (viii) $^{197}_{79}Au + ^{28}_{14}Si \rightarrow ^{203}_{85}At + ^{15}_{8}O$

Section 4.5

Solved Problems

1. Calculate the Q-value for the reaction

$$^{2}H + ^{3}H \rightarrow ^{4}He + ^{1}n$$

Given:

$$M(^{1}n) = 1.00866501 \text{ amu}$$
$$M(^{2}H) = 2.014102 \text{ amu}$$
$$M(^{3}H) = 3.016049 \text{ amu}$$
$$M(^{4}He) = 4.002603 \text{ amu}$$

Solution: Q-value is given by

$$Q = [M(^{2}H) + M(^{3}H) - M(^{4}He) - M(^{1}n)]$$
$$= [2.014102 + 3.016049 - 4.002603 - 1.00866501]$$
$$= 0.01888299 \text{ amu}$$
$$= 0.01888299 \times 931.49 = 17.6 \text{ MeV}$$

Therefore, the Q-value for this reaction is 17.6 MeV.

2. Prove that ^{252}Cf is unstable and decays by α-emission as under:

$$^{252}Cf \rightarrow ^{248}Cm + ^{4}He$$

Given:

$$M(^{252}Cf) = 252.081621 \text{ amu}$$
$$M(^{248}Cm) = 248.072343 \text{ amu}$$
$$M(^{4}He) = 4.002603 \text{ amu}$$

Solution: Let us calculate the Q-value for this decay

$$Q = [M(^{252}Cf) - M(^{248}Cm) - M(^{4}He)]$$
$$= [252.081621 - 248.072343 - 4.002603]$$
$$= 0.006675 \text{ amu}$$
$$= 0.006675 \times 931.49 = 6.21 \text{ MeV}$$

Since Q-value for this reaction is +6.21 MeV, therefore, ^{252}Cf will decay by α-emission.

3. Find out the Q-value for the reaction

$$^{208}Pb(^{56}Fe,\ ^{54}Fe)^{210}Pb$$

Given:

$$M(^{208}Pb) = 207.976641 \text{ amu}$$
$$M(^{56}Fe) = 55.934939 \text{ amu}$$
$$M(^{210}Pb) = 209.984178 \text{ amu}$$
$$M(^{54}Fe) = 53.939612 \text{ amu}$$

Also find the threshold for this reaction.

Solution: In this reaction the projectile is ^{56}Fe and the reaction is

$$^{208}Pb + {}^{56}Fe \rightarrow {}^{210}Pb + {}^{54}Fe$$

the Q-value for this reaction is given by

$$Q = [M(^{208}Pb) + M(^{56}Fe) - M(^{210}Pb) - M(^{54}Fe)] \text{ amu}$$
$$= [207.976641 + 55.934939 - 209.984178 - 53.939612] \text{ amu}$$
$$= -0.001221 \text{ amu} = -0.01221 \times 931.49 \text{ MeV} = -11.37 \text{ MeV}$$

Therefore, the Q-value for this reaction is -11.37 MeV.

Threshold for this reaction is

$$E_{th} = -Q\frac{M(\text{projectile}) + M(\text{target})}{M(\text{target})}$$

or

$$= -Q\frac{M(^{56}Fe) + M(^{208}Pb)}{M(^{208}Pb)}$$

In this relation atomic masses can be replaced by respective mass numbers. The error introduced by this is negligible.

Therefore,

$$E_{th} = 11.37\frac{56 + 208}{208}$$
$$= 14.43 \text{ MeV}$$

So, the threshold for the reaction is 14.43 MeV.

4. When a proton captures a neutron to produce a deuteron nucleus, a γ-ray of energy 2.230 MeV is released based on the following equation:

$${}^{1}_{1}H + {}^{1}_{0}n \rightarrow {}^{2}_{1}H + \gamma$$

Calculate the mass of neutron. Given:

$$\text{Mass of } {}^{1}_{1}H = 1.008142 \text{ amu}$$
$$\text{Mass of } {}^{2}_{1}H = 2.014735 \text{ amu}$$

Solution: Total mass of the reactants = Total mass of the products
Let us take all masses in MeV
Therefore,

$$1.008142 \times 931.47 + m \times 931.47 = 2.014735 \times 931.47 + 2.230$$

where m is the mass of neutron.

$$931.47 \times m = 939.8613$$

$$m = \frac{939.8412}{931.47} = 1.008987 \text{ amu}$$

5. Find out whether the following reaction is exoergic or endoergic.

$$^{6}\text{Li} + {}^{1}n \rightarrow {}^{4}\text{He} + {}^{3}\text{H}$$

Given:

$$M(^{6}\text{Li}) = 6.0151234 \text{ amu}$$
$$M(^{1}n) = 1.0086654 \text{ amu}$$
$$M(^{4}\text{He}) = 4.0026034 \text{ amu}$$
$$M(^{3}\text{H}) = 3.0160294 \text{ amu}$$

Solution: Q-value for this reaction is given by

$$\begin{aligned} Q &= [M(^{6}\text{Li}) + M(^{1}n) - M(^{4}\text{He}) - M(^{3}\text{H})] \\ &= [6.0151234 + 1.0086654 - 4.0026034 - 3.0160294] \\ &= 0.005156 \text{ amu} \\ &= 0.005156 \times 931.47 = 4.8 \text{ MeV} \end{aligned}$$

Therefore, the Q-value for this reaction is 4.8 MeV and the reaction is exoergic.

6. Prove that ^{252}Cf can undergo spontaneous fission as under:

$$^{252}\text{Cf} \rightarrow {}^{98}\text{Zr} + {}^{145}\text{Ce} + 9\,{}^{1}n$$

Given:

$$M(^{252}\text{Cf}) = 252.081621 \text{ amu}$$
$$M(^{98}\text{Zr}) = 97.912735 \text{ amu}$$
$$M(^{145}\text{Ce}) = 144.917230 \text{ amu}$$
$$M(^{1}n) = 1.008664916 \text{ amu}$$

Solution: Q-value for the spontaneous fission is given by

$$\begin{aligned} Q &= M(^{252}\text{Cf}) - M(^{98}\text{Zr}) - M(^{145}\text{Ce}) - 9 \times M(^{1}n) \\ &= (252.081621 - 97.912735 - 144.917230 - 9.0779850) \\ &= 0.1737009 \text{ amu} \\ &= 0.1737009 \times 931.47 = 161.8 \text{ MeV} \end{aligned}$$

Since Q-value is positive, so spontaneous decay of ^{252}Cf is possible.

7. Calculate the separation energy to remove one proton from ${}^{16}_{8}O$. Given:

$$m(H) = 1.007825 \text{ amu}$$
$$m({}^{15}_{7}N) = 15.000108 \text{ amu}$$
$$m({}^{16}_{8}O) = 16 \text{ amu}$$

Solution: Given reaction is

$${}^{16}_{8}O \rightarrow {}^{15}_{7}N + {}^{1}_{1}H$$

Q-value for this reaction is

$$\begin{aligned} Q &= M_O - M_N - M_H \\ &= 16 - 15.000108 - 1.007825 = -0.007933 \\ &= -0.007933 \times 931.47 \\ &= -7.4 \text{ MeV} \end{aligned}$$

Therefore, energy required to remove one proton from ${}^{16}_{8}O = 7.4$ MeV

8. The Q-value of the ${}^{23}Na(n, \alpha){}^{20}F$ reaction is -5.4 MeV. Determine the threshold energy of the neutrons for this reaction. Given:

$$\text{Mass of neutron} = 1.00866 \text{ amu}$$
$$\text{Mass of } {}^{23}Na = 22.99097 \text{ amu}$$

Solution: Given reaction is

$${}^{23}Na + n \rightarrow {}^{20}F + {}^{4}_{2}He$$
$$Q\text{-value} = -5.4 \text{ MeV}$$

$$\text{Therefore, threshold energy} = -Q\frac{M_{Na} + M_n}{M_{Na}}$$

$$= -5.4\,\frac{22.99097 + 1.00866}{22.99097}$$

$$= 5.64 \text{ MeV.}$$

9. Calculate the Q-value in MeV for the following nuclear reaction:

$${}^{4}_{2}He + {}^{14}_{7}N \rightarrow {}^{17}_{8}O + {}^{1}_{1}H$$

Given masses of

$${}^{4}_{2}He = 4.00388 \text{ amu}$$
$${}^{14}_{7}N = 14.00755 \text{ amu}$$
$${}^{17}_{8}O = 17.00452 \text{ amu}$$
$${}^{1}_{1}H = 1.00815 \text{ amu}$$

Solution: Q-value is given by the relation

$$Q = M_N + M_{He} - M_O - M_H$$

$$= 14.00755 + 4.00388 - 17.00452 - 1.00866$$

$$= -0.00124 \text{ amu}$$

$$= -0.00124 \times 931.47 = -1.15 \text{ MeV}$$

Therefore, Q-value = −1.15 MeV

10. Calculate the minimum energy of γ-rays necessary to disintegrate deuteron into a proton and neutron. Given masses of

$$^2_1\text{H} = 2.014735 \text{ amu}$$

$$^1_1\text{H} = 1.008142 \text{ amu}$$

$$^1_0n = 1.008987 \text{ amu}$$

Solution: The given reaction is

$$^2_1\text{H} + \gamma \rightarrow {}^1_1\text{H} + {}^1n$$

Let E_γ be the energy of γ-ray to start this reaction.
Total mass of the reactants = Total mass of the products
Let us take all masses in MeV
Therefore,

$$2.014735 \times 931.47 + E_\gamma = 1.008142 \times 931.47 + 1.008987 \times 931.47$$

or

$$E_\gamma = 2.2299 \text{ MeV}$$

Unsolved Problems

1. Find out whether the following reaction is exoergic or endoergic.

$$^{27}\text{Al} + {}^1\text{H} \rightarrow {}^{27}\text{Si} + {}^1n$$

Given:

$$M(^1\text{H}) = 1.00727647 \text{ amu}$$

$$M(^1n) = 1.008664916 \text{ amu}$$

$$M(^{27}\text{Al}) = 26.981539 \text{ amu}$$

$$M(^{27}\text{Si}) = 26.986704 \text{ amu}$$

[**Ans.** $Q = -6.10$ MeV, endoergic]

2. Find out the Q-value for the reaction

$$^{56}\text{Fe}\,(^{208}\text{Pb},\ ^{210}\text{Pb})\,^{54}\text{Fe}$$

Given:

$$M(^{208}\text{Pb}) = 207.976641 \text{ amu}$$

$$M(^{56}\text{Fe}) = 55.934939 \text{ amu}$$

$$M(^{210}\text{Pb}) = 209.984178 \text{ amu}$$

$$M(^{54}\text{Fe}) = 53.939612 \text{ amu}$$

Also find the threshold for this reaction.

[**Ans.** −11.37 MeV, 54.60 MeV]

3. (i) Calculate the Q-value for the decay of a neutron into a proton and an electron.

[**Ans.** 0.783 MeV]

(ii) Calculate the Q-value for the decay of a proton into a neutron and an electron.

[**Ans.** −1.805 MeV]

4. Calculate the Q-value for the reaction

$$^{27}Al + {}^{4}He \rightarrow {}^{30}Si + {}^{1}H$$

Given:

$$M(^{1}H) = 1.00727647 \text{ amu}$$
$$M(^{4}He) = 4.002603 \text{ amu}$$
$$M(^{27}Al) = 26.981539 \text{ amu}$$
$$M(^{30}Si) = 29.973772 \text{ amu}$$

[**Ans.** 2.88 MeV]

5. Find the minimum energy of γ-ray required to produce photodisintegration of ^{9}Be. Given:

$$M(^{9}Be) = 9.012182 \text{ amu}$$
$$M(^{4}He) = 4.002603 \text{ amu}$$
$$M(^{1}n) = 1.00866501 \text{ amu}$$

[**Ans.** 1.57 MeV]

6. Find the minimum kinetic energy in the lab system needed by an α-particle to cause the reaction $^{14}N(\alpha, p)^{17}O$. The masses of

$$m(^{14}N) = 14.00307 \text{ amu}$$
$$m(\alpha) = 4.003879 \text{ amu}$$
$$m(p) = 1.00783 \text{ amu}$$
$$m(^{17}O) = 17.00013 \text{ amu}$$

[**Ans.** 1.21 MeV]

7. Calculate the Q-value for the reaction

$$^{27}Al + {}^{1}H \rightarrow {}^{27}Si + {}^{1}n$$

Given:

$$M(^{1}H) = 1.00727647 \text{ amu}$$
$$M(^{1}n) = 1.008664916 \text{ amu}$$
$$M(^{27}Al) = 26.981539 \text{ amu}$$
$$M(^{27}Si) = 26.986704 \text{ amu}$$

[**Ans.** –6.10 MeV]

8. Find the Q-value and threshold for the reaction

$$^{239}U(^{4}He, 3\,^{1}n)^{239}Pu$$

Given:

$$M(^{238}U) = 238.050785 \text{ amu}$$
$$M(^{4}He) = 4.002603 \text{ amu}$$
$$M(^{239}Pu) = 239.052158 \text{ amu}$$
$$M(^{1}n) = 1.00866501 \text{ amu}$$

[**Ans.** 23.1 MeV, 23.5 MeV]

9. Calculate the threshold energy for $^{14}N(n, \alpha)^{11}B$ reaction. Given:

$$m(n) = 1.008983 \text{ amu}$$
$$m(\alpha) = 4.003879 \text{ amu}$$
$$m(^{14}N) = 14.007550 \text{ amu}$$
$$m(^{11}B) = 11.012811 \text{ amu}$$

[**Ans.** 0.157 MeV]

10. Find the Q-value for the nuclear reaction

$$^{9}_{4}\text{Be} + ^{4}_{2}\text{He} \rightarrow ^{12}_{6}\text{C} + ^{1}_{0}n + Q$$

Given masses of

$$^{9}_{4}\text{Be} = 9.01286 \text{ amu}$$
$$^{4}_{2}\text{He} = 4.00388 \text{ amu}$$
$$^{12}_{6}\text{C} = 12.000 \text{ amu}$$
$$^{1}_{0}n = 1.00866 \text{ amu}$$

[**Ans.** 7.52 MeV]

11. A tritium gas target ($^{3}_{1}$H) is bombarded with a beam of protons ($^{1}_{1}$H) of kinetic energy 3 MeV. Determine the Q-value of the following reaction:

$$^{1}_{1}\text{H} + ^{3}_{1}\text{H} \rightarrow ^{3}_{2}\text{He} + ^{1}_{0}n + Q$$

Given:

$$m(^{1}_{1}\text{H}) = 1.007276 \text{ amu}$$
$$m(^{3}_{1}\text{H}) = 3.016056 \text{ amu}$$
$$m(^{1}_{0}n) = 1.00866 \text{ amu}$$
$$m(^{3}_{2}\text{He}) = 3.016030 \text{ amu}$$

[**Ans.** –1.26 MeV]

12. A beam of α-particles of energy 7.3 MeV is used for $^{27}Al(\alpha, p)^{30}Si$ reaction. The protons emitted at an angle of 0° are found to have energy of 9.34 MeV. Find the Q-value of this reaction. [**Ans.** 2.22 MeV]

13. A deuteron beam of energy 1.51 MeV is used for $^{16}O(d, \alpha)\,^{14}N$ reaction. α-particles emitted at an angle of 90° are found to have energy of 3.42 MeV. Find the Q-value of this reaction. [**Ans.** 3.10 MeV]

Section 4.6

Unsolved Problems

1. What target isotope must be used to form the compound nucleus ^{28}Si, when the projectile is (i) neutron, (ii) proton, and (iii) alpha-particle?

[**Ans.** (i) ^{27}Si, (ii) ^{27}Al, (iii) ^{24}Mg]

2. List at least four nuclear reactions for making compound nucleus ^{36}Cl by bombarding different projectiles on stable nuclei.

3. When ^{27}Al is bombarded with protons, what compound nucleus is formed? List four possible decay modes of this compound nucleus assuming that sufficient energy exists with the compound nucleus for decay into the chosen mode.

4. What target isotope must be used to form the compound nucleus ^{24}Na, when the projectile is (i) n, (ii) p, (iii) α-particle and (iv) ^{3}He?
5. Devise three nuclear reactions that forms ^{24}Mg as a compound nucleus starting with stable target and stable projectile.
6. List at least three nuclear reactions for making compound nucleus ^{60}Co by bombarding different projectiles on stable nuclei.

Section 4.7

Solved Problems

1. ^{235}U loses about 0.1% of its mass when it undergoes fission.
   (i) How much energy is released when 1 kg of ^{235}U undergoes fission?
   (ii) One ton of TNT releases about 4 GJ when it is detonated. How many tons of TNT are equivalent in destructive power to a bomb that contains 1 kg of ^{235}U?

Solution: (i) 0.1% of 1 kg = 0.001 kg

Applying the relation $E = mc^2$, we get

$E = 0.001 \times (3 \times 10^8)^2 = 9 \times 10^{13}$ J

(ii) 4 GJ = 4×10^9 J of energy is produced by 1 t of TNT

$$9 \times 10^{13} \text{ J of energy requires TNT} = \frac{1}{4 \times 10^9} \times 9 \times 10^{13} = 22{,}500 \text{ t}$$

Therefore, the destructive power of bomb using 1 kg of ^{235}U = 22,500 t of TNT = 22.5 kt of TNT.

2. A beam of low energy neutrons, intensity 10^6s^{-1} traverses a foil of ^{235}U, thickness 0.15 kg m^{-2}. If the fission cross-section is $2 \times 10^{-26}\text{m}^2$, find the rate of fissions induced in the foil by the neutrons. Here, we neglect the reduction of the beam as it passes through the foil.

Solution: Thickness of the foil = 0.15 kg m^{-2}

This means that a foil of area S m^2 will have a mass of $S \times 0.15$ kg.

235 kg of ^{235}U contains = 6.023×10^{26} nuclei

$$S \times 0.15 \text{ kg will contain} = \frac{S \times 0.15 \times 6.023 \times 10^{26}}{235}$$

$$= S \times 3.844 \times 10^{23} \text{ nuclei}$$

Each nucleus presents an area = $2 \times 10^{-26}\text{m}^2$ to the beam of neutrons

So the total area presented = $S \times 3.844 \times 10^{23} \times 2 \times 10^{-26}\text{m}^2$.

$$= S \times 7.686 \times 10^{-3}$$

Thus, the fraction of the area S which is covered = 7.686×10^{-3}

This is the fraction of the beam removed by the fission process. Hence, the total induced fission rate in this fractional area is $7.686 \times 10^{-3} \times 10^6 = 7685 \text{ s}^{-1}$, when a beam of intensity 10^6s^{-1} is incident on that area.

3. The nuclide ^{256}Fm decays through spontaneous fission with a half-life of 158 min. If the energy released is about 220 MeV per fission, calculate the fission power produced by 1 μg of this isotope.

Solution: We have the fission rate

$$\frac{dN}{dt} = -N\lambda$$

Here, N is the number of nuclei present and λ is the decay constant.

Number of nuclei of ^{256}Fm present in 1 μg of sample is

$$= \frac{6.023 \times 10^{23}}{256} \times 10^{-6} = 2.35 \times 10^{15}$$

and decay constant $\lambda = \dfrac{\ln 2}{t_{1/2}} = \dfrac{\ln 2}{158 \times 60} = 7.31 \times 10^{-5}\,\text{s}^{-1}$

Therefore, fission rate $= 2.35 \times 10^{15} \times 7.31 \times 10^{-5}$

$= 1.72 \times 10^{11}$ fissions s^{-1}

Energy released in fission of one nucleus = 220 MeV

$= 220 \times 1.602 \times 10^{-13}$ J

$= 3.52 \times 10^{-11}$ J

Power released in fission of 1.72×10^{11} nuclei $= 3.52 \times 10^{-11} \times 1.72 \times 10^{11}$

= 6 W

4. The nuclide ^{252}Cf undergoes spontaneous fission with a half-life of 2.62 years. Energy released in this process is about 210 MeV. If 100 mg of this isotope is taken in a space-ship, calculate the power produced by this isotope.

Solution: We have the fission rate

$$\frac{dN}{dt} = -N\lambda$$

Here, N is the number of nuclei present and λ is the decay constant.

Number of nuclei of ^{252}Cf present in 100 mg of sample is

$$= \frac{6.023 \times 10^{23}}{252} \times 0.1 = 2.39 \times 10^{20}$$

and decay constant $\lambda = \dfrac{\ln 2}{t_{1/2}} = \dfrac{\ln 2}{2.62 \times 365 \times 24 \times 3600} = 8.39 \times 10^{-9}\,\text{s}^{-1}$

Therefore, fission rate $= 2.39 \times 10^{20} \times 8.39 \times 10^{-9}$

$= 2.01 \times 10^{12}$ fissions/s

Energy released in fission of one nucleus = 210 MeV

$$= 210 \times 1.602 \times 10^{-13} \text{ J}$$

$$= 3.36 \times 10^{-11} \text{ J}$$

Energy released in fission of 2.01×10^{12} nuclei $= 3.36 \times 10^{-11} \times 2.01 \times 10^{12}$

$$= 67.6 \text{ W}$$

5. The energy released during the nuclear explosion at Hiroshima has been estimated as equivalent to that released by 20,000 tons of TNT. Assume that 200 MeV is released when a ^{235}U nucleus absorbs a neutron and fissions and that 4.18×10^9 J is released during detonation of 1 ton of TNT. How many nuclear fissions occurred at Hiroshima, and what was the total decrease in mass?

Solution: Energy released in detonation of 1 ton of TNT $= 4.18 \times 10^9$ J

Energy released in detonation of 20,000 tons of TNT

$$= 4.18 \times 10^9 \times 20{,}000 \text{ J}$$

$$= 8.36 \times 10^{13} \text{ J}$$

Energy released in fission of 1 nucleus of ^{235}U $= 200$ MeV

$$= 200 \times 1.603 \times 10^{-13} \text{ J}$$

$$= 3.206 \times 10^{-11} \text{ J}$$

3.206×10^{-11} J of energy requires 1 fission

8.36×10^{13} J of energy will require $\dfrac{8.36 \times 10^{13}}{3.206 \times 10^{-11}} = 2.61 \times 10^{24}$ fissions

Total energy released in 2.61×10^{24} fissions $= 2.61 \times 10^{24} \times 200$ MeV

$$= 5.22 \times 10^{26} \text{ MeV} = 8.362 \times 10^{13} \text{J}$$

Using the relation $E = mc^2$, we get

$$m = \frac{E}{c^2} = \frac{8.362 \times 10^{13}}{(3 \times 10^8)^2} = 9.29 \times 10^{-4} \text{ kg} = 0.929 \text{ g}$$

$$= 929 \text{ mg}$$

Unsolved Problems

1. Fission of 1 atom of ^{235}U produces about 200 MeV of energy. Estimate the fission rate required to produce 2 watts of energy per second. **[Ans.** 6.25×10^{10} fissions/s].
2. Calculate the energy released if 0.2 kg of ^{235}U undergoes complete fission. **[Ans.** 3.9×10^{12} calories]
3. If each fission of ^{235}U releases 200 MeV, how many fissions occur per second to yield a power level of 2.0 GJ? How much mass is lost per day by the nuclear reactor operated at this power level? **[Ans.** 6.25×10^{19} fissions/s, 1.92 g/day]
4. The fissioning of ^{235}U atom yields 200 MeV of energy. How many atoms of ^{235}U are required to provide an amount of energy equal to that required to lift a housefly by 10 cm? Assume that the mass of the average fly is 5.0 mg. **[Ans.** 1.53×10^5]

5. Calculate the energy released in

(i) Spontaneous fission of ^{235}U to ^{147}La and ^{88}Br.

(ii) Thermal neutron induced fission of ^{235}U to ^{147}La and ^{89}Br.

Given:

$$M(^{235}\text{U}) = 235.043922 \text{ amu}$$
$$M(^{147}\text{La}) = 146.922780 \text{ amu}$$
$$M(^{88}\text{Br}) = 87.924070 \text{ amu}$$
$$M(^{89}\text{Br}) = 88.926400 \text{ amu}$$

[**Ans.** (i) 178.5 MeV, (ii) 184.4 MeV]

6. Calculate the energy released in the spontaneous fission of $^{238}_{92}$U into two nuclei, each $^{119}_{46}$Pd.

Given:

$$M(^{238}\text{U}) = 238.0507835 \text{ amu}$$
$$M(^{119}\text{Pd}) = 118.922680 \text{ amu}$$

[**Ans.** 190.5 MeV]

7. Calculate the energy released in the fission of $^{235}_{92}$U assisted by thermal neutrons into two nuclei, each $^{118}_{46}$Pd. Given:

$$M(^{235}\text{U}) = 235.043922 \text{ amu}$$
$$M(^{118}\text{Pd}) = 117.91890 \text{ amu}$$

[**Ans.** 200.1 MeV]

8. Upon combustion in air, methane gas undergoes the following chemical reaction:

$$CH_4 + 2O_2 \rightarrow CO_2 + 2H_2O$$

During this process about 9 eV/methane molecule energy is released. Find the relative energy released per unit mass for nuclear fission as against this chemical reaction.

[**Ans.** 7.6×10^6]

9. Assume that $^{235}_{92}$U undergoes fission by a thermal neutron producing $^{97}_{40}$Zr, ^{134}Te and some neutrons. (a) Determine the atomic number of Te from the above data. (b) How many neutrons are released? [**Ans.** 52, 5]

10. $^{235}_{92}$U undergoes photo fission by γ-rays of energy of 10 MeV. If the fission products are $^{90}_{36}$Kr, $^{142}_{56}$Ba and few neutrons, find the number of neutrons emitted in this reaction and write a complete equation for this process. [**Ans.** 3]

11. $^{235}_{92}$U undergoes photo fission by γ-rays of energy of 10 MeV. If the fission products are $^{90}_{36}$Kr, $^{142}_{56}$Ba and three neutrons, calculate the total energy of fission products. Given:

$$m(\text{U}) = 235.0439299 \text{ amu}$$

$$m(\text{Ba}) = 141.916457 \text{ amu}$$
$$m(\text{Kr}) = 89.919517 \text{ amu}$$
$$m(\text{n}) = 1.008665$$

[**Ans.** 179.5 MeV]

12. A 100 MW reactor consumes half its fuel in three years. How much ^{235}U does it contain? Given that fission of one nuclei yield energy = 200 MeV. [**Ans.** 232.4 kg]

13. Find the ratio of energy released when 1 mg of ^{233}U undergoes fission to the energy released when 1 mg of TNT explodes. Given that energy released in the explosion of 1 g of TNT is 2.5×10^{22}eV. [**Ans.** 2.06×10^7]

14. If an atom bomb of 10 MT explodes, find the total number of fissions that occur and also find the mass which is converted to energy. Assume that 200 MeV is released when a ^{235}U nucleus absorbs a neutron and fissions and that 4.18×10^9 J is released during detonation of 1 ton of TNT. How many nuclear fissions occurred at Hiroshima, and what was the total decrease in mass? [**Ans.** 1.305×10^{27} fissions, 464 g]

15. ^{236}U undergoes fission into ^{102}Mo and ^{131}Sn isotopes. Complete this reaction. How many neutrons are ejected in this reaction? How many neutrons will be ejected if the fission products are ^{88}Br and ^{140}La isotopes?

$$\left[\textbf{Ans.}\quad {}^{236}\text{U} + {}^1_0 n \rightarrow {}^{102}_{42}\text{Mo} + {}^{131}_{50}\text{Sn} + 4{}^1_0 n, \text{ 4 neutrons,}\right.$$
$$\left. {}^{236}\text{U} + {}^1_0 n \rightarrow {}^{88}_{35}\text{Br} + {}^{140}_{57}\text{La} + 9{}^1_0 n, \text{ 9 neutrons}\right]$$

16. If one of the fission fragments from the fission of ^{240}Pu is ^{90}Sr, what is the second fragment, if its mass is 142? How many free neutrons are produced?

$$\left[\textbf{Ans.}\quad {}^{240}\text{Pu} + {}^1_0 n \rightarrow {}^{90}_{38}\text{Sr} + {}^{142}_{56}\text{Ba} + 9{}^1_0 n, \text{ 9 neutrons}\right]$$

17. For the fission reaction

$$^{235}\text{U} + n \rightarrow {}^{142}_{55}\text{Cs} + {}^{90}_{37}\text{Rb} + 4{}^1_0 n$$

estimate the energy released per fission. Given $m_{\text{U}} = 235.0439231$ amu, $m_{\text{Cs}} = 141.924292$ amu, $m_{\text{Rb}} = 89.914809$ amu and $m_n = 1.008665$ amu. [**Ans.** 166.6 MeV]

18. Calculate the energy released in the symmetric fission of the nuclei with following A and Z values

(i) $A = 238$ and $Z = 92$
(ii) $A = 100$ and $Z = 44$

Given the masses ${}^{238}_{92}\text{U} = 238.050788$ amu, ${}^{119}_{46}\text{Pd} = 118.923100$ amu, ${}^{100}_{44}\text{Ru} = 99.904219$ amu and ${}^{50}_{22}\text{Ti} = 118.923100$ amu. [**Ans.** 190.6 MeV, –5.9 MeV]

19. One gram of ^{235}U upon fission by thermal neutrons releases 2.29×10^4kW · h of energy. Find the mass of coal required to produce an equivalent amount of energy in a thermal power plant. Given that on burning of 1 kg of coal produces 3.5×10^4 kJ of energy. [**Ans.** 2355 kg]

Section 4.8

Solved Problems

1. Calculate the energy liberated if 10 kg of ^{2}H undergoes the following fusion reaction:

$$5\,{}^2_1\text{H} \rightarrow {}^3_2\text{He} + {}^4_2\text{He} + {}^2_1\text{H} + 2\,{}^1_0n + 25\text{ MeV}$$

Solution: 2 kg of ^{2}H contains $= 6.023 \times 10^{26}$ atoms

$$10\text{ kg of }^2\text{H contains} = \frac{6.023\times 10^{26}}{2} \times 10 = 3.01 \times 10^{27}\text{ atoms}$$

Now,

5 atoms of ^{2}H on fusion liberate energy = 25 MeV

$$3.01 \times 10^{27}\text{ atoms on fusion liberate energy} = \frac{25}{5} \times 3.01 \times 10^{27} = 1.51 \times 10^{28}\text{ MeV}$$

Therefore, 10 kg of ^{2}H in fusion will liberate energy $= 1.51 \times 10^{28}$ MeV $= 2.42 \times 10^{15}$ J.

2. In their old age, heavy stars obtain part of their energy by the reaction:

$${}^4_2\text{He} + {}^{12}_6\text{C} \rightarrow {}^{16}_8\text{O}$$

How much energy does each such event give off?

Solution: Total mass of the reactants = 4.002603 + 12.000000 = 16.002603 amu

Total mass of the product = 15.994915 amu

Difference in the mass $= 7.688 \times 10^{-3}$ amu

Energy produced $= 7.688 \times 10^{-3} \times 931.49 = 7.16$ MeV.

3. Consider a gas of atoms undergoing fusion. Calculate the temperature required to overcome the Coulomb barrier and the fusion energy released if the gas consists of (a) ^{10}B and (b) ^{24}Mg.

Solution:

(a) Let us estimate the height of the Coulomb barrier. It is given by the relation

$$V_{\text{Coul}} = \frac{Z_1 Z_2 e^2}{r}$$

Here, r is the separation between two nuclei at the point of closest approach. It is given by the sum of radii of two ^{10}B nuclei. The radius of each of the nucleus can be estimated using $r = 1.2A^{1/3}$f. Therefore,

$$r = 1.2 \times 10^{1/3} + 1.2 \times 10^{1/3} = 5.17\text{ f}$$

Coulomb barrier can be written as

$$V_{\text{Coul}} = \frac{e^2}{\hbar c} \times \frac{Z_1 Z_2 \times \hbar c}{r}$$

$$\frac{e^2}{\hbar c} = \frac{1}{137} = \alpha\text{ (fine structure constant)}$$

Substituting various values, we get

$$V_{\text{Coul}} = \frac{1}{137} \times \frac{5 \times 5 \times 197.5 \text{ MeV f}}{5.17 \text{ f}}$$

$$= 6.97 \text{ MeV}$$

$$= 1.12 \times 10^{-12} \text{ J}$$

In order to calculate the temperature required to overcome the Coulomb barrier, we equate this energy to thermal energy as

$$\frac{3}{2} KT = E = V_{\text{Coul}}$$

where K is Boltzmann's constzant and T is absolute temperature.
or

$$\frac{3}{2} \times 1.38 \times 10^{-23} T = 1.12 \times 10^{-12}, \text{ which gives}$$

$$T = 5.4 \times 10^{10} \text{ K}$$

Similar calculations are performed for the case of ^{24}Mg fusing with ^{24}Mg as under.

(b) Let us estimate the height of the Coulomb barrier. It is given by the relation

$$V_{\text{Coul}} = \frac{Z_1 Z_2 e^2}{r}$$

Here, r is the separation between two nuclei at the point of closest approach. It is given by the sum of radii of two ^{24}Mg nuclei. The radius of each of the nucleus can be estimated using $r = 1.2A^{1/3}$ f. Therefore,

$$r = 1.2 \times 24^{1/3} + 1.2 \times 24^{1/3} = 6.92 \text{ f}$$

Coulomb barrier can be written as

$$V_{\text{Coul}} = \frac{e^2}{\hbar c} \times \frac{Z_1 Z_2 \times \hbar c}{r}$$

$$\frac{e^2}{\hbar c} = \frac{1}{137} = \alpha \text{ (fine structure constant).}$$

Substituting various values, we get

$$V_{\text{Coul}} = \frac{1}{137} \times \frac{12 \times 12 \times 197.5 \text{ MeV f}}{6.92 \text{ f}}$$

$$= 29.99 \text{ MeV}$$

$$= 4.80 \times 10^{-12} \text{ J}$$

In order to calculate the temperature required to overcome the Coulomb barrier, we equate this energy to thermal energy as

$$\frac{3}{2} KT = E = V_{\text{Coul}},$$

where K is Boltzmann's constant and T is absolute temperature.

or

$$\frac{3}{2}\times 1.38\times 10^{-23}T = 4.80 \times 10^{-12}, \text{ which gives}$$

$$T = 23.2 \times 10^{10} \text{ K}$$

4. Suppose that the sun consists entirely of hydrogen and that the dominant energy-releasing reaction is

$$4{}^1_1\text{H} \rightarrow {}^4_2\text{He} + 2{}^0_1e + 2\nu + \gamma$$

If the total power output of the sun is assumed to remain constant at 3.9×10^{26}W, how long will it take for all of the hydrogen to be burned up? Take the mass of the sun as 2.0×10^{30} kg.

Solution: Total number of hydrogen atoms in the sun $= \dfrac{2.0\times 10^{30}\,\text{kg}}{1.67\times 10^{-27}\,\text{kg/atom}}$

$$= 1.2 \times 10^{57} \text{ atoms}$$

The given reaction has a mass difference

$$= 4 \times 1.007825 - 4.002603 - 2 \times 0.000549 = 0.027599 \text{ amu}$$

which in terms of energy is $0.027599 \times 931.47 = 25.7$ MeV

Since each fusion reaction consumes 4 hydrogen atoms, the total energy available in the sun

$$= \frac{1.2\times 10^{57}}{4}\times 25.7 \text{ MeV}$$

$$= 7.71 \times 10^{57} \text{ MeV}$$

$$= 1.24 \times 10^{45} \text{ J}$$

3.9×10^{26} J of energy is released in 1s.

1.24×10^{45} J of energy will be released in $\dfrac{1.24\times 10^{45}}{3.9\times 10^{26}} = 3.18\times 10^{18}$ s.

$$= 1.008 \times 10^{11} \text{ years}$$

Thus, this gives an estimated lifetime of about 100.8 billion years.

5. Calculate the mass defect and Q-values for the fusion reactions

$$d + d \rightarrow {}^3\text{He} + n$$

$$d + d \rightarrow {}^3\text{H} + p$$

Assuming these occur with the deuterons at rest, find the kinetic energies of the outgoing particles in each case. Given $m_p = 1.007825$ amu, $m_n = 1.008665$ amu, $m(^2\text{H}) = 2.014102$ amu, $m(^3\text{H}) = 3.016049$ amu and $m(^3\text{He}) = 3.016029$ amu.

Solution: We have

$$d + d \rightarrow {}^3\text{He} + n$$

Mass defect is given by the relation

$$\text{Mass defect} = 2 \times m_d - m(^3\text{He}) - m_n$$

Substituting various given masses, we get

$$\text{Mass defect} = 2 \times 2.014102 - 3.016029 - 1.008665$$
$$= 0.00351 \text{ amu}$$

and

$$Q\text{-value} = \text{mass defect (in amu)} \times 931.47 \text{ MeV}$$
$$= 0.00351 \times 931.47 \text{ MeV}$$
$$= 3.27 \text{ MeV}$$

$$d + d \rightarrow {}^3\text{H} + p$$

Mass defect is given by the relation

$$\text{Mass defect} = 2 \times m_d - m(^3\text{H}) - m_p$$

Substituting various given masses, we get

$$\text{Mass defect} = 2 \times 2.014102 - 3.016049 - 1.007825$$
$$= 0.00433 \text{ amu}$$

and

$$Q\text{-value} = \text{mass defect (in amu)} \times 931.47 \text{ MeV}$$
$$= 0.00433 \times 931.47 \text{ MeV}$$
$$= 4.03 \text{ Mev}$$

Assuming the initial state deuterons are essentially at rest then the final state kinetic energy is equal to Q. By applying the conservation of momentum it can be seen that the share of the kinetic energy that each particle has is inversely proportional to its mass. Thus, for these reactions, the heavier particle takes one quarter while the lighter particle takes three quarters of the total kinetic energy (Q).

Unsolved Problems

1. Calculate the energy librated in the following reaction:

$$p + {}^{11}\text{B} \rightarrow 3\,{}^4_2\text{He}$$

Given:

$$M(p) = 1.007825 \text{ amu}$$
$$M(^4\text{He}) = 4.002603 \text{ amu}$$
$$M(^{11}\text{B}) = 11.009305 \text{ amu}$$

[**Ans.** 8.68 MeV]

2. ^{2}H and ^{3}H undergoes the following fusion reaction:

$$^2\text{H} + {}^3\text{H} \rightarrow {}^4_2\text{He} + {}^1_0n + 17.6 \text{ MeV}$$

Calculate the amounts of ^{2}H and ^{3}H required to produce 10^{16} J of energy.

[**Ans.** 11.8 kg, 17.7 kg]

3. Upon combustion in air, methane gas undergoes the following chemical reaction:

$$CH_4 + 2O_2 \rightarrow CO_2 + 2H_2O$$

During this process about 9 eV/methane molecule energy is released. In sun, 4^1_1 H atoms undergoes fusion as under:

$$4^1_1\text{H} \rightarrow {}^4_2\text{He} + 2\,{}^0_{+1}e + 2\nu_e + 26.73 \text{ MeV}$$

Calculate the relative energy released per unit mass for nuclear fusion as against this chemical reaction. **[Ans.** 5.9×10^7]

4. The energy released during the explosion of hydrogen bomb has been estimated as equivalent to that released by 50 mega tons of TNT. Assume that the following reaction takes place in the bomb

$$4^1_1\text{H} \rightarrow {}^4_2\text{He} + 2^0_1 e + 2\nu$$

Find the weight of hydrogen required to produce the required power output. Also, find the weight of hydrogen converted to energy. Given that 4.18×10^9 J is released during detonation of 1 ton of TNT, $m_p = 1.007276$ amu, $m_{He} = 4.0015$ amu and $m_2 = 0.00054858$ amu. **[Ans.** 353 kg, 2.32 kg]

5. In 2006, total world wide energy consumption was about 6×10^{20} J per year. Suppose this entire energy is derived from the reaction

$$3^2_1\text{H} \rightarrow {}^4_2\text{He} + {}^1_1\text{H} + {}^1_0 n$$

How many deuteron atoms would be required to generate this energy? What would be the weight of these atoms? Given $m_d = 2.0141022$ amu, $m_a = 4.002603$ amu, $m_n = 1.00866501$ amu and $m_p = 1.00727647$ amu. **[Ans.** 5.07×10^{32}, 1684 tons]

6. In a fusion reaction six deuterons fuse to give 2 ^{4}He nuclei, two protons and two neutrons. Total energy released in this process is 43.2 MeV. Calculate the energy released by fusion of 2 g of deuterons. **[Ans.** 4.34×10^{24} MeV]

REVIEW QUESTIONS

Short Answer Type

1. Define nuclear reaction cross-section. What are its units and dimensional formula?
2. What is nuclear reaction cross-section?
3. What is differential cross-section of a nuclear reaction?
4. What are stripping and pickup reactions?
5. What is spalliation nuclear reaction?
6. What is nuclear transmutation?
7. Name and explain the physical quantities that are not conserved in a nuclear reaction.
8. What are the quantities which remain conserved in a nuclear reaction?
9. How do you support the formation of compound nucleus in a nuclear reaction?
10. What is half-thickness of a material?
11. Differentiate between nuclear decay process and a nuclear reaction.
12. For a nuclear reaction ${}^{24}_{12}\text{Mg}(\alpha, n)\,{}^{27}_{14}\text{Si}$, how do you think, we can calculate the Q-value from the masses?

13. What does the X in the following reaction represent? Also write about the compound nucleus formed.

$$^{14}_{7}\text{N}(X, p)\,^{17}_{8}\text{O}$$

14. What is meant by Q-value of a nuclear reaction?
15. Can a nuclear reaction be induced by photons? Explain giving an example.
16. Name various types of nuclear reactions.
17. Define the differential and total scattering cross-sections.

Long Answer Type

1. What is kinematics of nuclear reaction? What is the Q-value and its significance?
2. What is the Q-value of a nuclear reaction? Derive an expression for Q-value of a nuclear reaction in terms of masses and kinetic energies of incident particle, product particle and nuclei.
3. What is a compound nucleus? Who proposed it? What is its significance?
4. Define nuclear cross-section. Derive an expression for number of surviving particles, if a number of particles is incident on a slab of certain area, thickness and number density.
5. What is a nuclear reaction? Define its cross-section and Q-value. Derive an expression for the Q-value of the reaction $X(a, b)Y$ in terms of the kinetic energy of particles.
6. Write notes on:
   (i) Nuclear reaction cross-section.
   (ii) Compound nucleus and assumptions.
7. Define the threshold energy for a nuclear reaction and derive an expression for it for the reaction $X(x, y)Y$ in terms of the Q-value and masses of the participating nuclei. Write the conditions for exothermic and endoergic nuclear reactions.
8. Explain the terms nuclear reaction cross-section and differential cross-section. Derive an expression for nuclear reaction cross-section. Has the nuclear reaction cross-section any relation to the target area?
9. What is a nuclear reaction? Discuss various conservation laws in nuclear reactions with illustrative examples.
10. Define and calculate the threshold energy of an endoergic nuclear reaction that proceeds through formation of a compound nucleus.
11. What is a nuclear reaction? Give two examples of nuclear reactions produced in each case by α-particle, proton and neutron.
12. Discuss possible reactions that may occur when high energy charged particle approaches nucleus.
13. What are nuclear reactions? State conservation laws obeyed in such reactions. What is meant by Q-value of a reaction?
14. Discuss by giving examples the different types of nuclear reactions.
15. Explain compound nucleus theory of nuclear reactions. What do you mean by cross-section of a nuclear reaction?

Chapter 5

Interaction of Radiations with Matter

5.1 INTRODUCTION

Nuclear radiations can be divided into three categories:

1. Charged particles like electrons, protons, deuterons, alpha-particles and heavy ions.
2. Electromagnetic radiations like X- and gamma rays.
3. Neutral particles like neutrons.

In this chapter, we shall first discuss the interaction of heavy charged particles with matter and then interaction of electrons will be discussed. Interaction of electrons with matter is much more complicated phenomenon compared to interaction of protons and other heavy charged particles with matter. Then, we shall discuss the interaction of electromagnetic radiations with matter.

Neutrons being neutral particles can easily penetrate the nucleus and induce nuclear reaction. This interaction is not discussed here.

5.2 ENERGY LOSS BY HEAVY CHARGED PARTICLES

When a heavy charged particle moves in a material, it undergoes collisions with atom as a whole present in the material and also collides with electrons present in the atom. Since there are large number of electrons present in an atom (e.g. in Pb, there are 82 electrons per Pb atom), so most of the time incident particle collides with an electron. The collisions with atom as a whole are very rare and in the present calculations, we neglect these collisions. Further, in the present derivation, we assume that the energy of the incident particle is low, so that no nuclear reaction is possible.

The collision between a heavy charged particle and an electron can be treated classically. In this type of collision, the incident heavy charged particle keeps on moving in a straight line, while the electron after collision is deflected by a large angle.

Classically in each collision with an electron, the incident charged particle loses energy, ΔE, and is given as

$$\Delta E = \frac{4m}{M}E$$

where m is the mass of the electron, E is the kinetic energy and M is the mass of the incident particle.

Let us assume that a 5 MeV α-particle collides with an electron. The energy lost by α-particle in a single collision is

$$\Delta E = \frac{4 \times 0.511\ \text{MeV}}{4 \times 938\ \text{MeV}} \times 5\ \text{MeV}$$
$$\approx 2.7\ \text{keV}$$

Total number of collisions required to completely stop α-particle is

$$\text{Total number of collision} = \frac{5000\ \text{keV}}{2.7\ \text{keV/collision}}$$
$$\approx 1852\ \text{collisions}$$

This simple calculation shows that though the amount of energy lost per collision is small, but the number of collisions is very large. After a collision, incident particle keeps on moving in a straight line. As the incident charged particles move in a straight line, their numbers remain the same, however their energy keeps on decreasing continuously. After travelling certain thickness in matter, energy reduces to zero and incident particles just stop.

The average distance travelled by the incident charged particle in a material before it comes to rest is called its mean range.

Let us derive a relation for the energy lost by a charged particle per unit path length in a given material.

Consider a collision of an incident heavy particle of charge Ze at B with an electron of charge e placed at A as shown in Figure 5.1. Assume that the velocity of heavy charged particle in matter is v and is moving in the $\hat{r}$ direction. Let b be the impact parameter or the perpendicular distance of the heavy incident particle from the electron.

At any instant of time, assume the distance between the charge Ze and the electron be r. The Coulomb force acting on electron due to charge Ze is

$$\vec{F} = \frac{Ze^2}{4\pi\varepsilon_0 r^2}\hat{r}$$

Two components of this force along x and y directions are:

$$F_x = \frac{Ze^2}{4\pi\varepsilon_0 r^2}\cos\varphi$$

$$F_y = \frac{Ze^2}{4\pi\varepsilon_0 r^2}\sin\varphi$$

where φ is the angle between $\hat{r}$ and x-axis. The F_y component of the force imparts momentum to the electron; the value of momentum p is given by

$$p = \int_{-\infty}^{\infty} F_y dt$$

$$= \int_{-\infty}^{\infty} \frac{Ze^2}{4\pi\varepsilon_0 r^2}\sin\varphi\, dt \qquad (5.1)$$

while x component of the force cancels out.

The limits of $-\infty$ to ∞ over time are taken as we have said earlier, we are considering the problem at any instant of time. This instant of time could be at a time much earlier or much later. Now,

$$x = vt \quad \text{and} \quad x = b \cot\varphi \text{ (from Figure 5.1), so } vt = b\cot\varphi$$

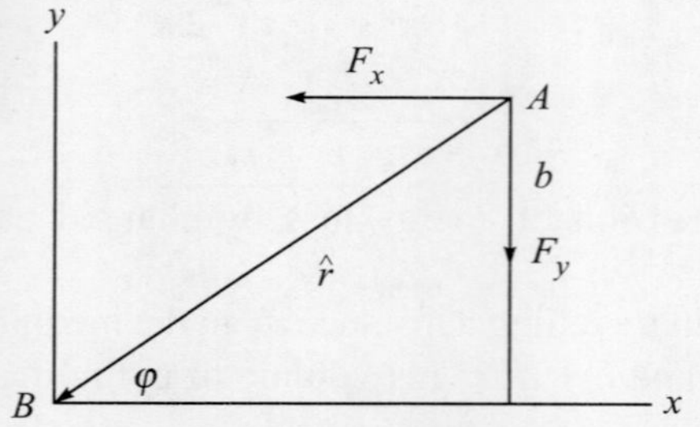

Figure 5.1 Interaction of charged particles with electrons.

Therefore,

$$v\,dt = -b\,\text{cosec}^2\varphi\,d\varphi$$

or

$$dt = -\frac{b}{v}\text{cosec}^2\varphi\,d\varphi$$

Also from Figure 5.1

$$r = \frac{b}{\sin\varphi}$$

Substituting for r and dt in Eq. (5.1), we have

$$p = \int_{\pi}^{0} \frac{Ze^2}{4\pi\varepsilon_0 \dfrac{b^2}{\sin^2\varphi}}\left(-\frac{b}{v}\right)\text{cosec}^2\varphi\,\sin\varphi\,d\varphi$$

The limits of t are from $-\infty$ to ∞ and the corresponding limits of φ are from π to 0. The above equation can also be written as

$$p = \int_{0}^{\pi} \frac{Ze^2\sin\varphi}{4\pi\varepsilon_0 bv}\,d\varphi \tag{5.2}$$

We know that

$$\int_0^{\pi} \sin\varphi\,d\varphi = \left|-\cos\varphi\right|_0^{\pi} = 2$$

Substituting the value of integral in Eq. (5.2), we get the expression for momentum transfer by the charged particle to an electron as

$$p = \frac{2Ze^2}{4\pi\varepsilon_0 bv}$$

The kinetic energy transferred to the electron of mass m by the charged particle is

$$\text{Kinetic energy } E = \frac{p^2}{2m}$$
$$= \frac{4Z^2e^4}{16\pi^2\varepsilon_0^2 b^2 v^2}\frac{1}{2m}$$
$$= \frac{Z^2e^4}{8\pi^2\varepsilon_0^2 b^2 v^2 m}$$

Now, we work out expression of energy loss by charged particle when it moves a unit distance in the medium.

As shown in Figure 5.2, in travelling a distance dx in the medium, the charged particle covers the volume lying between b and $b + db$. This volume of cylindrical shell is

$$\pi[(b+db)^2 - b^2]\,dx = 2\pi b\,db\,dx$$

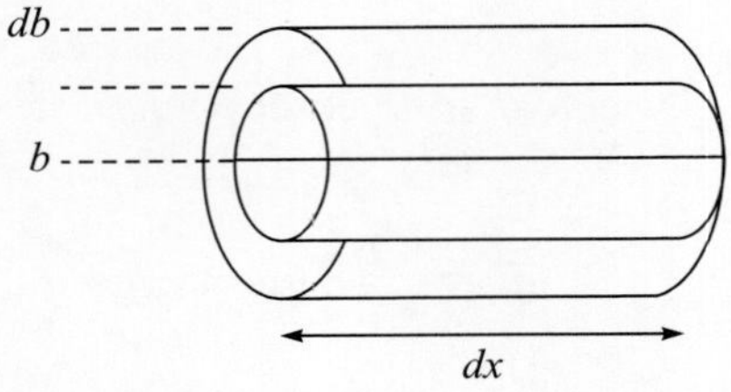

Figure 5.2 Volume of the spherical shell covered by a moving charged particle.

where we have neglected the term db^2 as $db << b$. If n is the electron density, then the number of electrons in this shell is $2\pi n b\,db\,dx$. Thus, the kinetic energy loss $(-dE)$ due to interactions with electrons is

$$dE = -\frac{Z^2e^4}{8\pi^2\varepsilon_0^2 b^2 v^2 m} 2n\pi\, b\,db\,dx$$

or

$$\frac{dE}{dx} = -\frac{Z^2ne^4}{4\pi\varepsilon_0^2 mv^2}\frac{db}{b} \tag{5.3}$$

If we want to calculate total energy loss per unit distance in the impact parameter range b_{min} to b_{max}, then we have to integrate Eq. (5.3) for b. Thus, Eq. (5.3) becomes

$$-\frac{dE}{dx} = \frac{Z^2e^4n}{4\pi\varepsilon_0^2 mv^2}\int_{b_{min}}^{b_{max}}\frac{db}{b}$$

or

$$-\frac{dE}{dx} = \frac{Z^2e^4n}{4\pi\varepsilon_0^2 mv^2}[\ln(b_{max}) - \ln(b_{min})] \tag{5.4}$$

$-\dfrac{dE}{dx}$, energy loss per unit path length is also known as *specific energy loss* or *stopping power* of the medium for a given charged particle.

Evaluation of b_{min} and b_{max}

One question which arises is as to why we do not take $b_{min} = 0$ and $b_{max} = \infty$. These values give absurd values for $-\frac{dE}{dx}$ in Eq. (5.4). Reasons for this absurdity become clear as we calculate b_{min} and b_{max}.

Calculation of b_{min}

The minimum value of b that is meaningful in this problem is wavelength of electron as viewed by the passing charged particle. This wavelength is given by

$$\lambda = b_{min} = \frac{h}{mv}$$

Calculation of b_{max}

The maximum value of b is the distance in which maximum energy can be transferred and corresponding time of interaction is of the order of the time period of the electron in its orbit. The time of interaction = b/v and time period of the electron can be calculated by solving time-dependent Schrödinger equation and its value is h/I. Here, I is the energy of the electron orbit and is more commonly known as *ionization potential*. Equating the two times, we have

$$\frac{b_{max}}{v} = \frac{h}{I}$$

or

$$b_{max} = \frac{hv}{I}$$

Substituting these values, the logarithm term in Eq. (5.4) becomes

$$\ln\left(\frac{b_{max}}{b_{min}}\right) = \ln\left(\frac{mv^2}{\langle I\rangle}\right)$$

where $\langle I\rangle$ is the average potential energy of all the electrons in the atom. This is the result obtained by classical argument. A complete calculation based on quantum mechanical arguments gives

$$\ln\left(\frac{b_{max}}{b_{min}}\right) = \ln\left(\frac{2mv^2}{\langle I\rangle}\right)$$

Substituting this in Eq. (5.4), we get the classical relation for the energy lost by a heavy charged particle as

$$-\frac{dE}{dx} = \frac{Z^2e^4n}{4\pi\varepsilon_0^2mv^2}\left[\ln\left(\frac{2mv^2}{\langle I\rangle}\right)\right] \tag{5.5}$$

This relation was first derived by Bohr and is known as *Bohr's relation for stopping power.* Complete relativistic calculations were performed by Bohr and Bethe and they modified Bohr's stopping power relation as

$$-\frac{dE}{dx} = \frac{Z^2 e^4 n}{4\pi\varepsilon_0^2 mv^2}\left[\ln\left(\frac{2mv^2}{\langle I\rangle}\right) - \ln\left(1 - \frac{v^2}{c^2}\right) - \frac{v^2}{c^2}\right] \tag{5.6}$$

This relation is known as *Bohr–Bethe expression for stopping power*. This is the rate at which a charged particle loses energy when it passes through matter. Average ionization potential $\langle I\rangle$ is about 13 eV times Z, the number of electrons of the atom of the medium in which charged particle is moving. For very light atoms, this value is slightly higher. An empirical relation for average ionization potential $\langle I\rangle$ is found as

$$\langle I\rangle = 9.1\, Z(1 + 1.9\, Z^{-2/3})$$

The experimentally observed rate of energy loss of charged particles in the middle energy range is in good agreement with the values derived using Eq. (5.6). However, this agreement is not very good in all energy ranges due to the following reasons:

- If the velocity of the incident particle is large, comparable with velocity of light c, the normally spherical fields become distorted, shrinking in the direction of motion of the charged particle and expanding laterally. This effect leads to an increase in the rate of energy loss for very high-energy particles.
- If a particle moves in a medium with velocity larger than the velocity of light in that medium, it emits visible radiations, known as *Cerenkov radiations*. This effect reduces the energy loss rate.
- As incident particle slows down to a speed of the same order as that of the outer orbital atomic electrons, it picks up less tightly bound electrons of the medium. This reduces the effective charge of the incident particle and, therefore, energy loss rate.
- When velocity of the incident ion is so small that $2mv^2$ becomes less than I, the first term (in square bracket) in Eq. (5.6) becomes negative, therefore, $\left(-\frac{dE}{dx}\right)$ becomes negative, i.e. it is gaining energy, which is not possible.

5.3 INTERACTION OF ELECTRONS WITH MATTER

As already stated energy loss of electrons in matter is a much more complicated phenomenon than the energy loss by heavy charged particles. Some of the reasons for this difference are as under:

- Because of smaller mass, electrons when collides with the atomic electrons of the absorber, suffer deflections through much larger angles. Therefore, the path of an electron in matter is very irregular and not at all as straight as the path of heavy charged particles.

- Electrons emitted in β-decay have a continuous distribution of energy, whereas α-particles and other heavy ions are generally monoenergetic.
- The electron detected after crossing the material could be the incident electron or the electron from the material as these are identical. In performing theoretical calculations, this phenomenon must be taken into account.
- Generally kinetic energy of β-particles is comparable or greater than the rest mass energy of electrons (0.511 MeV). This makes it necessary to employ relativistic treatment of the collision process.
- Because the kinetic energy of electrons is in relativistic region and when such electrons move in a material, they are accelerated and deaccelerated in the electric fields of nucleus and electrons. During acceleration and deacceleration these relativistic electrons emit radiations called *Bremsstrahlung radiations*. This mode of energy loss is not possible for low energy heavy charged particles. However, if heavy charged particles have kinetic energy comparable to their rest mass energy (for proton this value is 938 MeV), this mode of energy loss becomes important.

Out of several mechanisms for energy loss of electrons in a material, two processes are most important:

1. Energy loss by ionization of the material by incident electrons. This energy loss is denoted by $\left.\frac{dE}{dx}\right)_c$.

2. Energy loss due to Bremsstrahlung. This energy loss is denoted by $\left.\frac{dE}{dx}\right)_r$. This is also known as radiative term.

The calculations for these two energy losses involve advanced quantum mechanics, correction due to collisions between identical particles, etc. The final expressions for these energy losses are as under:

$$\left.\frac{dE}{dx}\right)_c = \frac{2\pi N_0 Z\rho}{mc^2\beta^2 A}\left[\ln\frac{T(T+mc^2)}{2I^2mc^2}\beta^2 + F\right]$$

where Z, A and ρ are the atomic number, atomic weight and density of stopping medium. N_0 is Avogadro's number, m is the electron mass, $\beta = \frac{v}{c}$, where v is the velocity of the electron and T is the kinetic energy of the electron and F is

$$F = \left[(1-\beta^2) - 2(2\sqrt{1-\beta^2} - 1 + \beta^2)\ln 2 - \frac{1}{8}(1-\sqrt{1-\beta^2})^2\right]$$

and

$$\left.\frac{dE}{dx}\right)_r = \frac{Z^2 N_0 (T+mc^2)\rho}{137 m^2 c^4 A}\left[4\ln\frac{2(T+mc^2)}{mc^2} - \frac{4}{3}\right]$$

The total energy loss is the sum of these two energy losses, or

$$\left.\frac{dE}{dx}\right)_T = \left.\frac{dE}{dx}\right)_c + \left.\frac{dE}{dx}\right)_r$$

To estimate the relative contributions of the two energy loss processes, we have

$$\frac{dE/dx)_r}{dE/dx)_c} = \frac{T + mc^2}{mc^2}\frac{Z}{1600} \tag{5.7}$$

The radiative term is significant at high energy and in heavy Z materials.

The differences between the stopping of heavy charged particles and electrons in an absorber are shown in Table 5.1.

TABLE 5.1 Difference between stopping of heavy charged particles and that of electrons

S.No.	*Stopping of heavy charged particles*	*Stopping of electrons*
1.	Their kinetic energy is small compared to the rest mass energy, so no relativistic effects are needed.	Their kinetic energy is larger or comparable to the rest mass energy, so relativistic effects are to be applied.
2.	In a single collision with electrons, they lose very small fraction of energy.	In a single collision with electrons, they lose large fraction of energy.
3.	Because of small energy loss in a single collision, they move almost in a straight line.	Because of violent collision with absorber electrons, they get deflected by large angles and hence their path is zigzag.
4.	Because their kinetic energy is small compared to the rest mass energy, their energy loss mechanism is mostly by ionization and excitation.	Because their kinetic energy is large compared to their rest mass energy, the predominant energy loss mechanism is radiation (Bremsstrahlung).
5.	α-particles emitted in α-decay have discrete energy, so energy loss mechanism is simple.	β-particles emitted in β-decay have continuous energy distribution, so the energy loss mechanism is quite complex.
6.	No exchange phenomenon is required in heavy ion collisions with atomic electrons.	As electrons incident on absorber and electrons present in absorber are identical, so we have to take into account the exchange phenomenon in electron-electron collisions.

5.4 RANGE OF CHARGED PARTICLES

As stated earlier, heavy charged particles lose small fraction of energy in each collision and suffer a large number of collisions before they are finally stopped by the medium. During this motion they move in a straight line in the forward direction.

The average distance travelled by a charged particle before it comes to rest is called the *range of the charged particle*.

To derive an expression for the range of charged particle, let E_i be the kinetic energy at the time of entry in the medium and assume that it loses dE energy in travelling a distance dx. So, the range R is

$$R = \int_{E_i}^{0} dx = -\int_{0}^{E_i} dx$$

$$R = \int_{0}^{E_i} \left(-\frac{dE}{dx}\right)^{-1} (-dE) = \int_{0}^{E_i} \left(\frac{dE}{dx}\right)^{-1} dE$$

Substituting for $\left(-\frac{dE}{dx}\right)$ for velocity $v << c$, we have from Eq. (5.5)

$$R = \int_{0}^{E_i} \left[\left(\frac{Z^2 e^4 n}{4\pi\varepsilon_0^2 mv^2}\right) \ln \frac{2mv^2}{h\nu}\right]^{-1} dE$$

In the above expression except for Z and v all other parameters are constant for a medium. So,

$$R \propto \int_{0}^{E_i} \left(\frac{Z^2}{v^2}\right)^{-1} dE$$

where we have taken $\ln \frac{2mv^2}{h\nu}$ as a constant as variation of $\ln v^2$ is very small compared to v^2

Now, $E = \frac{1}{2}Mv^2$. Therefore,

$$R \propto \int_{0}^{E_i} \left(\frac{Z^2 M}{2E}\right)^{-1} dE$$

or

$$R \propto \frac{E^2}{MZ^2}$$

This formula is fairly accurate. For protons of energy up to 200 MeV, the empirical relation for range in air is given as

$$R = \left(\frac{E}{9.3}\right)^{1.8} \times 10^2 \text{ cm} \tag{5.8}$$

where E is the energy in MeV.

5.5 INTERACTION OF GAMMA RAYS WITH MATTER

If gamma ray passes through an absorber of thickness Δx, its intensity I changes. The change in intensity ΔI, is proportional to the intensity I of the γ-rays falling on the absorber and thickness of the absorber. The change in intensity is given by

$$\Delta I \propto - I\Delta x$$

or

$$\Delta I = - \mu I\Delta x \tag{5.9}$$

Negative sign indicates that as x increases, I decreases. Here μ is constant of proportionality and is known as *absorption coefficient* or *attenuation coefficient*. For a given absorber, μ depends upon the energy of gamma rays. Equation (5.9) can be written as

$$\frac{\Delta I}{I} = -\mu\, \Delta x$$

which on integration gives

$$\ln(I) = -\mu\, x + c \tag{5.10}$$

when $x = 0$, $I = I_0$. Here I_0 is the initial intensity of γ-rays falling on the absorber.

From Eq. (5.10)

$$c = \ln(I_0)$$

Substituting for c in Eq. (5.10), we get

$$\ln\left(\frac{I}{I_0}\right) = -\mu\, x$$

This equation can also be written as

$$I = I_0 e^{-\mu\, x} \tag{5.11}$$

Absorption coefficient μ is also called *linear absorption coefficient.* Sometimes we also use the term mass absorption coefficient μ_m, which is defined as

$$\mu_m = \frac{\mu}{\rho} \text{ where } \rho \text{ is the density of the absorber.}$$

Since in Eq. (5.11) μx must be a dimensionless quantity, therefore, if x is expressed in metres, then μ will have the dimensions of metre^{-1}.

From Eq. (5.11), it is clear that intensity of γ-rays decreases exponentially and never becomes zero. Interaction of γ-rays is markedly different from that of charged particles. γ-rays have greater penetrating power and different absorption laws.

As stated earlier, charged particles suffer a large number of collisions before finally stopped by the medium. In γ-rays and thin absorber, each photon is removed individually in a single event. The number of photons removed in thickness Δx is proportional to Δx and also proportional to the number of photons reaching Δx. This dependence leads to Eq. (5.11).

In Eq. (5.11), if $x = \frac{1}{\mu}$, then we get

$$I = I_0 e^{-1}$$

or

$$\boxed{I = \frac{I_0}{e}}$$

Therefore, linear attenuation coefficient is also defined as reciprocal of that thickness of the material which reduces the intensity of the incident γ-rays to $\frac{1}{e}$ of its original value.

5.5.1 Radiation Length

Reciprocal of linear attenuation coefficient is known as *radiation length*. It is also known as *absorption length*. The radiation length is also defined as the thickness of the material which reduces the intensity of the incident γ-rays to $\frac{1}{e}$ of its original value. If the units of attenuation coefficient μ are metres^{-1}, then the units of radiation length are metres.

5.5.2 Half-thickness

Half-thickness of an absorber is that thickness which reduces the intensity of γ-ray beam by one-half of its original value.

From Eq. (5.11)

$$I = \frac{I_0}{2}$$

$$\therefore \qquad \frac{I_0}{2} = I_0 e^{-\mu x}$$

or

$$e^{\mu x} = 2$$

Taking $\log_e$ on both sides, we get

$$\mu x_{1/2} = \ln 2$$

where $x_{1/2}$ is half-thickness, i.e. the thickness which reduces the intensity of γ-rays by one-half of its original value. This can also be written as

$$\boxed{\mu = \frac{0.6931}{x_{1/2}}} \tag{5.12}$$

5.5.3 Experimental Determination of Attenuation Coefficient

Experimentally, linear attenuation coefficient is determined using Eq. (5.11), which can be written as

$$\frac{I}{I_0} = e^{-\mu x}$$

or

$$-\mu x = \ln\left(\frac{I}{I_0}\right)$$

This relation is a straight line between thickness x and $\ln\left(\frac{I}{I_0}\right)$. Slope of this straight line gives μ.

The apparatus used to determine the value of attenuation coefficient is shown in Figure 5.3. A radioactive source say ^{137}Cs, (S) which emits γ-rays of energy 662 keV, is enclosed in a thick block of lead with one very fine opening on one side as shown in Figure 5.3. From this arrangement we get a very fine beam of γ-rays, which passes through a lead collimator A. This fine beam of γ-rays passes through the absorber B for which attenuation coefficient is to be measured. Any scattered photons from the absorber B are again stopped by another lead collimator C. Only beam of γ-rays attenuated and transmitted through the absorber reaches the heavily shielded γ-ray detector D.

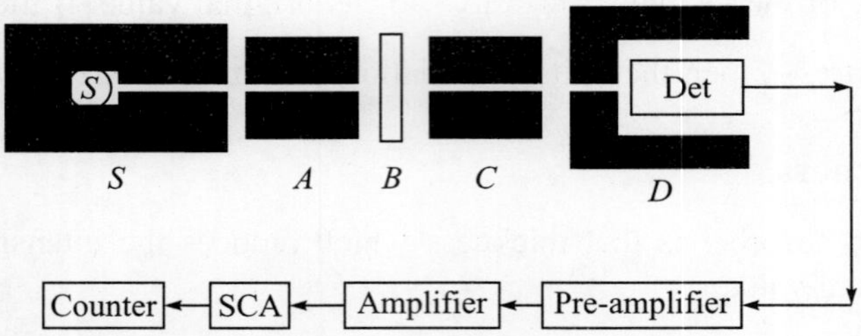

Figure 5.3 Block diagram of the apparatus used for determination of attenuation coefficients.

The experiment is first performed without putting any absorber at position B. The counts recorded in the γ-ray spectrometer give I_0. Then by putting absorbers of different thicknesses at position B, corresponding counts I are recorded. A plot between $\ln\left(\frac{I}{I_0}\right)$ versus thickness of the absorber x is shown in Figure 5.4 for Al and Pb absorbers. The slope of this line gives the value of μ, which in this case is 0.18 cm^{-1} for Al and 0.94 cm^{-1} for Pb.

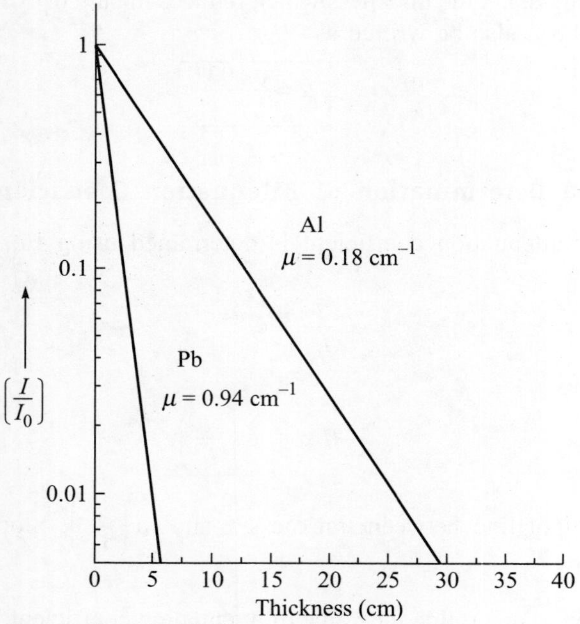

Figure 5.4 Plot of $\ln\left(\frac{I}{I_0}\right)$ versus thickness of the absorber for Al and Pb. This plot is on semi-log scale. Therefore, 'ln' on the y-axis has been omitted.

The ^{137}Cs source can also be replaced by other radioactive sources like ^{65}Zn, ^{24}Na, etc. to measure attenuation coefficient at different energies. Similarly, different materials can be placed at position B to determine attenuation coefficient for different materials.

Experiments have been performed using different absorbers and γ-rays of different energies. It has been found that

- Attenuation coefficient μ depends upon the nature of absorber. μ is larger for heavier elements like Pb, Bi, etc. and is smaller for lighter elements like C, Al, etc.
- Attenuation coefficient μ depends upon the energy of γ-rays. It is larger for low energy γ-rays and is smaller for high energy γ-rays.

Three processes are mainly responsible for the absorption of γ-rays. These are:

(i) Photoelectric effect
(ii) Compton scattering
(iii) Pair production.

The probability of each process is expressed as absorption coefficient. The total absorption coefficient is sum of absorption coefficients of the three processes. The various absorption coefficients depend upon the energy of the γ-rays as well as nature of absorbing material. Thus, the absorption of γ-rays cannot be expressed in a single formula. Each partial absorption coefficient is expressed as a function of energy for a given material and tables have been prepared for different materials. The total absorption coefficient μ of a given material is expressed by the formula

$$\mu(E) = \mu_{pe}(E) + \mu_{\sigma}(E) + \mu_{pp}(E) \tag{5.13}$$

where indices *pe*, σ and *pp* stand for photoelectric effect, Compton effect and pair production respectively.

Now we discuss these effects in detail.

5.6 PHOTOELECTRIC EFFECT

In 1887, Hertz observed that a metallic surface emits electrons when a photon of low wavelength is incident on it. He further observed that when ultraviolet light falls on zinc plate (which is electrically neutral), the plate becomes positively charged, i.e. electrons are emitted by the plate. This phenomenon of emission of electrons by a metallic plate when photons of suitable wavelength falls on it, is known as *photoelectric effect*. Prof. Albert Einstein got Noble Prize in 1916 for the explanation of photoelectric effect.

Further experiments give us the following results:

1. A minimum frequency of photons is needed for the emission of electrons from a metallic surface. This energy associated with the minimum frequency is known as the *work function of the material*.
2. The number of photoelectrons emitted by the material depends upon the intensity of the incident photons.

3. If the frequency of the incident photon is higher than the work function, then the difference between the incident energy and the work function becomes the kinetic energy of the photoelectrons, i.e.

$$\frac{1}{2}mv^2 = h\nu - h\nu_0 \tag{5.14}$$

where m is the mass of the electrons, v is the velocity of the electron, $h\nu$ is the energy of the incident photon and this has to be greater than $h\nu_0$, the minimum energy required to eject the electron from the surface. $h\nu_0 = w$, the work function ν_0 is called threshold frequency and the corresponding wavelength λ_0 is called threshold wavelength.

The photoelectric effect is observed from the inner shell electrons with high probability and is predominant with high Z materials.

Why Photoelectric Effect is Not Possible with Free Electrons?

Let us assume a completely non-relativistic case and assume photoelectric effect is taking place with free electrons. Electron is initially at rest. So, its kinetic energy and momentum are zero. Incident photon has kinetic energy $h\nu$ and momentum $h\nu/c$. After photoelectric effect, incident photon is completely absorbed and electron moves with a kinetic energy $1/2mv^2$ and momentum mv.

Applying law of conservation of momentum, we get

$$\frac{h\nu}{c} + 0 = mv$$

or

$$h\nu = mvc \tag{5.15}$$

Similarly, applying law of conservation of energy, we get

$$h\nu + 0 = \frac{1}{2}mv^2$$

or

$$h\nu = \frac{1}{2}mv^2 \tag{5.16}$$

Solving Eqs. (5.15) and (5.16) for v, we get

$$v = 2c$$

This means that electron will move with a velocity which is twice the velocity of light. This is in contradiction with the theory of relativity, which states that no material particle can travel with velocity greater than velocity of light in vacuum. Hence, it is not possible to have photoelectric effect with free electron. This holds true also for the relativistic case. We must have a bound electron, so that momentum and energy conservation laws are not violated.

Experimental verification of the photoelectric effect discussed above is not simple. It requires electromagnetic radiations of energy higher than the binding energy of electrons in K, L, ... shells. This energy is of the order tens of eV to tens of keV. This requires a source of electromagnetic radiations emitting either UV radiations or X-rays. However, we consider only those electrons which are very weakly bound to the atom or are almost free. These electrons can be removed by supplying energy equal to work function of the material. The work function of most of the materials is of the order of few eV. This energy can be provided by visible light. Below we present different experiments based on the above discussion.

5.6.1 Experimental Verification of Photoelectric Effect

The apparatus used to verify photoelectric effect is shown in Figure. 5.5. It consists of an evacuated quartz tube. The tube contains two metal electrodes. The electrodes are connected to a variable power supply. One of the electrodes is exposed to light acts as an anode. When anode is exposed to light, it ejects photoelectrons. Some of the photoelectrons that emerge from anode have enough energy to reach the cathode despite of its negative polarity and constitute the current in the circuit. The electrons moving with lower energy are repelled by the cathode and will not reach the cathode. When potential between the two electrodes is increased, no more photoelectrons arrive at the cathode and current in the circuit drops to zero. This extinction voltage corresponds to the maximum kinetic energy of photoelectrons.

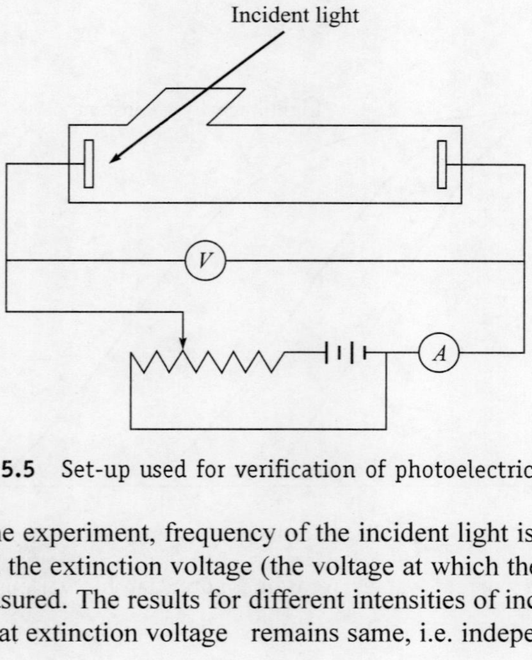

Figure 5.5 Set-up used for verification of photoelectric effect.

In the first part of the experiment, frequency of the incident light is kept constant. Intensity of the light is varied and the extinction voltage (the voltage at which the current in the ammeter becomes zero) V_0 is measured. The results for different intensities of incident light are shown in Figure 5.6. It is found that extinction voltage remains same, i.e. independent of the intensity of light.

In the second part of the experiment, intensity of the incident light is kept constant and the frequency of the incident light is varied. The results of the experiment are shown in Figure 5.7. It is found that as the frequency of the light increases, extinction voltage also increases. In other words, the kinetic energy of photoelectrons is directly proportional to the frequency of incident light.

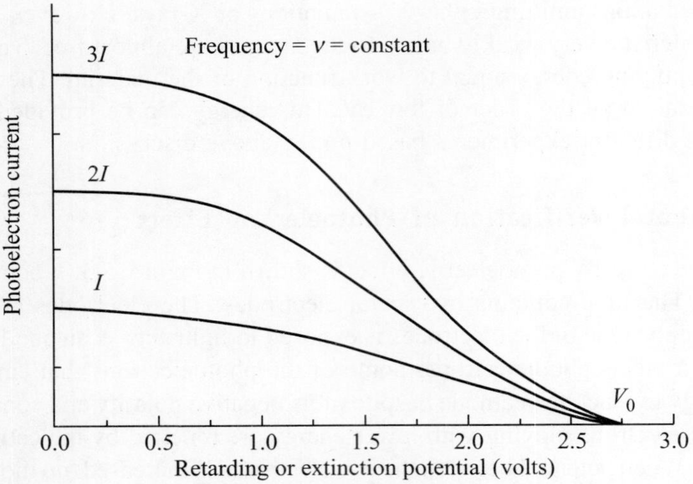

Figure 5.6 Photoelectric current versus retarding potential for various intensities of incident light.

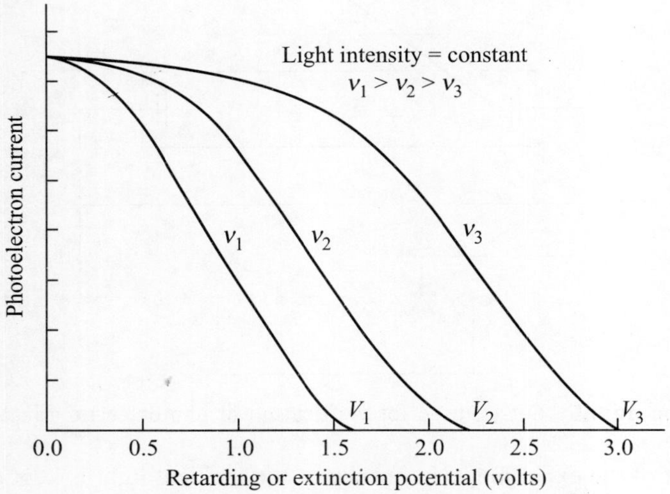

Figure 5.7 Photoelectric current versus retarding potential for various frequencies of incident light.

These two experiments verify basic aspects of photoelectric effect. It has been shown that

- Probability of photoelectric effect $\propto Z^{4.5}$.
- Probability of photoelectric effect $\propto \dfrac{1}{h\nu_0}$, if $h\nu_0 > m_0c^2$ and $\propto \dfrac{1}{(h\nu_0)^3}$, if $h\nu_0 < m_0c^2$.

5.7 COMPTON EFFECT

Compton in 1923 was able to demonstrate that if a beam of monochromatic X-rays or light of high frequency was incident on an element of low Z, such as carbon, the scattered radiations consisted of two components, one component was of same frequency as the incident radiation and the second component was of lower frequency.

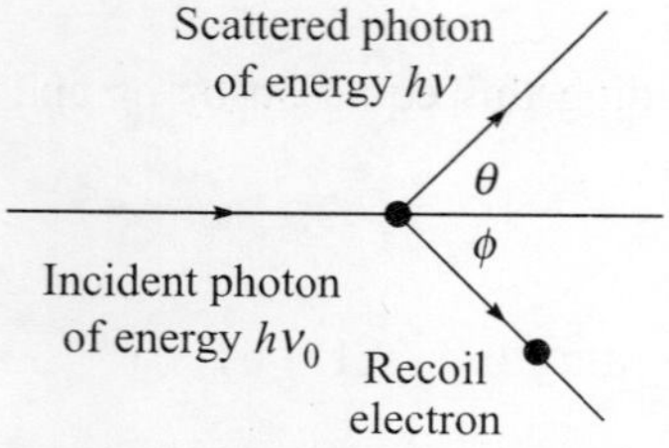

Figure 5.8 Schematic representation of Compton effect.

Compton assumed that scattering process between the incident photon and electron (assumed to be free and at rest) is an elastic collision. So, the kinetic energy and momenta are conserved. After scattering, the recoil electron takes energy from the incident photon. Thus, the energy of the incident photon is reduced or scattered photon has a lower frequency or longer wavelength. Schematically the scattering process is shown in Figure 5.8.

Compton effect studies are generally carried out with gamma rays having relatively high energies, so the recoil energy of the Compton scattered electrons is also high. Since rest mass energy of the electrons is only 0.511 MeV, so relativistic effects have to be considered. Thus, the mass of the electron m moving with velocity v is given by

$$m = \frac{m_0}{\sqrt{1 - \frac{v^2}{c^2}}}$$

where m_0 is the mass of the electron at rest and c is the velocity of light.

Applying the law of conservation of energy to the Compton scattering process, we get the following equation

$$m_0c^2 + h\nu_0 = h\nu + mc^2 \tag{5.17}$$

Since momentum is a vector quantity, applying the law of conservation of momentum along the x-axis, we have

$$\frac{h\nu_0}{c} = \frac{h\nu}{c} \times \cos\theta + mv \times \cos\phi \tag{5.18}$$

Similarly, applying the law of conservation along the y-axis, we get

$$0 = \frac{h\nu}{c} \times \sin\theta - mv \times \sin\phi \tag{5.19}$$

where $\frac{h\nu_0}{c}$ is the initial momentum of the gamma ray and $\frac{h\nu}{c}$ is the momentum of the scattered gamma ray.

Multiplying Eqs. (5.18) and (5.19) by c, we get

$$h\nu_0 - h\nu \times \cos\theta = mvc \times \cos\phi \tag{5.20}$$

$$h\nu \times \sin\theta = mvc \times \sin\phi \tag{5.21}$$

Squaring Eqs. (5.20) and (5.21) and adding, we have

$$m^2v^2c^2 = h^2 (\nu_0^2 + \nu^2 - 2\nu\nu_0 \times \cos\theta) \tag{5.22}$$

Dividing this equation by c^2 and using the relation $\lambda = \frac{c}{\nu}$, one obtains

$$m^2v^2 = \frac{h^2}{\lambda_0^2} + \frac{h^2}{\lambda^2} - \frac{2h^2 \times \cos\theta}{\lambda\lambda_0} \tag{5.23}$$

Rewriting Eq. (5.17) as

$$mc^2 = m_0c^2 - h(\nu - \nu_0)$$

Squaring this equation, we get

$$m^2c^4 = m_0^2c^4 + h^2(\nu - \nu_0)^2 - 2m_0c^2h(\nu - \nu_0) \tag{5.24}$$

Subtracting Eq. (5.22) from Eq. (5.24), we get

$$m^2c^2 (c^2 - v^2) = -2h^2(1 - \cos\theta)\nu\nu_0 + m_0^2c^4 - 2h(\nu - \nu_0)m_0c^2$$

Using the relation $m = \dfrac{m_0}{\sqrt{1 - \dfrac{v^2}{c^2}}}$, Eq. (5.24) becomes

$$\frac{m_0^2c^2(c^2 - v^2)}{1 - \dfrac{v^2}{c^2}} = -2h^2\nu\nu_0 (1 - \cos\theta) + m_0{}^2c^4 - 2h(\nu - \nu_0)\, m_0c^2$$

Simplifying this equation, we have

$$2h(\nu_0 - \nu)\, m_0c^2 = 2h^2\, \nu\nu_0 (1 - \cos\theta)$$

Cancelling $2h$ and dividing by $\nu\nu_0$, we get

$$\frac{\nu_0 - \nu}{\nu\nu_0} = \frac{h}{m_0c^2}(1 - \cos\theta)$$

or

$$\frac{1}{\nu} - \frac{1}{\nu_0} = \frac{h}{m_0c^2}(1 - \cos\theta) \tag{5.25}$$

Using the relation $\lambda = \frac{c}{\nu}$, we get

$$\frac{\lambda}{c} - \frac{\lambda_0}{c} = \frac{h}{m_0c^2}(1 - \cos\theta)$$

or

$$\boxed{\lambda - \lambda_0 = \Delta\lambda = \frac{h}{m_0c}(1 - \cos\theta)} \tag{5.26}$$

This change in wavelength is known as *Compton shift*. As is clear from the above equation, this shift depends upon θ only, as h, m_0 and c are constants. It is independent of incident energy.

Different cases of Compton shift can be considered for different values of θ:

1. If $\theta = 0°$, $\cos\theta = 1$, and $\Delta\lambda$ becomes 0, which means, that there is no scattering of the incident photon as Compton shift is 0.
2. If $\theta = 90°$, $\cos\theta = 0$, then

$$\Delta\lambda = \frac{h}{m_0 c}.$$

Substituting the constants $h = 6.625 \times 10^{-34}$ J s, $m_0 = 9.1 \times 10^{-31}$ kg, $c = 3 \times 10^8$ m/s, we get

$$\Delta\lambda = 2.42 \times 10^{-12} \text{ m} = 0.0242 \text{ Å}$$

3. If $\theta = 180°$, then $\cos\theta = -1$ and $\Delta\lambda = \dfrac{2h}{m_0 c} = 0.0484$ Å

$\dfrac{h}{m_0 c}$ is known as *Compton wavelength* and its value is 2.42×10^{-12} m.

Energy of Scattered Photon

Rewriting Eq. (5.25)

$$\frac{1}{h\nu} = \frac{1}{h\nu_0} + \frac{1}{m_0 c^2}(1 - \cos\theta)$$

or

$$\frac{1}{h\nu} = \frac{m_0 c^2 + (1 - \cos\theta)h\nu_0}{h\nu_0 m_0 c^2}$$

Taking inverse

$$h\nu = \frac{h\nu_0 m_0 c^2}{m_0 c^2 + (1 - \cos\theta)h\nu_0}$$

Dividing numerator and denominator on the right-hand side by $m_0 c^2$, we have

$$h\nu = \frac{h\nu_0}{1 + \dfrac{h\nu_0}{m_0 c^2}(1 - \cos\theta)}$$

Substituting $\dfrac{h\nu_0}{m_0 c^2} = \alpha$, we get

$$h\nu = \frac{h\nu_0}{1 + \alpha(1 - \cos\theta)} \tag{5.27}$$

This relation for $h\nu$ gives us the energy of the scattered photon.

Special cases:

- For $\theta = 0°$, $h\nu = h\nu_0$, i.e. the energy of the scattered photon is same as the energy of the incident photon. In this case Compton effect is not taking place.
- For $\theta = \dfrac{\pi}{2}$, $h\nu = \dfrac{h\nu_0}{1 + \alpha}$. Since $\alpha > 0$, so energy of the scattered photon is always be less than that of incident photon i.e. Compton effect is taking place

- For $\theta = \pi$, i.e. incident photon is back-scattered. In this case

$$h\nu = \frac{h\nu_0}{1+2\alpha}$$

As energy of the incident photon is increased, the energy of the back-scattered photon converges to 0.255 MeV.

Energy of Recoil Electrons

From Eq. (5.17), we have
Kinetic energy of the recoil electron $T = h\nu_0 - h\nu$.
Substituting for $h\nu$ from Eq. (5.27), we get

$$T = h\nu_0 - \frac{h\nu_0}{1+\alpha(1-\cos\theta)}$$

or

$$T = h\nu_0\left[1 - \frac{1}{1+\alpha(1-\cos\theta)}\right] \tag{5.28}$$

Special cases:

- For $\theta = 0°$, $\quad T = 0$
- For $\theta = \frac{\pi}{2}$, $\quad T = h\nu_0 \frac{\alpha}{1+\alpha}$
- For $\theta = \pi$, $\quad T = h\nu_0 \frac{2\alpha}{1+2\alpha}$

5.7.1 Experimental Verification of Compton Effect

The set-up used for verification of Compton effect is shown in Figure 5.9. In this set-up, source of radiations is a radioactive source (S) of ^{137}Cs, which emits gamma rays of 662 keV. These gamma rays after collimation (A) are made to fall on an aluminium cylinder (C). Aluminium cylinder acts as a scatterer. The Compton scattered gamma rays go in different directions. The number of photons and their energy at different angles are measured by a scintillation spectrometer (this spectrometer is discussed in Chapter 7). For different angles we observe two gamma rays, one corresponding to the incident energy of 662 keV (elastically scattered) and the other is corresponding to the Compton scattered gamma ray. This is shown in Figure 5.10. In the top part at 0°, only one γ-ray at 662 keV is present,

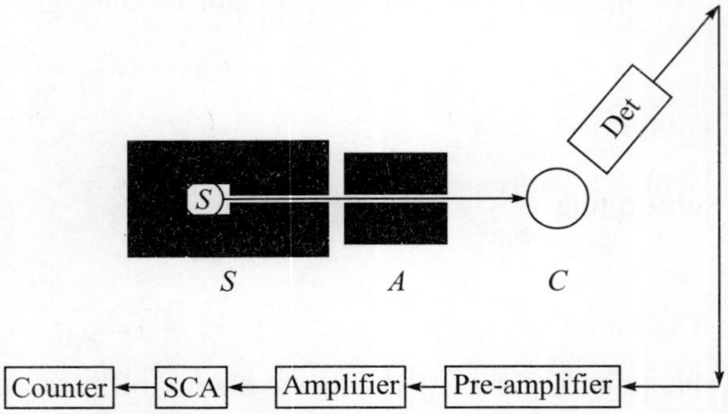

Figure 5.9 Experimental set-up to verify Compton effect.

corresponding to unshifted peak. In the other parts of the figure, two peaks are present, one corresponding to unshifted component and the other due to Compton scattering. The energy of this Compton scattered gamma ray can be calculated from Eq. (5.27).

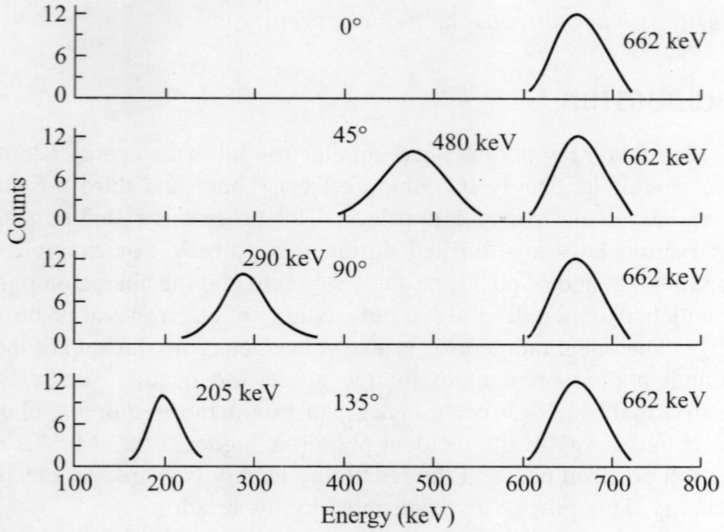

Figure 5.10 Compton scattered 662 keV γ-rays of ^{137}Cs at different angles.

The probability of Compton effect depends upon the atomic number of the scatterer and the energy of the incident gamma rays. This variation is as under:

- Probability of Compton effect $\propto Z$
- Probability of Compton effect $\propto \dfrac{1}{h\nu_0}$

Compton Effect with Visible Light

Compton shift is given by

$$\lambda - \lambda_0 = \Delta\lambda = \frac{h}{m_0 c}(1 - \cos\theta)$$

It will be maximum at $\theta = 180°$ and is

$$\lambda - \lambda_0 = \Delta\lambda = \frac{2h}{m_0 c} = \frac{2 \times 6.626 \times 10^{-34}}{9.1 \times 10^{-31} \times 3 \times 10^{8}}$$

$$\lambda - \lambda_0 = \Delta\lambda = 0.0485 \times 10^{-10}\ \text{m}$$

This small change in wavelength of visible light is not possible to detect with any optical instrument. Therefore, Compton effect cannot be observed with visible light. Similarly, ultraviolet radiation which extends from ~ 4000×10^{-10} m to ~ 10×10^{-10} m can not be used for observing Compton effect.

Further, Compton effect takes place from free electrons. The binding energy of electrons in an atom generally varies from ~ 13 eV *to* ~ 80 keV. *The energy of visible light is of the order of* 2 eV *and that of ultraviolet light it is of the order of* 10 eV. *For such low energies, electrons in atom are bound. For X-rays or γ-rays the binding energy of electrons in an atom is negligibly small and the electrons can be treated as free electrons.*

5.8 PAIR PRODUCTION

We have already seen that γ-ray photons give an electron all of its energy (Photoelectric effect) or transfer only a part of its energy (Compton effect). There is a third possibility in which a photon materializes into an electron and a positron. This process is called *pair production*.

All the conservation laws are fulfilled during this process. For example, the sum of the charge of electron ($q = -e$) and of positron ($q = +e$) is zero and the charge on photon is also zero. The total energy which also includes the rest mass energy of electron and positron is equal to the γ-ray photon energy. The linear momentum is also conserved (with the help of the nucleus, which carries away enough photon momentum for the process to occur). The rest mass energy of electron and positron is 0.511 MeV each, hence pair production requires a photon energy of at least 1.022 MeV. If the energy of the incident photon is higher, 1.022 MeV is consumed in the formation of electron positron pair and the remaining energy is shared by electron and positron as their kinetic energy. This process is represented by the relation

$$\gamma \rightarrow e^- + e^+$$

Dirac Theory of Pair Production

The creation of electron–positron pair was explained by Dirac in 1931 by developing relativistic quantum mechanical equation for electrons.

One of the puzzling features of the Dirac equation was that for a free electron, it predicted not only a continuum of energy states greater than its rest mass energy but also a continuum of negative energy states. This theory is discussed below in a very simplified way. Relativistically the total energy E of a particle of rest mass m_0 and momentum p is given by

$$E^2 = m_0^2c^4 + p^2c^2 \tag{5.29}$$

or

$$E = \pm\sqrt{m_0^2c^4 + p^2c^2}$$

Thus

$$E \geq m_0c^2$$

or

$$E \leq -m_0c^2$$

Dirac interpreted that an electron may have total positive energy from $+m_0c^2$ to ∞ or a negative energy from $-m_0c^2$ to $-\infty$ as shown in Figure 5.11. We see only electrons in positive energy state. Dirac proposed that the negative energy states (called Dirac sea) are completely filled with electrons and Pauli exclusion principle forbids any transition from positive energy state to negative energy state. Electron from negative energy state cannot make a transition to positive energy state by itself as it requires 1.022 MeV energy.

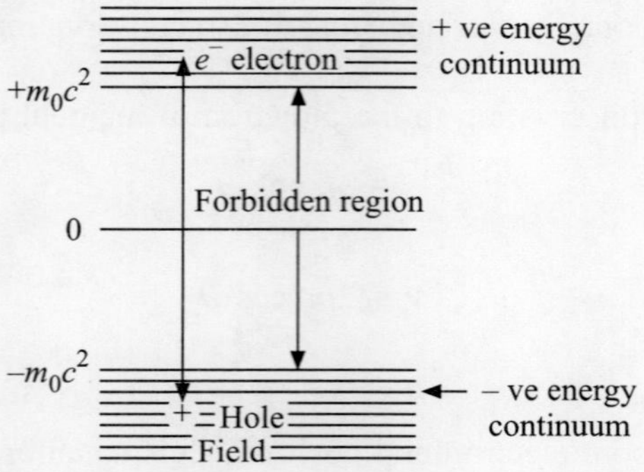

Figure 5.11 Dirac diagram for electron hole (positron) production.

However, if a photon of energy greater than 1.022 MeV or $2m_0c^2$ falls on an electron in negative energy region, it raises the electron to positive energy state. This results in a "hole" in the negative energy state. This hole in the negative energy sea is supposed to behave like a positron.

In quantum electrodynamics (QED), the final theory of electro-magnetic interactions developed independently by Feynman, Schwinger and Tomonaga in 1948, where electron–positron creation was properly described. According to this theory, '*a positron can be considered as an electron moving backward in time*'.

5.8.1 Positron Annihilation

An inverse process occurs when a positron comes closer to an electron under the influence of their opposite charges. Both electron and positron vanish simultaneously and two γ-rays of energy > 0.511 MeV moving in opposite directions appear.

$$e^- + e^+ \to \gamma + \gamma$$

This process in which a positron captures an electron and both annihilate into two γ-rays of energy moving in opposite directions is known as *positron annihilation* or *pair annihilation*.

The two γ-rays move in opposite directions to conserve linear momentum. For positron annihilation presence of nucleus (for momentum conservation) is not required. Electron and positron can annihilate in vacuum also.

Pair Production Cannot Occur in Vacuum

Let us assume that pair production is taking place in empty space. An incident photon of energy $h\nu$ and momentum $h\nu/c$ gets annihilated in empty space and forms e^- and e^+ pair. The vector diagram of momenta involved when photon is converted into electron–positron pair is shown in Figure 5.12. Applying energy conservation, we get

$$h\nu = 2mc^2 \tag{5.30}$$

where m is the moving mass of moving electron or positron.

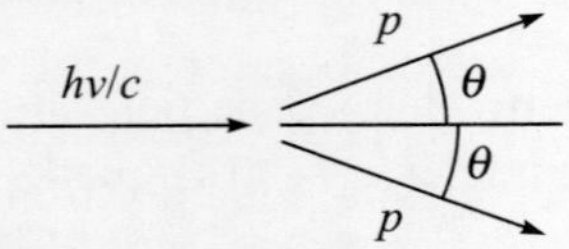

Figure 5.12 Momentum conservation in pair production.

The angles of emission of e^- and e^+ are equal due to conservation of linear momentum along perpendicular direction.

Applying momentum conservation in the direction of incident photon, we get

$$\frac{h\nu}{c} = 2p \cos \theta \tag{5.31}$$

or

$$h\nu = 2pc \cos \theta$$

and

$$p = mv \tag{5.32}$$

where v is the velocity of both electron and positron. This equation can also be written as

$$h\nu = 2mc^2 \frac{v}{c} \cos \theta \tag{5.33}$$

As $\frac{v}{c} < 1$ and $\cos \theta \le 1$, therefore,

$$h\nu < 2mc^2 \tag{5.34}$$

This contradicts with the law of energy conservation ($h\nu = 2mc^2$). Therefore, it is not possible for pair production to take place in vacuum as it violates energy, momentum conservation laws. Presence of third body—the nucleus is a must which carries away part of the momentum of the incident photon, thereby satisfying both the laws.

NUMERICAL PROBLEMS

Section 5.2

Solved Problem

1. A proton of energy 10 MeV is moving in Pb. How much energy does it lose in a collision with an electron? How many collisions are required to completely stop it?

Solution: Energy lost in a single collision with an electron $= \frac{4m}{M}E$

$$= \frac{4 \times 0.511 \times 10}{938}$$

$$= 0.02179 \text{ MeV}$$

$$= 21.79 \text{ keV}$$

Total number of collisions required to completely stop proton $= \frac{10{,}000 \text{ keV}}{21.79 \text{ keV/collision}}$

$$= 459 \text{ collisions}$$

2. The scattering of an energetic charged particle in matter is mostly due to interactions with (a) electron (b) nuclei (c) quarks

Solution: The incident charged particle undergoes Coulomb interaction with both electrons and nuclei. The collisions with electrons are numerous, but the momentum transfer is small and therefore suffers very small scattering. Only collisions with the nuclei result in appreciable scattering of the incident particle. Therefore, scattering of energetic charged particles is due to collisions with nuclei.

Unsolved Problems

1. An oxygen ion of energy 10 MeV is moving in Pb. What is the energy lost by oxygen ion in a single collision with electron? How many such collisions are required to stop the oxygen ion?

   [**Ans.** 1.36 keV, 7342 collisions]

2. A proton of energy (i) 2 MeV, (ii) 5 MeV is moving in Al. How much energy does it lose in a collision with an electron?

   [**Ans.** 4.36 keV, 10.90 keV]

Section 5.3

Solved Problem

1. Compare the radiation losses with the ionization losses for 0.9756, 9.756 and 97.56 MeV β-particles in Pb.

Solution: We have from Eq. (5.7)

$$\frac{dE/dX)_{\text{radiation}}}{dE/dX)_{\text{ionization}}} = \frac{Z(T + m_0c^2)}{m_0c^2 \times 1600}$$

Therefore, for β-particles of energy 0.9756 MeV

$$\frac{dE/dX)_{\text{radiation}}}{dE/dX)_{\text{ionization}}} = \frac{82(0.9756 + 0.511)}{0.511 \times 1600} = 0.15$$

Similarly, for β-particles of energy 9.756 MeV

$$\frac{dE/dX)_{\text{radiation}}}{dE/dX)_{\text{ionization}}} = \frac{82(9.756 + 0.511)}{0.511 \times 1600} = 1.03$$

and for β-particles of energy 97.56 MeV

$$\frac{dE/dX)_{\text{radiation}}}{dE/dX)_{\text{ionization}}} = \frac{82(97.56 + 0.511)}{0.511 \times 1600} = 9.84$$

Unsolved Problems

1. Compare the radiation losses with the ionization losses for 20.0 MeV β-particles in Al and Pb. What conclusion you draw from the results?

   [**Ans.** 0.33, 2.06]

2. Calculate the β-energy for which radiation and ionization losses are equal in Pb. Repeat the calculations for Al.

[Ans. 10.48 MeV, 63.40 MeV]

3. For $Z = 10, 20, 30, 40, 50, 60, 70, 80$ and 90, what are the values of β-energies for which radiation and ionization losses are equal? Plot E versus Z for this case.

[Ans. 81.3, 40.4, 26.7, 19.9, 15.8, 13.1, 11.1, 9.7, 8.6]

Section 5.5

Solved Problems

1. In an absorption experiment with 1.14 MeV γ-radiation from ^{65}Zn, it is found that 20 cm of Al reduced the beam intensity to 3%. Calculate the half-value thickness of Al for this radiation.

Solution:
Given:

$$\frac{I}{I_0} = \frac{3}{100}$$

Thickness of Al = 20 cm = 0.2 m. We have from Eq. (5.11)

$$I = I_0 e^{-\mu x}$$

or

$$\frac{I}{I_0} = e^{-\mu x}$$

or

$$\frac{3}{100} = e^{-0.2\times\mu}$$

or

$$\frac{100}{3} = e^{0.2\times\mu}$$

Taking $\log_e$ on both sides, we get

$$3.50656 = 0.2\mu$$

which gives

$$\mu = 17.533 \text{ m}^{-1}$$

Half-value thickness is given by

$$x_{1/2} = \frac{0.6931}{\mu}$$

$$= \frac{0.6931}{17.533}$$

$$= 0.0395 \text{ m}$$

2. Linear absorption coefficient of Pb for 1 MeV γ-rays is 0.75 cm^{-1}. Calculate thickness of lead required to reduce the intensity of γ-rays to $\frac{1}{100}$ times.

Solution:
Given:

$$\frac{I}{I_0} = \frac{1}{100} \quad \text{and} \quad \mu = 0.75 \text{ cm}^{-1}$$

We have from Eq. (5.11)

$$I = I_0 e^{-\mu x}$$

or

$$\frac{I}{I_0} = e^{-\mu x}$$

$\Rightarrow$

$$\frac{I_0}{I} = e^{\mu x}$$

or

$$\frac{100}{1} = e^{0.75 \times \mu}$$

Taking $\log_e$ on both sides, we get

$$4.6052 = 0.75x$$

or

$$x = 6.579 \text{ cm}$$

3. Half-thickness of an absorber of γ-rays is 5 mm. What is the percentage loss of intensity of these γ-rays in travelling 20 mm thickness of the absorber?

Solution:
Given:

$$x_{1/2} = 5 \text{ mm}$$

We have from Eq. (5.12)

$$x_{1/2} = \frac{0.6931}{\mu}$$

or

$$\mu = \frac{0.6931}{5} \text{ mm}^{-1}$$

or

$$\mu = 0.13862 \text{ mm}^{-1}$$

$$I = I_0 e^{-\mu x}$$

or

$$\frac{I}{I_0} = e^{-\mu x}$$

$$= e^{-0.13862 \times 20}$$

$$\frac{I}{I_0} = 0.0625$$

$$I = 0.0625 I_0$$

Thus, the intensity of γ-rays will reduce to 6.25% of its original value.

4. ^{24}Na emits two γ-rays of energy 1378 keV and 2754 keV respectively. Calculate the relative intensities of the two γ-rays after they have passed through 55.0 g/cm^2 of Pb (ρ = 11.0 g/cm^3). The linear absorption coefficients are 0.5 cm^{-1} for the 2754 keV and 0.6 cm^{-1} for the 1378 keV γ-rays.

Solution: The thickness of Pb sheet $= \dfrac{55.0 \text{ g/cm}^2}{11.0 \text{ g/cm}^3} = 5.0$ cm

The attenuation of the two γ-rays is given by

$$\frac{I_1}{I_0} = e^{-\mu_1 x}$$

and

$$\frac{I_2}{I_0} = e^{-\mu_2 x}$$

Therefore, on dividing these two relations, we get

$$\frac{I_1}{I_2} = \frac{e^{-\mu_1 x}}{e^{-\mu_2 x}}$$

$$= e^{(\mu_2 - \mu_1)x}$$

or

$$\frac{I_1}{I_2} = e^{(0.6-0.5)\times 5.0}$$

$$= e^{0.5}$$

$$= 1.65$$

5. Find the half-thickness for β-particles from ^{210}Bi in Al. The end point energy of β-particles from ^{210}Bi is 1.17 MeV. Given absorption coefficient μ of Al at this energy is 17.8 cm^2/g.

Solution: Half-value thickness is defined as that thickness of absorber which reduces the intensity to one-half of its original value. The relations for calculating half-value thickness given for γ-rays are also applicable to β-particles. Therefore, from Eq. (5.11)

$$\frac{I}{I_0} = \frac{1}{2} = e^{-\mu x_{1/2}}$$

Therefore,

$$x_{1/2} = \frac{\log 2}{\mu}$$

$$= \frac{0.693}{17.8}$$

$$= 0.03893 \frac{\text{g}}{\text{cm}^2}$$

In order to convert this thickness from $\frac{\text{g}}{\text{cm}^2}$ to usual units cm, let us divide it by density of Al (= 2.7 g/cc). Therefore,

$$x_{1/2} = \frac{0.03893\ \text{g/cm}^2}{2.7\ \text{g/cc}}$$

$$= 0.0144\ \text{cm}$$

6. A beam of antineutrinos from a reactor is incident on earth. Find the absorption length, i.e. the distance in which the beam is reduced by a factor $1/e$ of its original value. Given density of earth = density 5000 kg m^{-3}, cross section 10^{-19} barn per electron. (Ignore interactions with nuclei and assume that Z/A is approximately a half.)

Solution: As antineutrinos interact with electrons present in the earth, let us first estimate the number of electrons present in 1 m^3 of soil.

The weight of 1 m^3 of earth is equal to its density = 5000 kg.

The number of nucleons (protons + neutrons) present in 1 m^3 of earth is calculated by dividing the weight of the earth (5000 kg) by mass of nucleon 1.674×10^{-27} kg.

Therefore, the number of nucleons present in 1 m^3 of earth = $\frac{5000}{1.674 \times 10^{-27}}$.

The number of protons present $= \frac{1}{2}$ of total nucleons present

$$= \frac{1}{2} \times \frac{5000}{1.674 \times 10^{-27}}$$

This number must be equal to the number n of electrons present as the earth is electrically neutral.

Therefore,

$$n = \frac{1}{2} \times \frac{5000}{1.674 \times 10^{-27}}$$

Now we have

$$x = \lambda = \frac{1}{n\sigma}$$

Substituting various values, we get

$$x = \frac{1}{\frac{5000}{2 \times 1.674 \times 10^{-27}} \times 10^{-19} \times 10^{-28}}\ \text{m}$$

$$= 6.7 \times 10^{16}\ \text{m} = 6.7 \times 10^{13}\ \text{km}$$

It is worth mentioning here that this thickness is approximately equal to the distance of nearest star (Alpha Centauri) from the earth. This distance is approximately equal to about 4.35 light years.

7. The thickness of Cd sheet required to reduce the intensity of slow neutrons to 2% of its original value is 0.03 cm. Find the neutron absorption cross section of Cd. Given density of Cd = 8.7 g/cm^3, its atomic weight = 112.

Solution: The number of Cd atoms per unit volume is

$$n = \frac{\rho}{A} N_{av}$$

$$= \frac{8.7}{112} \times 6.023 \times 10^{23}$$

$$= 4.679 \times 10^{22}\ \text{cm}^{-3}$$

The required thickness t of Cd foil is given by

$$I = I_0\, e^{-n\sigma t}$$

or

$$t = \frac{1}{n\sigma} \ln \frac{I_0}{I}$$

or

$$\sigma = \frac{1}{nt} \ln \frac{I_0}{I}$$

$$\sigma = \frac{1}{4.7 \times 10^{22} \times 0.03} \ln \frac{1}{0.02}$$

$$= 2.7745 \times 10^{-21}\ \text{cm}^2$$

$$= 2774.5\ \text{barn}$$

Unsolved Problems

1. How much Pb is required to reduce the intensity of 500 keV γ-ray to 50% of its initial intensity? Given: μ for Pb = 164 m^{-1}. **[Ans.** 0.0042 m]
2. How much air column is required to reduce the intensity of 500 keV γ-ray to 1% of its initial intensity? Given: μ for air = 0.0112 m^{-1}. **[Ans.** 411.2 m]
3. Dentists commonly use 50 kV X-ray machines generating X-rays of 50 keV. Estimate the thickness of Pb sheet that absorbs 99.99% of 50 keV X-rays. Given linear attenuation coefficient for X-rays of 50 keV in Pb is 93.2 cm^{-1}. **[Ans.** 0.99 mm]
4. A beam of neutrons is incident upon a thick layer of cadmium. Find the absorption length, i.e. the distance in which the beam is reduced by a factor $1/e$ of its original value. Given density of cadmium = 8650 kg m^{-3}, and neutron absorption cross section = 24,500 barn.

   [Ans. 8.88 μm]
5. A beam of protons is incident upon a thick layer of lead. Find the absorption length, i.e. the distance in which the beam is reduced by a factor $1/e$ of its original value. Given density of lead = 11,340 kg m^{-3}, and proton absorption cross section = 15.7 barn per atom. **[Ans.** 1.93 cm]

6. How much energy does an electron of 1 GeV lose after traversing two radiation length?

[**Ans.** 0.845 GeV]

7. Find the thickness of Cd sheet required to reduce the intensity of slow neutrons to 2% of its original value. Given density of Cd = 8.7 g/cm^3, its atomic weight = 112, absorption cross-section = 3000 barn.

[**Ans.** 0.028 cm]

Section 5.6

Solved Problems

1. Work function of Cs is 1.9 eV. If light of wavelength 2500 Å falls on it, what is the maximum velocity of the ejected photoelectrons?

Solution: We have

$$h\nu = h\frac{c}{\lambda}$$

or

$$h\nu = \frac{6.626 \times 10^{-34} \text{ J s} \times 3 \times 10^8 \text{ m/s}}{2500 \times 10^{-10} \text{ m}}$$

$$= 7.95 \times 10^{-19} \text{ J}$$

or

$$= \frac{7.95 \times 10^{-19}}{1.602 \times 10^{-19}}$$

$$= 4.96 \text{ eV}$$

$$E = 4.96 - 1.9 = 3.06 \text{ eV}$$

Let velocity of photoelectrons be v. Therefore,

$$\frac{1}{2}mv^2 = E$$

or

$$v = \sqrt{\frac{2E}{m}}$$

$$= \sqrt{\frac{2 \times 3.06 \times 1.602 \times 10^{-19}}{9.1 \times 10^{-31}}}$$

$$= 1.04 \times 10^6 \text{ m/s}$$

Therefore, velocity of photoelectrons = 1.04×10^6 m/s.

2. Ultraviolet light of intensity 2 W/m^2 and wavelength 250 nm is directed at a surface coated with sodium.

(i) What is the maximum kinetic energy of the photoelectrons emitted by the surface?

(ii) If 0.20% of the incident photons produce photoelectrons, how many electrons are emitted per second, if the surface area of the exposed surface is 2.0 cm^2? Given work function (φ) of sodium is 2.3 eV.

Solution:

(i) Since 1 nm = 10^{-9} m. Therefore, 250 nm = 250×10^{-9} m

Therefore, energy of 250 nm photons $E = h\nu = h\dfrac{c}{\lambda}$

$$= \frac{(6.626 \times 10^{-34}\ \text{Js}) \times (3 \times 10^{8}\ \text{m/s})}{250 \times 10^{-9}\ \text{m}}$$

$$= 7.951 \times 10^{-19}\ \text{J}$$

$$= \frac{7.951 \times 10^{-19}\ \text{J}}{1.602 \times 10^{-19}\ \text{J/eV}}$$

$$= 4.96\ \text{eV}$$

Maximum kinetic energy $E_{\text{max}} = h\nu - \varphi = 4.96 - 2.30 = 2.66$ eV.

(ii) Photons energy in joules $E_p = 4.96 \times 1.602 \times 10^{-19} = 7.95 \times 10^{-19}$ J

Hence, number of photons reaching the surface per second $= n_p = \dfrac{E/t}{E_p} = \dfrac{(P/A)(A)}{E_p}$

$$= \frac{(2.00\ \text{W/m}^2) \times (2.0 \times 10^{-4}\ \text{m}^2)}{7.95 \times 10^{-19}\ \text{J/photon}}\ \text{photons/s}$$

Therefore, $n_p = 5.03 \times 10^{14}$ photons/s

The rate at which photoelectrons are emitted $= \dfrac{0.2}{100} n_p$

$$= \frac{0.2}{100} \times 5.03 \times 10^{14}$$

$$= 1.01 \times 10^{12}$$

$$= 1.01 \times 10^{12}\ \text{photoelectrons/s}$$

3. The threshold wavelength for photoelectric emission in platinum is 190 nm. What wavelength of light must be used in order for electron with a maximum kinetic energy of 1.1 eV be ejected?

Solution: We have $\varphi = \dfrac{hc}{\lambda}$, substituting various values

$$\varphi = \frac{6.626 \times 10^{-34}\ \text{J}\,s}{1.602 \times 10^{-19}\ \text{J/eV}} \times \frac{3 \times 10^{8}\ \text{m/sec}}{190 \times 10^{-9}\ \text{m}} = 6.53\ \text{eV}$$

We have from Eq. (5.14), where $h\nu_0 = \varphi$

$$E_{\text{max}} = h\nu - \varphi$$

or

$$h\nu = 6.53 + 1.1 = 7.63\ \text{eV}$$

or

$$\frac{hc}{\lambda} = 7.63 \text{ eV}$$

or

$$\lambda = \frac{hc}{7.63 \text{ eV}}$$

Substituting various values, we get

$$\lambda = \frac{6.626 \times 10^{-34} \text{ J s} \times 3 \times 10^{8} \text{ m/s}}{7.63 \times 1.602 \times 10^{-19} \text{ J}}$$

$$= 1.626 \times 10^{-7} \text{ m}$$

Therefore, $\lambda \approx 163$ nm

Unsolved Problems

1. The work function of platinum is 64 eV. When it is exposed to light of wavelength 160 nm, the maximum kinetic energy of ejected photoelectrons is 1.36 eV. Find Planck's constant from this data.

[**Ans.** 6.62×10^{-34} J s]

2. A metal surface is exposed by 9.8×10^{14} Hz light emits photoelectrons with maximum kinetic energy of 0.60 eV. The same surface when exposed to 15.2×10^{14} Hz light ejects electrons with maximum kinetic energy of 2.15 eV. Find work function of the metal and also calculate Planck's constant from this data.

[**Ans.** 1.90 eV, 6.63×10^{-34} J s]

3. A metal surface is illuminated by 8.5×10^{14} Hz light emits electrons whose maximum kinetic energy is 0.52 eV. The same surface when illuminated by 12.0×10^{14} Hz light emits electrons of 1.97 eV. From this data find Planck's constant and the work function of the metal. [**Ans.** 6.64×10^{-34} J s, 3.0 eV]

4. Gamma rays of energy 100 keV falls on platinum surface, whose work function is 6.4 eV. Find the maximum kinetic energy and velocity of the ejected electrons.

[**Ans.** 99.99 keV, 1.88×10^{8} m/s]

5. Work function of cesium is 1.9 eV. If ultraviolet light of frequency 10^{16} Hz falls on it, find the maximum kinetic energy of the photoelectrons and the stopping potential.

[**Ans.** 39.51 eV, 39.51 V]

6. What is the maximum wavelength of light that will cause ejection of photoelectrons from potassium? What will be the maximum kinetic energy of the photoelectrons if light of wavelength 210 nm falls on it. Given work function of potassium = 2.2 eV.

[**Ans.** 564 nm, 3.71 eV]

7. Light of wavelength 250 nm falls on a metal surface. Its work function is 2.26 eV. Find the extinction voltage, i.e. the retarding voltage at which the photoelectric current becomes zero. Also find the speed of the fastest photoelectrons.

[**Ans.** 2.7 V, 9.75×10^{5} m/s]

Section 5.7

Solved Problems

1. X-rays with $\lambda = 1$ Å are scattered from carbon block. The scattered radiation is viewed at 65° to the incident beam. What is Compton shift? Given: $h = 6.62 \times 10^{-34}$ J s, $c = 3 \times 10^8$ m/s and $m_e = 9.11 \times 10^{-31}$ kg.

Solution:
Given:

$$\theta = 65°$$
$$h = 6.62 \times 10^{-34} \text{ J s}$$
$$c = 3 \times 10^8 \text{ m/s}$$
$$m_e = 9.11 \times 10^{-3} \text{ kg}$$

We have from Eq. (5.26)

$$\lambda - \lambda_0 = \Delta\lambda = \frac{h}{m_0 c}(1 - \cos\theta)$$

$$= \frac{6.62 \times 10^{-34}}{9.11 \times 10^{-31} \times 3 \times 10^8}(1 - \cos 65°)$$

$$\lambda - \lambda_0 = \Delta\lambda = 1.40 \times 10^{-12} \text{ m}$$

2. A γ-ray of wavelength 1.87 pm (= 661 keV) undergoes Compton scattering through 135° from a free electron. Calculate the wavelength of the scattered γ-ray.

Solution:
Given:

$$\lambda_0 = 1.87 \text{ pm} = 1.87 \times 10^{-12} \text{ m and } \theta = 135°$$

We have from Eq. (5.26)

$$\lambda - \lambda_0 = \frac{h}{m_0 c}(1 - \cos\theta)$$

or

$$\lambda = \lambda_0 + \frac{h}{m_0 c}(1 - \cos\theta)$$

Substituting various values, we get

$$\lambda = 1.87 \times 10^{-12} + \frac{6.626 \times 10^{-34}}{9.1 \times 10^{-31} \times 3 \times 10^8}(1 - \cos 135°)$$

or

$$\lambda = 6.00 \times 10^{-12} \text{ m} = 6.00 \text{ pm } (1 \text{ pm} = 10^{-12} \text{ m}).$$

Therefore, wavelength of scattered γ-ray is equal to 6.00 pm (= 165 keV).

3. A beam of X-rays is scattered by a target at 90°. The scattered beam has a wavelength of 3.8 pm. What is the wavelength of X-rays in the incident beam?

Solution:
Given:

$$\lambda = 3.8 \text{ pm} = 3.8 \times 10^{-12} \text{ m and } \theta = 90°.$$

We have from Eq. (5.26)

$$\lambda - \lambda_0 = \frac{h}{m_0 c}(1 - \cos\theta)$$

or

$$\lambda_0 = \lambda - \frac{h}{m_0 c}(1 - \cos\theta)$$

Substituting various values, we get

$$\lambda_0 = 3.8 \times 10^{-12} - \frac{6.626 \times 10^{-34}}{9.1 \times 10^{-31} \times 3 \times 10^{8}}(1 - \cos 90°)$$

or

$$\lambda_0 = 1.37 \times 10^{-12} \text{ m} = 1.37 \text{ pm}$$

Therefore, wavelength of scattered γ-ray is equal to 1.37 pm.

4. A X-ray photon of frequency 3.2×10^{19} Hz collides with an electron and gets scattered through 60°. Find the frequency of scattered photon.

Solution:
Given:

$$\nu_0 = 3.2 \times 10^{19} \text{ Hz and } \theta = 60°$$

We have from Eq. (5.25)

$$\frac{1}{\nu} - \frac{1}{\nu_0} = \frac{h}{m_0 c^2}(1 - \cos\theta)$$

Therefore,

$$\frac{1}{\nu} = \frac{1}{\nu_0} + \frac{h}{m_0 c^2}(1 - \cos\theta)$$

Substituting various values, we get

$$\frac{1}{\nu} = \frac{1}{3.2 \times 10^{10}} + \frac{6.626 \times 10^{-34}}{9.1 \times 10^{-31} \times (3 \times 10^{8})^2}(1 - \cos 60°)$$

or

$$\nu = 2.47 \times 10^{19} \text{ Hz}$$

5. 1.836 MeV γ-ray from ^{88}Y undergoes Compton scattering from free electron through 180°. Find the energy of scattered photon. Also find the recoil energy of the electron.

Solution:
Given:

$$h\nu_0 = 1.836 \text{ MeV and } \theta = 180°.$$

We have from Eq. (5.26)

$$\lambda - \lambda_0 = \frac{h}{m_0 c}(1 - \cos\theta)$$

or

$$\frac{c}{v} - \frac{c}{v_0} = \frac{h}{m_0 c}(1 - \cos\theta)$$

or

$$\frac{1}{hv} - \frac{1}{hv_0} = \frac{1}{m_0 c^2}(1 - \cos\theta)$$

or

$$\frac{1}{hv} = \frac{1}{hv_0} + \frac{1}{m_0 c^2}(1 - \cos\theta)$$

$$m_0 c^2 = 511 \text{ keV for electron}$$

Therefore,

$$\frac{1}{hv} = \frac{1}{1836} + \frac{1}{511}(1 - \cos 180°)$$

or

$$hv = 224 \text{ keV}$$

Therefore, energy of the scattered photon $hv = 224$ keV
Energy of recoil electron = 1836 – 224 = 1612 keV.

6. At what scattering angle will incident 200 keV X-ray leave the scatterer with energy of 180 keV?

Solution:

$$hv_0 = 200 \text{ keV} \quad \text{and} \quad hv = 180 \text{ keV}$$

$$\frac{1}{hv} = \frac{1}{hv_0} + \frac{1}{m_0 c^2}(1 - \cos\theta)$$

Substituting various values, we get

$$\frac{1}{180} = \frac{1}{200} + \frac{1}{511}(1 - \cos\theta)$$

which gives

$$\theta = 44°$$

Unsolved Problems

1. Find the Compton wavelength for (i) electron, and (ii) proton.

   [**Ans.** 2.427 pm, 0.00132 pm]

2. A beam of X-rays having wavelength 56.2 pm is scattered through different angles, i.e. 30, 60, 90, 120 and 150 degrees. Find the Compton shift in wavelengths of the scattered photons at these angles and plot a graph between the shift and angles.

3. A γ-ray photon of energy 1332 keV is scattered through 90° by a free electron. Calculate the energy of the scattered photon and recoil electron.

   [**Ans.** 369 keV, 963 keV]

4. A γ-ray photon of frequency 6.66×10^{20} Hz collides with an electron and is scattered through 90°. Find the frequency of scattered γ-ray photon.

[**Ans.** 1.04×10^{20} Hz]

5. A photon having initial frequency of 1.65×10^{19} Hz collides with an electron and the frequency of the scattered photon is 1.15×10^{19} Hz. How much kinetic energy is imparted to the electron during collision?

[**Ans.** 20.7 keV]

6. What is the energy of the Compton scattered photons at 90°, when $E_\gamma >> m_e c^2$?

[**Ans.** 511 keV]

7. A γ-ray photon of initial frequency 2×10^{20} Hz collides with a free electron at rest and is scattered through 60°. Find the frequency of the scattered γ-ray. The electron Compton wavelength is 2.4×10^{-12} m. [**Ans.** 1.11×10^{-20} Hz]

8. Find the Compton wavelength of a proton at rest.

[**Ans.** 1.32×10^{-13} cm]

Section 5.8

Solved Problem

1. A γ-ray of energy 5.0 MeV undergoes pair production. Electron and positron formed move in opposite directions with equal kinetic energy. Find the kinetic energy of electron and positron.

Solution: Rest mass of electron = 0.511 MeV
Rest mass of positron = 0.511 MeV

Total energy consumed in creating electron and positron pair = 0.511 + 0.511 = 1.022 MeV

Energy left = 5.0 – 1.022 = 3.978 MeV

This energy will be shared equally by electron and positron.

$$\text{Each will carry kinetic energy} = \frac{3.978}{2} = 1.989 \text{ MeV}$$

Unsolved Problem

1. A proton and antiproton annihilates forming 2 γ-ray photons each moving in opposite directions. Find the energy of the γ-ray photons.

[**Ans.** 938.28 MeV]

REVIEW QUESTIONS

Short Answer Type

1. Why cannot visible light be used to demonstrate Compton effect?
2. What is mass attenuation coefficient?
3. What is pair production?

4. An electron and proton have the same energy. Which is more energetic?
5. Why cannot UV light be used to demonstrate Compton effect?
6. What is the wavelength of 1.02 MeV photon?
7. How does momentum remain conserved in electron–positron pair production?
8. Compare the positron emission and the electron capture.
9. Name at least four processes which contribute to the dissipation of energy of electrons passing through matter.
10. What is internal conversion? How is it different from photoelectric effect?
11. Differentiate the energy loss by electrons from the energy loss by heavy charged particles while passing through matter.
12. What is half-thickness of the material?
13. What is Compton shift in wavelength when scattering angle is π?
14. Write a note on stopping power.
15. What is Bremsstrahlung? How does pair production take place?
16. What is the unit of mass absorption coefficient?
17. What is the SI unit of linear absorption coefficient?
18. Is it possible for a photon to transfer its entire energy to the electron in Compton scattering? Explain.
19. Define threshold frequency in photoelectric effect.
20. Define Compton effect.
21. State different processes by which γ-rays interact with matter.
22. Define photoelectric effect.
23. Photoelectric effect occurs from bound electrons. Why?

Long Answer Type

1. What do you understand by absorption of γ-rays by matter? Derive an expression when γ-rays are made to pass through a thin plate. Also discuss half-thickness of the absorber.
2. What is Compton effect? Derive an expression for the change in wavelength of scattered photon. Show that change in wavelength in Compton scattering is independent of wavelength of incident photon. What is Compton wavelength?
3. What is Compton effect? Prove that Compton shift depends upon angle of scattering only. Can we use visible light to observe Compton effect? Why?
4. Find the expression for absorption of γ-rays in matter. What are absorption coefficient, mass absorption coefficient and half-thickness of the material?
5. What are pair production and electron–positron annihilation processes?
6. A charged particle having charge Ze, mass M and kinetic energy E is passing through a certain medium having n electrons per unit volume. If the motion of the particle is along x-direction, derive an expression for the rate of loss of energy. State the Bethe–Bloch modifications and write the Bethe–Bloch formula. What is meant by stopping power of the medium? On what factors does it depend?
7. Discuss the interactions of heavy charged particles with matter. Obtain the Bethe–Bloch formula and hence obtain the result of the range of charged particle.
8. Explain pair production and discuss its theory.

9. Derive the Bethe–Bohr formula for loss of energy due to ionization.

10. Discuss various processes that can take place when γ-radiation is absorbed by matter.

11. Describe Bohr's formula for energy loss of a heavy charged particle moving through matter.

12. How does a heavy charged particle interact with matter? Derive Bohr's formula for stopping power.

13. Obtain the classical formula for energy loss of heavy charged particles through matter and then introduce relativistic corrections to obtain Bethe–Bloch formula.

14. What is attenuation coefficient? How can it be measured experimentally? Explain.

15. Discuss the interaction of heavy charged particle with matter. How does it differ from interaction of lighter particles like electrons with matter?

16. Derive the Bethe–Bohr formula for energy loss of electrons.

17. Write notes on

(i) Bremsstrahlung

(ii) Dirac theory of pair production.

18. How does classical theory fail to explain photoelectric effect?

19. What do you understand by absorption of γ-rays by matter? Also give the concept of absorption coefficient.

20. Explain the passage of γ-radiation through matter. What is the mass absorption coefficient?

21. Write a note on Compton scattering.

22. Determine the minimum value of impact parameter in the interaction of heavy charged particle with matter.

23. Derive an expression for the stopping power of a heavy charged particle in matter.

24. Explain the interaction of γ-rays with matter. Name the processes responsible for γ-ray absorption.

25. What is photoelectric effect? Discuss the significance of Einstein's equation.

26. How do γ-rays lose energy in travelling through matter? Discuss relative importance of various processes involved.

27. Describe a method of measurement of absorption coefficient of γ-rays in a homogeneous material. What is the relative importance of photoelectric effect, Compton scattering and pair production with respect to E and Z.

28. Write notes on radiation length and positron annihilation.

29. Derive the wavelength shift formula in Compton effect.

30. What are stopping power, impact parameter and range of charged particles? Give the Bethe–Bohr formula for the specific ionization loss of a charged particle.

Chapter 6

Particle Accelerators

6.1 INTRODUCTION

A particle accelerator is an instrument used to increase the kinetic energy of charged particles such as electrons, protons, alpha-particles and other heavy ions. The relation of a particle accelerator to a nuclear scientist is the same as that of a microscope to a biologist or a telescope to an astronomer.

Sometimes we use an instrument without realizing the basic principle upon which the instrument is based. Suppose you are asked a question, have you ever seen and used a particle accelerator? Most of us will immediately say 'NO'. But in actual practice we are almost daily using a particle accelerator in our homes and laboratories. In our homes, we daily use Television. Similarly, in our laboratories, we use Cathode Ray Oscilloscope. The picture tube in the television and cathode ray tube in the oscilloscope are nothing but small-scale or mini particle accelerators. These picture tubes and accelerators have many common features. Both require a source of charged particles (electrons from a hot filament in picture tubes and ionized atoms from an ion source in case of accelerator). In both cases, an electric field to accelerate particles (~ 10^4 V in a picture tube and 10^7 V in accelerator) is required. Focusing electrodes to focus the beam, deflectors to aim the beam in a desired direction are required in both the cases. In accelerator beam is allowed to fall on a selected target, whereas in picture tube, beam is allowed to fall on the screen. In both these devices, i.e. accelerator and picture tube, all the components are housed in a chamber with high vacuum. This avoids the beam from scattering with air molecules.

Rutherford for the first time demonstrated that nitrogen nucleus can be modified by bombarding it with α-particles emitted by radioactive materials like Thorium or Radium. During that period, it was felt that more energetic projectiles produce changes in atomic nuclei. The kinetic energy of these particles should be higher so as to overcome the repulsive Coulomb force between positively charged nucleus and positively charged projectile.

It was in 1932 that J.D. Cockcroft and E.T.S. Walton built first particle accelerator, capable of accelerating protons to 400 keV and these 400 keV protons could induce an artificial nuclear reaction in ^{7}Li.

6.2 COCKCROFT AND WALTON ACCELERATOR

The first accelerator was built in 1932 by J.D. Cockcroft and E.T.S. Walton in Cavendish laboratory at the University of Cambridge. This accelerator was capable of producing potential difference of 400 kV and thereby accelerating protons to 400 keV.

6.2.1 Principle

The basic principle of this accelerator is that when a positively charged particle having a charge q units, is left near a point, where the potential is positive (V_0 volts), the particle repelled towards the point at (say) ground potential, its kinetic energy will be qV_0 eV. For example, if $V_0 = 400$ kV, and the particle is proton, its kinetic energy will be 400 keV and if the particle is He atom with both the electrons removed, its kinetic energy will be 2×400 keV = 800 keV.

6.2.2 Construction

It consists of voltage multiplier circuit. The accelerator, originally designed by Cockcroft and Walton, contained voltage quadrupler as shown in Figure 6.1. It consists of capacitors C_1, C_2, C_3 and C_4 having equal capacity and four diodes, D_1, D_2, D_3 and D_4. This arrangement is connected to the secondary of a high-voltage step up transformer. The voltage in the secondary coil of the transformer varies as $V(t) = V_0 \sin \omega t$, where V_0 is of the order of 100 kV.

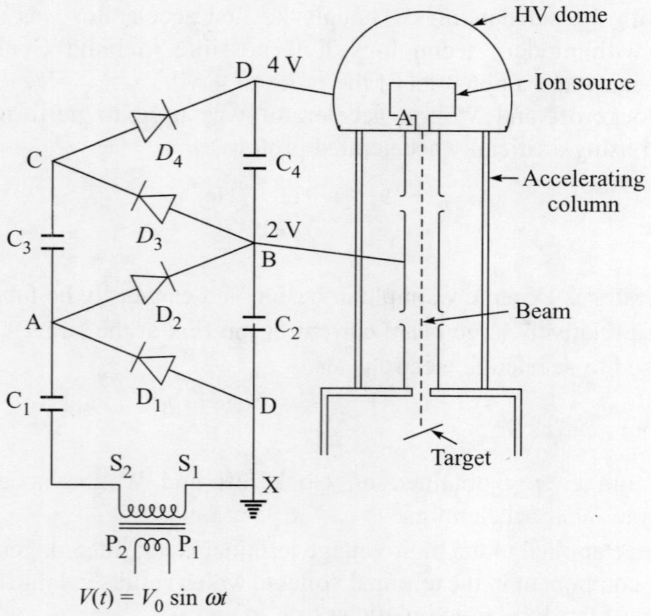

Figure 6.1 Cockcroft and Walton accelerator.

In the circuit the point X is always at ground potential and the upper point D is at highest potential which is connected to the high-voltage dome, which is spherical hollow conductor. In this high-voltage dome there is an ion source, which produces ions (by ionization) which are to be accelerated. Suppose we wish to accelerate protons, then ion source ionizes hydrogen gas and for α-particles, the gas to be ionized is ^{4}He.

6.2.3 Working

First let us consider capacitors C_1 and C_2; and diodes D_1 and D_2. Suppose during the first-half cycle, lower end S_1 of the transformer is positive and upper end S_2 of transformer is negative. This makes diode D_1 forward biased and diode D_2 reverse biased. Due to this capacitor C_1 is charged to peak voltage V_0. Since the voltage at the secondary of the transformer is varying between maximum value $+V_0$ to minimum value $-V_0$, therefore, the net voltage at point A varies between 0 and $2V_0[V_0 + (-V_0)]$ to $[V_0 + (+V_0)]$.

During the second-half of ac cycle, diode D_2 becomes forward biased and conducts, while diode D_1 is reverse biased. Therefore, point B reaches $2V_0$ and the potential difference between B and X is $2V_0$. Continuing the same argument the potential difference between point D and X is $4V_0$. If $V_0 = 100$ kV, then this potential difference is 400 kV. This circuit is known as *voltage multiplier* or *cascade rectifier*. This high-voltage is applied to a hemispherical dome, known as *high-voltage terminal or simply terminal.* The particles to be accelerated are produced here by ionizing suitable gas with the help of an ion source. The resultant positive charged particles or ions are accelerated towards the target kept at ground potential. The particles are accelerated in an evacuated tube to avoid collisions and scattering from air molecules.

Higher voltages on the high-voltage terminal are limited by its inability to hold its potential without sparking with the surroundings. Though the first accelerator was able to stand about 400 kV only, but with modern technology it is possible to build Cockcroft and Walton accelerators, which can stand a potential of the order of 4 MV.

The earliest Cockcroft and Walton accelerator was used to perform the first nuclear disintegration of ^{7}Li using artificially accelerated protons.

$$p + {}^7\text{Li} \rightarrow {}^4\text{He} + {}^4\text{He}$$

6.2.4 Advantages

1. This accelerator is extremely simple in design and can easily be fabricated.
2. It provides a relatively large beam current or ion flux at the target.
3. It can be used to accelerate electrons also.

6.2.5 Limitations

1. The maximum energy obtained on Cockcroft and Walton accelerators is low in comparison to other accelerators.
2. As the voltage applied to the high-voltage terminal is not filtered, generally there is large ripple or ac component in the terminal voltage, which results in large energy spread in the final energy of the accelerated particles.
3. The ion source is in the high-voltage area. If we wish to make some adjustments in the ions source like replacing the burnt out filament, etc., we have to reduce the high-voltage to zero, make adjustments and then increase the voltage. This entire operation takes many hours.

Some of these limitations have been taken care of in another accelerator known as *Van de Graaff accelerator*.

6.3 VAN DE GRAAFF ACCELERATOR

This accelerator was developed by Robert J. Van de Graaff in 1931 at Massachusetts Institute of Technology, USA and was initially capable to stand a potential difference of 5 million volts but later on it was upgraded to withstand a potential difference of 7 million volts.

6.3.1 Principle

Van de Graaff accelerator is based on the principle that if a charged conductor is brought in internal contact with another conductor, which is hollow, all of its charge gets transferred to the hollow conductor, no matter how high the potential is on the later. Therefore, by successively adding charge, the potential of the hollow conductor can be raised to any desired value.

6.3.2 Construction

It consists of a hollow spherical-shaped conductor, which is mounted on a long insulating support. The hollow conductor is also known as *high-voltage terminal* or *terminal* or *dome*. An insulating belt made up of silk, rubber or plastics, etc., passes over two frictionless pulleys, P_1 and P_2. Pulley P_1 is mounted on the grounded end of the structure, while second pulley P_2 is enclosed within spherical-shaped high-voltage terminal as shown in Figure 6.2. C_1 and C_2 are two sharp pointed combs. Comb C_1, called *spray comb*, charges the belt with charge sprayed by its sharp needles, close to the belt at the ground pulley. This comb is connected to a power supply that raises its potential to a few tens of kilovolts.

The comb C_2 is called charge collection comb and is placed near the belt inside the high-voltage terminal or dome and is connected to the inner side of the dome. An ion source is placed inside the dome and is connected to the accelerating tube A. On the other side of the accelerating tube, the target is placed in the vacuum. The entire accelerating tube is highly evacuated. This is to make sure that accelerated particles do not collide with air molecules and get scattered. Entire structure which includes high-voltage terminal, pulleys, belt, etc. is enclosed in a steel pressure tank (not shown in Figure 6.2) in

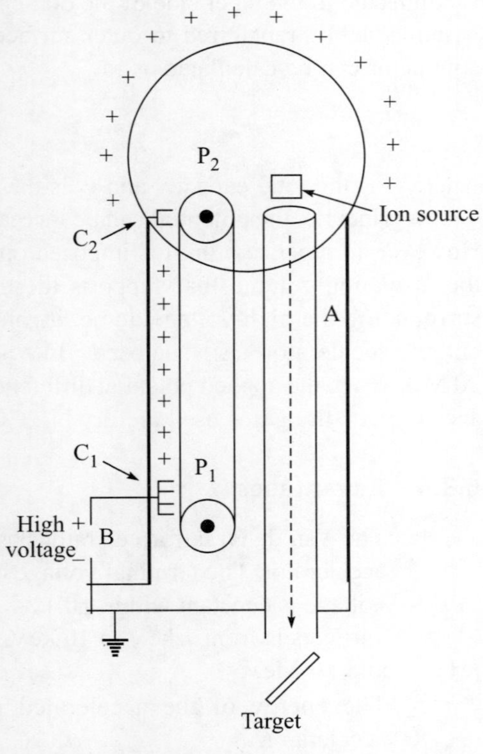

Figure 6.2 Schematic diagram of Van de Graaff accelerator.

which SF_6 gas is filled at pressure of the order of 10–20 atmospheres. SF_6 gas is chosen as it has much higher breakdown voltage compared to air. With SF_6 gas accelerators have been built where high-voltage terminal can withstand potential differences of the order of 20–30 MV.

6.3.3 Working

A dc potential of 5 kV to 50 kV is applied at a point B, the positive end is connected to the comb C_1 and negative end is grounded. The belt is run at high speed with the help of motor. The comb C_1 is connected to high-voltage supply, an intense electrostatic field is set up near the sharp points of the comb. This field produces positive and negative ions in the air. This discharge of air is also known as *Corona discharge*. (The electric field gradient at the surface of the conductor depends upon its radius of curvature. When a conductor having charge is a sharply pointed edge like comb, the radius of curvature of the pointed edge is very small. The charge density at the tip will be very large. So the pointed edge leaks the charge in the form of a discharge known as *corona discharge*. It sprays charge on the moving belt.) Sharp points of the comb C_1 attracts negative electrons and repel positive ions towards the fast-moving belt. The positive ions get deposited on the belt or belt becomes positively charged. As the positively charged belt moves up in the dome, it picks up electrons from the comb C_2 and the positive charge on the belt gets neutralized. As the comb C_2 is connected to the inner side of the dome, the inner side of the dome gets positive charge, which is immediately transferred to outer surface of the dome. The resulting potential V on the outer conductor can be calculated as

$$V = \frac{Q}{C}$$

where C is the total capacity and Q is the total charge on the conductor.

In principle, the potential can be increased without limits as we add more and more charge Q. However, in practice a limit is imposed on the potential by the electric breakdown (sparking) of the insulating column that supports the dome or of the surrounding atmosphere. When air is surrounding the high-voltage dome, it can stand voltages up to about 2–5 MV. However, if the entire accelerator is enclosed in SF_6 gas, it can stand voltages up to about 20 MV. When the applied potential difference is low, few hundreds of kV, as in small Van de Graaff accelerators, the gases used are dry N_2, CO_2, etc.

6.3.4 Advantages

1. The Van de Graaff accelerator has one enormous advantage over the Cockcroft Walton accelerator. The terminal voltage of Van de Graaff accelerator is highly stable. Terminal voltage is constant within ±0.1%. This means that the spread in the energy of accelerated particles is from 1 keV to 10 keV, when the energy of accelerated particles is between 1 and 10 MeV.
2. The energy of the accelerated particles is higher compared to Cockcroft Walton accelerator.

6.3.5 Limitations

1. One major limitation of Van de Graaff accelerator is its low-current output. The beam current in this accelerator is of the order of μA.
2. The ion source in this accelerator is also in the high-voltage area like that of Cockcroft Walton accelerator.

6.4 TANDEM ACCELERATOR

Modified form of Van de Graaff accelerator which eliminates the problem of placing ion source in high-voltage area, is known as *Tandem accelerator*. Tandem accelerator is two-stage accelerator compared to Van de Graaff accelerator, which is a single stage accelerator.

6.4.1 Principle

It works on the principle that negatively charged ions are accelerated towards positive potential and in the positive potential region, if we remove few electrons from these negatively charged ions, thus making them positively charged, they further get accelerated towards ground potential area.

6.4.2 Construction and Working

In tandem accelerator, the high-voltage terminal is in the centre of pressure tank, which is filled with SF_6 gas at a pressure of 10–20 atmospheres. In an ion source positive ions are produced at the ground level at one end of the accelerator as shown in Figure 6.3. These positive ions are accelerated through 100 kV to 400 kV. These accelerated ions are allowed to pass through charge exchange canal containing vapours of Cs metal (Cs metal contains lot of loosely unbound electrons). While passing through charge exchange canal, 1 to 4% of positive ions picks up two electrons and become negatively charged ions with unit negative charge. For example, protons become H^- and O^+ is converted to O^-. It is highly improbable that positively charged ions pick up three electrons to become doubly negatively charged. The beam then passes through first analyzing magnet. Here, negative ions are separated and are injected into the accelerator. Negative ions are accelerated towards the positive terminal, which is raised to high potential by a charging belt and when they reach the positive terminal, their kinetic energy becomes V eV, where V is the voltage (in volts) on the terminal. At the high-voltage terminal, negative ions are allowed to pass through a gas at low pressure (known as *stripper gas canal*) or through extremely thin carbon foils (known as *stripper foil canal*). While passing through stripper, negative ions lose most of the electrons and become positively charged ions. These positively charged ions are repelled by the positive terminal. After passing through second analyzing magnet which analyze the beam with respect to its energy, or chooses a beam of particular energy only, reaches a switching magnet. The switching magnet puts the beam either in beam line 1 or 2, etc., where the target is placed. For heavy ions this 2nd stage is more important. For example, if we are accelerating ^{107}Ag ions and assume that the terminal is at 15 MV. In the first stage, Ag ions are singly charged, and gain energy equal to 15 MeV. In the stripper foil, suppose 11 electrons are removed from negatively charged Ag ions, so Ag ions acquire 10-unit positive charge. These ions in the second stage gain energy equal to $10 \times 15 = 150$ MeV. So, the total energy of Ag ions reaching the target is $15 + 150 = 165$ MeV. However, if we start with negatively charged hydrogen ions, in the first stage, they gain energy = 15 MeV and in the second stage they further gain energy = 15 MeV only. So, the total energy of protons is $15 + 15 = 30$ MeV.

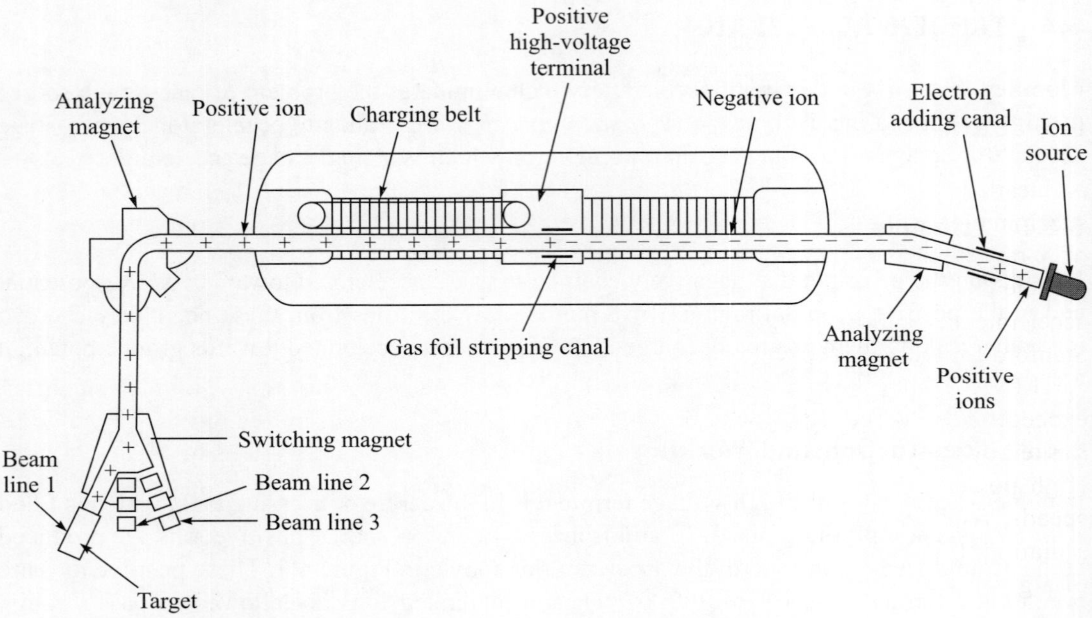

Figure 6.3 Detailed set-up of tandem accelerator.

6.4.3 Advantages

1. The ion source is at ground potential so any adjustment in the ion source can easily be made without reducing the terminal voltage.
2. Terminal voltage is used twice for accelerating the particles. For example, protons gain twice the energy for same terminal voltage as compared to Van de Graaff accelerator.

6.4.4 Limitations

1. The beam current is reduced further compared to Van de Graaff accelerator. In tandem accelerator beam currents are of the order of few nanoamperes.
2. The total energy gained by the particle is low compared to other accelerators. This limit again comes from the high-voltage on the terminal.

The accelerators discussed above cannot be used to accelerate particles to very high energies because of the breakdown in the electrical insulation and sparking in the air. This difficulty was overcome by the discovery of an altogether different principle for accelerating particles. Using the principle of resonance, particles are accelerated in steps and in each step; particles acquire additional energy by application of a relatively small voltage. This technique has led to the development of two types of accelerators:

- Linear accelerator LINAC or (drift tube accelerator), and
- Cyclic accelerator or cyclotron.

6.5 LINEAR ACCELERATOR (LINAC) OR DRIFT TUBE ACCELERATOR

A linear particle accelerator (also called a LINAC) is an electrical device for the acceleration of subatomic particles. This type of particle accelerator has many applications, from the generation of X-rays in hospitals, to an injector into higher energy accelerators at a dedicated experimental particle physics laboratory. The design of a LINAC depends on the type of particle that is being accelerated: electron, proton or ion. They range in size from a cathode-ray tube to the 2-mile long Stanford Linear Accelerator Center in California.

LINACs of appropriate design are capable of accelerating heavy ions to energies exceeding those available in cyclic accelerators (to be discussed in Section 6.7), which are limited by the strength of the magnetic fields required to maintain the ions on a curved path. High power LINACs are also being developed for production of electrons at relativistic speeds, required since fast electrons travelling in an arc lose energy through synchrotron radiation; this limits the maximum power that can be imparted to electrons in a synchrotron of a given size.

6.5.1 Principle

Charged particles can be accelerated by allowing them to fall through a suitable potential difference.

6.5.2 Construction

Linear accelerator consists of a number of hollow metallic cylindrical electrodes of increasing length, arranged in a straight line as shown in Figure 6.4. Alternate electrodes, i.e., the first, third, fifth, etc. are joined to one terminal A and the second, fourth, sixth, etc. to other terminal B of a high frequency generator. In this arrangement alternate electrodes carry opposite electrical potentials. For example, in a particular half-cycle of the oscillation all the odd-numbered electrodes would be positive whereas those with even numbers would be negative. In the next half-cycle, the potentials are reversed.

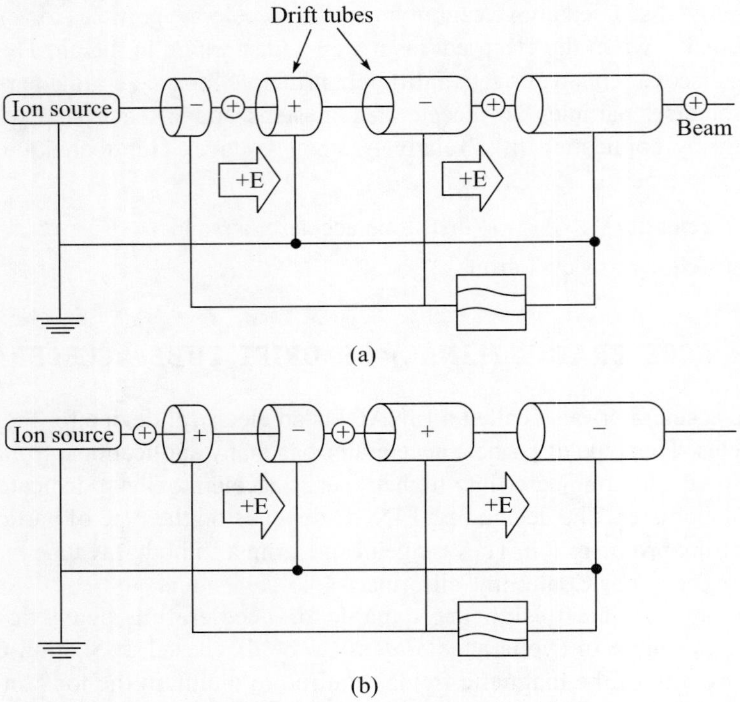

Figure 6.4 (a) the first drift tube is negatively charged. It means beam of +ve ions will be accelerated, (b) the first drift tube is positively charged. It means beam of +ve ions will not be accelerated.

6.5.3 Working

Suppose positively charged ions from the ion source S move from left to right along the common axis of the cylindrical electrodes. While passing through electrode 1, ions receive no acceleration, since they are moving in a field-free region. However, in traversing the gap between the 1st and 2nd electrodes, the ions are in the region in which there is a difference of potential. If the 1st electrode is positive and 2nd is negative, the positively charged ions are accelerated while crossing the gap. The positive ions then enter the 2nd electrode, and travel through it at a constant speed but at a higher velocity than in the 1st electrode. The length of this electrode is such that just as the ions reach the gap between 2nd and 3rd electrodes, the polarity of electrodes gets reversed, i.e. 2nd electrode becomes positive and 3rd becomes negative. This causes the positive ions to get additional acceleration in this gap also.

By making successive electrodes increasingly longer, to compensate for the increasing speed of the positive ions, it is possible to keep the ions in phase with the oscillating frequency. The positive ions gain energy each time they go from one electrode to the next.

If q is the charge on the ion, V_0 (volts) is the peak value of the ac frequency applied to the electrodes, then in each gap ions acquire energy $= qV_0$ eV. If there are n such gaps, then

$$\text{Energy acquired in } n \text{ gaps} = nqV_0 \text{ eV}. \tag{6.1}$$

The total kinetic energy acquired in n gaps $= \frac{1}{2}mv_n^2$.

Equating these two energies

$$\frac{1}{2}mv_n^2 = nqV_0$$

or

$$v_n = \sqrt{\frac{2nqV_0}{m}} \tag{6.2}$$

In these calculations, we have assumed that $v_n << c$.

Equation (6.2) shows that in order to get a high-energy beam

- the peak voltage V_0 of the oscillator should be higher.
- the number of gaps n should be as large as possible.

If f is the frequency of the oscillating potential, the time duration of a half-cycle is

$$= \frac{1}{2f}$$

The time ions take in passing through a nth cylinder of length L_n is

$$= \frac{L_n}{v_n}$$

where v_n is the velocity of the ions in nth cylinder.

For synchronization, we must have

$$\frac{L_n}{v_n} = \frac{1}{2f}$$

or

$$L_n = \frac{v_n}{2f}$$

Substituting for v_n from Eq. (6.2)

$$L_n = \frac{1}{2f}\sqrt{\frac{2nqV_0}{m}} \tag{6.3}$$

In 1931, Sloan and Lawrence built the first linear accelerator and using this accelerator they accelerated Hg ions to 2.85 MeV with peak oscillating voltage of 42 kV only.

Linear accelerators generally require very big set-ups. For example, LINAC at Los Almos in USA accelerating protons to 800 MeV is 805 metres long.

6.5.4 Advantages

1. Requirement of generating very high-voltages ~ million volts range is avoided in these accelerators.

2. They are economical for obtaining very high-energy particle beams.
3. They provide well-collimated beam of accelerated ions.

6.5.5 Limitations

1. They are inconveniently long in size.
2. They require extremely high frequency oscillators.

6.6 WAVE-GUIDE ACCELERATORS

With drift tube accelerators discussed in Section 6.5, it is not possible to accelerate electrons.The reason for this is that in drift tube accelerators, lower frequencies ~ 300 MHz are used. At these frequencies, power utilization will be very poor and the length of the drift tubes would be impossibly long. Further, if very high frequencies > 300 MHz are used to provide power to the electron, the outer tube of the linear accelerator or cavity would be so narrow that it could not accommodate the drift tubes.

In order to accelerate electrons beyond ~ 2 MeV, alternate version of linear accelerator was based on wave-guides (a wave-guide is a hollow cylindrical metallic conducting tube). These accelerators are known as *wave-guide accelerators*. In such accelerators drift tubes are completely eliminated. These accelerators consist of a circular wave-guide loaded with metallic annular discs as shown in Figure 6.5. An electromagnetic wave travels along the axis of the tube. In order to reduce the phase velocity of the wave, which is normally higher than the velocity of light, to that of the electrons, discs are spaced at about one-third wavelength intervals. The electrons can be imagined as riding on the advancing wavefront of the electromagnetic wave as surfer riders ride on a water wave.

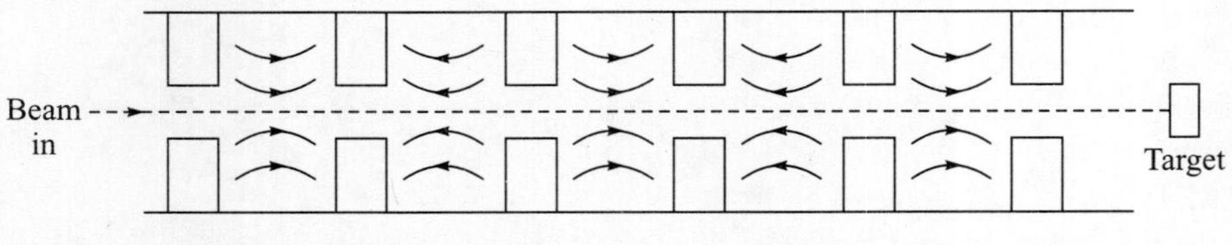

Figure 6.5 Resulting electric field in a disc loaded circular wave-guide accelerator.

Many electron linear accelerators based on this principle have been developed. The accelerator at Stanford University and that in the Russia produce 400 MeV electrons and the length of the accelerators are 79 and 110 metres respectively. A 20 GeV electron accelerator operational at Stanford University has a length of about 3 km.

6.7 MAGNETIC RESONANCE ACCELERATORS OR CYCLOTRONS

Because of the breakdown of electric insulation and sparking, particles cannot be accelerated to very high energies in electrostatic accelerators like Van de Graaff or Tandem accelerators.

Similarly, at very high energies, the length of linear accelerators becomes abnormally long (tens of kilometres). In 1930 Lawrence and Livingston suggested an alternative way of accelerating particles using a magnetic field to make the charged particles follow a spiral path of increasing radius. A beam of charged particles makes perhaps hundreds of revolutions, through the device and in each revolution particles receive two small energy increments until the particle energy reaches the desired value. This type of accelerators were initially called *magnetic resonance accelerators* but later named as *Cyclotrons*. In 1930, Lawrence and Livingston built the first experimental cyclotron, which was able to accelerate protons to 80 keV using an oscillator of peak voltage of only 2000 volts. The diameter of magnet pole pieces was 4 inch. Later on in 1932, another cyclotron was built which could accelerate protons to 1.2 MeV. It had a magnet with pole faces 11 inches in diameter. The diameter of the pole faces of the cyclotron is generally used to describe the size of the cyclotron. For example, if the diameter of poles of a magnet of a cyclotron is 88 inches, it would be referred as an "88-inch cyclotron". As we will see later in this section, larger the diameter of the pole face of the cyclotron magnet, greater would be the energy of the accelerated particles.

6.7.1 Principle

A positively charged particle can be accelerated to high energy with the help of an oscillating electric field, by making the particle to cross the same electric field time and again with the use of a strong magnetic field.

6.7.2 Construction

A cyclotron consists of two D-shaped hollow semicircular chambers D_1 and D_2 called "Dees" because of their shapes. These dees are placed with their diametric edges parallel to each other and are slightly separated from each other as shown in Figure 6.6. These dees are connected to the terminals of an alternating high frequency (10–20 MHz) and high voltage with peak value of 10–50 kV. This arrangement makes one dee positive and other negative during one half-cycle and vice versa in the next half-cycle. An ion source S of positive ions, such as protons, deuterons, α-particles, etc. is placed in the central region of the gap between the dees D_1 and D_2. This entire set-up, i.e. dees, ion source, etc. are placed in a highly evacuated chamber. The dees are completely insulated from this chamber. A uniform magnetic field is applied perpendicular to the cross-sectional area of the dees by placing them between the pole faces of a large electromagnet as shown in Figure 6.7.

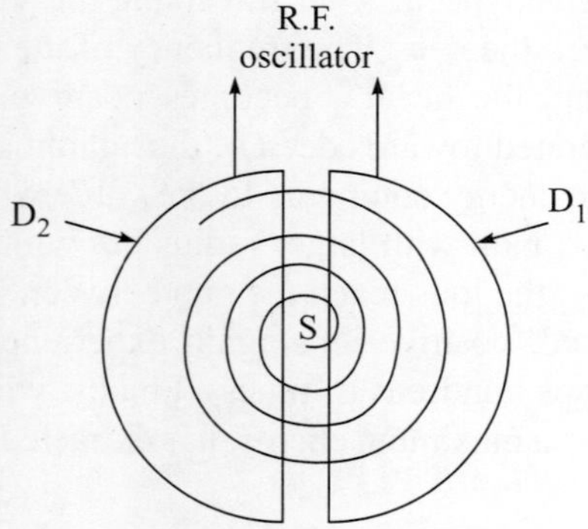

Figure 6.6 Cyclotron consisting of two dees placed in magnetic field.

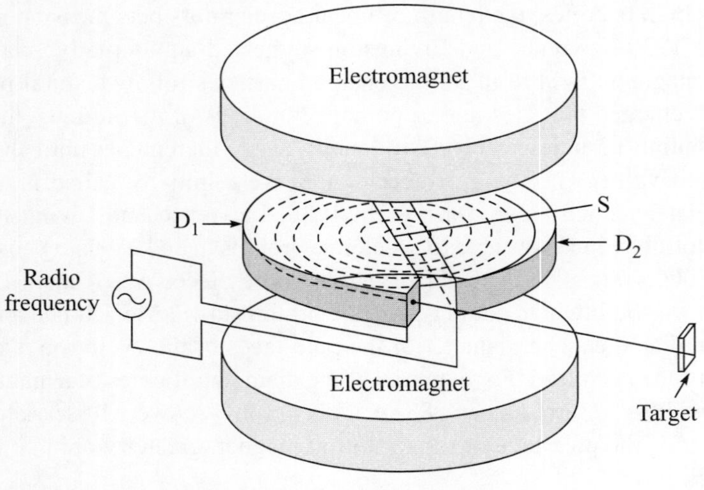

Figure 6.7 Detailed diagram of cyclotron.

6.7.3 Working

Ion source S produces positive ions within the gap between dees D_1 and D_2. At that particular instance, D_1 is positive and D_2 is negative. The positive ions are accelerated towards dee D_2 and it enters the dee D_2. As there is no electric field inside the dees, the ions are subjected to perpendicular magnetic field. The perpendicular magnetic field acts on the ions and the positive ions move in a circular path. Once inside the dee, they experience no acceleration and move with a constant speed. After traversing the semicircular path inside the dee D_2, they return to the gap between the dees. The frequency of the oscillator is adjusted in such a way that when ions reach the gap, the dee D_2 becomes positive and D_1 becomes negative. Now the positive ions are accelerated towards dee D_1, thus gaining in kinetic energy. The ions enter the dee D_1 with higher kinetic energy compared to the value when it was in dee D_2. In the dee D_1, they move again in a circular path with larger radius but with constant speed. After covering the semicircle inside the dee D_1, the ions reach the gap between the dees. At this instance again the polarity of the dees is reversed, positive ions again experience acceleration and this process is repeated many times (perhaps hundreds of times). Finally, when the positive ions reach the periphery of the dees after gaining a maximum energy, it is extracted out of the dees by means of high-voltage deflecting plates.

6.7.4 Theory

Let m be the mass of the positive ions to be accelerated, q is the charge on the ion. The ion moves in a semicircle of radius r with velocity v in a perpendicular magnetic field B.

$$\text{Magnetic force acting on the ion} = qvB$$

This force provides the necessary centripetal force $\dfrac{mv^2}{r}$ to the ions. Therefore,

$$\frac{mv^2}{r} = qvB$$

or

$$r = \frac{mv}{qB} \tag{6.4}$$

According to this equation, if m, q and B remain constant, then

$$r \propto v$$

As positive ions gain energy, i.e. v increases, the ions move in a semicircle of larger and larger radius. Therefore, the path of the ions is spiral in the dees.

If t is the time taken by the ions to complete a semicircular path, then

$$t = \frac{1}{2}\frac{2\pi}{\omega} = \frac{\pi}{\omega}$$

where ω is the angular frequency and $\omega = \frac{v}{r}$.

Therefore,

$$t = \frac{\pi r}{v}$$

Substituting r from Eq. (6.4)

$$t = \frac{\pi m}{Bq} \tag{6.5}$$

Time T to completing a full circular path is given by

$$T = 2t = \frac{2\pi m}{Bq}$$

Again if m, q and B are constant, then t, the time taken by the ion to complete the semicircular path inside the dee, is independent of the velocity of the ion, radius of the semicircular path and radius of dees or the size of pole pieces of the magnet. Hence, the frequency of the oscillations required to keep the ion in phase is given by

$$f = \frac{1}{T} = \frac{Bq}{2\pi m} \tag{6.6}$$

where f is known as cyclotron frequency or cyclotron resonance frequency and it is the frequency of ac oscillator also.

If $R = r_{max}$ is the radius of the last orbit from which the ions are extracted out of the dees and v_{max} is the velocity of the ions in the last orbit, then from Eq. (6.4)

$$R = \frac{mv_{max}}{qB}$$

or

$$v_{max} = \frac{BqR}{m} \tag{6.7}$$

Correspondingly, the maximum kinetic energy of the ions is

$$E_{max} = \frac{1}{2}mv_{max}^2$$

Substituting v_{max} from Eq. (6.7)

$$E_{max} = \frac{1}{2}\frac{mB^2q^2R^2}{m^2}$$

or

$$E_{max} = \frac{1}{2}\frac{B^2q^2R^2}{m} \tag{6.8}$$

The maximum kinetic energy acquired by the ions can also be calculated by multiplying the charge on the ion q, the peak value of the ac voltage V and the total number of circles N, completed by the ion before it is extracted out of the dees. Therefore,

$$E_{max} = Nq7V$$

Substituting E_{max} from Eq. (6.8)

$$\frac{1}{2}\frac{B^2q^2R^2}{m} = NqV$$

or

$$R = \frac{1}{B}\sqrt{\frac{2mV}{q}}\,N^{1/2} \tag{6.9}$$

From Eq. (6.8), it is clear that in order to accelerate particles to higher energy, we can

- increase B, i.e. make magnets with higher and higher pole strength. But increasing B beyond a certain limit is not possible as most of the materials saturate at a particular magnetic field.
- increase R, i.e. build pole pieces of larger and larger diameter so that the particles are extracted at larger R.

6.7.5 Advantages

1. Cyclotron is much smaller in size compared to linear accelerators.
2. No high voltages (like in Van de Graaff accelerator) are required. Only low-voltage ac oscillator (10–50 kV peak value) is required.
3. Cyclotron can deliver tens of microamperes of current at the target.

6.7.6 Limitations

1. It has been estimated that the cost of building larger cyclotrons scales roughly as the cube of the energy. For example, the cost of 500 MeV cyclotron is about US\$ 10^8. To build a cyclotron of 5 GeV is beyond the means of most of the countries.
2. As the energy of ions increases, relativistic effects come into picture. Because of the relativistic effects, mass of the ions m increases. m was assumed to be constant in Eqs. (6.4) and (6.5). Now it becomes a function of the velocity of the ions. This causes the ions to go out of phase with the applied ac oscillations and ions instead of accelerating may start de-accelerating. The upper limit of energy in case of protons is about 25–40 MeV.

3. Cyclotron cannot be used to accelerate electrons. The rest mass energy of electrons is very small (0.511 MeV); therefore, even for very low energies, mass of electrons does not remain constant, and they go out of phase with the applied ac voltage.

It was shown that allowance could be made for the effect of the increase of mass of the ions moving at high speed so as to keep them in phase with the oscillating potential. Two methods for this compensation are possible.

One method is to leave the magnetic field unchanged, but to decrease the frequency of the oscillating potential in accordance with Eq. (6.6). As the mass of the particle increases fm is kept constant. Cyclotrons based on this principle are known as *Synchrocyclotrons* or *Frequency Modulated Cyclotrons*.

The other method is to increase the magnetic field B in proportion to the mass, so that B/m remains constant. Due to this, the time period t in Eq. (6.5) remains unaffected, by the increase of mass. Accelerators based on this principle are known as *Sector Focusing or Azimuthally Varying Field* (*AVF*) Cyclotrons.

6.8 BETATRON

It is an accelerator for accelerating electrons or beta particles. There are definite problems associated with accelerating electrons by Van de Graaff and Cyclotron accelerators. The former machine could accelerate electrons up to a few MeV, while in case of latter machine relativistic effect becomes prominent. D.W. Kerst in 1940 built a new accelerator called *betatron*, which could accelerate electrons up to 250 MeV.

6.8.1 Principle

This machine uses the concept of electromagnetic induction as the accelerating force. It employs time varying magnetic field, which gives rise to induced electric field on the electrons moving in a fixed circular orbit and these electrons are accelerated. Here electrons move in a circular orbit of fixed radius. The rate of increase of magnetic flux (ϕ) is very slow compared to the frequency with which electrons are orbiting. In each orbit electrons are accelerated.

6.8.2 Construction

It consists of a doughnut-shaped chamber as shown in Figure 6.8, which is placed between the pole pieces of an electromagnet. The chamber is highly evacuated and the electrons with certain kinetic energy are injected into a circular orbit by an electron gun. By application of suitable electric and magnetic fields, electrons are forced to move in a constant radius orbit as shown by dotted line in Figure 6.8. When they attain full energy, they are deflected with the help of magnetic field on the target. This magnet and target are not shown in Figure 6.8 for making it simple. The electromagnet is powered by an alternating current. The inner layer of the chamber is coated with a thin layer of silver to avoid surface charge accumulation.

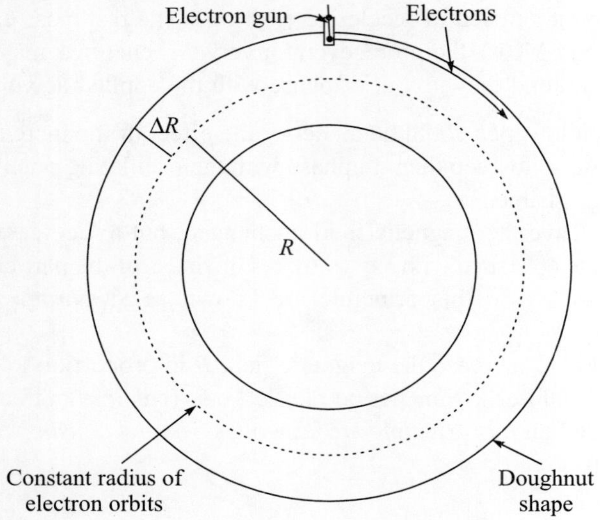

Figure 6.8 Schematic diagram of constant radius of electron orbits in a doughnut shaped betatron.

6.8.3 Working

The electrons are injected in the doughnut of the betatron during the first-quarter of a cycle in which magnetic field linking electron orbit is rising. Now to have electrons in a fixed orbit of radius r_0, a relation between the magnetic field at the orbit B and the total magnetic flux ϕ has to be derived.

An ac current having a frequency in the range 60–100 Hz powers the electromagnets. This generates a slow-varying field at the electron orbit of fixed radius.

The momentum of the electron $p = r_0 e B_0$

$$\text{The force acting on the electron} = \frac{d}{dt}(p) = \frac{d}{dt}(r_0 e B) = r_0 e \frac{dB}{dt} \tag{6.10}$$

Induced e.m.f. (ε) in the electron orbit is equal to work done by the unit charge in going around the orbit of radius r_0, i.e.

$$\varepsilon = \oint \vec{E} \cdot \vec{dl}$$

where E is the electric field which accelerates the electrons and from Faraday's laws ε is given by rate of change of magnetic flux.

$$\varepsilon = 2\pi r_0 E = \frac{d\phi}{dt} \tag{6.11}$$

The force on the electron is

$$eE = \frac{dp}{dt} \tag{6.12}$$

Combining Eqs. (6.10), (6.11) and (6.12)

$$\frac{d\phi}{dt} = 2\pi r_0^2 \frac{dB_0}{dt} \tag{6.13}$$

Integrating with respect to t, for ϕ varying from 0 to ϕ_0 and B varying from 0 to B_0 we get

$$\phi = 2\pi r_0^2 B_0 \tag{6.14}$$

This equation must be satisfied during the entire accelerating period, if the electrons have to be in the same orbit.

If B' is the average magnetic field over the whole area of the orbit, then

Total magnetic flux $$\phi = \pi r_0^2 B' \tag{6.15}$$

Comparing Eqs. (6.14) and (6.15), we get

$$B_0 = \frac{1}{2} B' \tag{6.16}$$

This is known as *betatron condition.*

6.8.4 Average Energy per Orbit

The total flux ϕ varies as $\phi_0 \sin \omega t$ so the average kinetic energy gained per orbit E' in the first-quarter of the cycle, i.e.

$$\frac{T}{4} = \frac{\pi}{2\omega}$$

is given by the integral of the product of electron charge and induced e.m.f. over this time span. Mathematically, it is written as

$$E' = \frac{e\phi_0}{\pi/2\omega} \int_0^{\pi/2} \frac{d}{dt}(\sin \omega t)dt$$

or $$E' = \frac{2e\omega\phi_0}{\pi} \tag{6.17}$$

Now $$\phi_0 = 2\pi r_0^2 B_0$$

So, the average energy gained in one orbit is $E' = 4e\omega r_0^2 B_0$. (6.18)

6.8.5 Calculation of Final Energy of Electrons

Distance travelled by the electron in $T/4$ seconds = Velocity × time = $\frac{v\pi}{2\omega}$ (6.19)

where v is the velocity of the electron, which is fairly close to velocity of light.

Distance travelled by the electron in one orbit= $2\pi r_0$.

$$\text{Number of orbits that electron makes} = \frac{\text{Total distance travelled by the electron}}{\text{Distance travelled by the electron in one orbit}}$$

$$\text{Number of orbits } N_0 = \frac{v}{4\omega r_0} \tag{6.20}$$

Combining Eqs. (6.18) and (6.20),

$$\text{Total energy} = ver_0B_0 \tag{6.21}$$

As an illustration, we will discuss how betatron produces electrons with high energy.

Suppose electrons with energy 70 keV are introduced in the doughnut. The corresponding speed of the electrons comes out to be $\approx 2 \times 10^{10}$ cm/s. The radius of the orbit is 50 cm and if the electromagnet is powered by an ac frequency of 60 Hz and magnetic field at the orbit is 1 T, let us calculate:

1. Total distance travelled in $\frac{T}{4}$ seconds.
2. Number of orbits.
3. Average energy per orbit.
4. Total energy.

1. **Total distance travelled in $\frac{T}{4}$ seconds:** Using Eq. (6.19)

$$\text{Distance} = \frac{2 \times 10^{10} \pi}{2 \times 2\pi \times \nu}$$

$$= \frac{2 \times 10^{10}}{2 \times 2 \times 60}$$

$$= 8.33 \times 10^7 \text{ cm}$$

$$= 8.33 \times 10^5 \text{ m}$$

2. **Number of orbits:** Using Eq. (6.20)

$$\text{Number of orbits} = \frac{2 \times 10^8}{4 \times 2 \times \pi \times \nu \times r_0}$$

$$= \frac{2 \times 10^8}{4 \times 2 \times \pi \times 60 \times 0.5}$$

$$= 2.65 \times 10^5 \text{ orbits}$$

3. **Average energy per orbit:** Using Eq. (6.18)

$$\text{Average energy per orbit} = 4 \times 1.6 \times 10^{-19} \times 2 \times \pi \times \nu \times 0.5^2 \times 1$$

$$= 4 \times 1.6 \times 10^{-19} \times 2 \times \pi \times 60 \times 0.5^2 \times 1$$

$$= 6.03 \times 10^{-17} \text{ J} \approx 375 \text{ eV}$$

4. **Total energy:** Using Eq. (6.21)

$$\text{Total energy} = ver_0B_0$$

$$= 2 \times 10^8 \times 1.6 \times 10^{-19} \times 0.5 \times 1$$
$$= 1.6 \times 10^{-11} \text{ J}$$
$$= 100 \text{ MeV}$$

6.9 SYNCHROCYCLOTRONS OR FREQUENCY MODULATED CYCLOTRONS

A synchrocyclotron is simply a cyclotron with the accelerating supply frequency decreasing as the particles become relativistic and begin to lag behind. Although in principle they can be scaled up to any energy they are not built any more as the synchrotron is a more versatile machine at high energies.

6.9.1 Principle

Synchrocyclotron is based on the principle that loss of resonance at high velocities, where there is an appreciable increase in the mass of the particle due to relativistic effects, can be compensated by decreasing the applied ac oscillating frequency.

6.9.2 Construction

The basic design of a synchrocyclotron is similar to that of a cyclotron. As shown in Figure 6.9, a synchrocyclotron consists of a single dee rather than two dees as in a conventional (low energy) cyclotron. Fixed magnetic field is applied with the help of pole pieces of an electromagnet, in a plane perpendicular to the plane of dee. The frequency of oscillating electric field applied is made to decrease continuously instead of keeping it constant so as to maintain the resonance with ion frequency. One terminal of the oscillating electric potential varying periodically is applied to the dee and the other terminal is earthed. The protons or deuterons to be accelerated are made to move in circles of increasing radii. The acceleration of particles takes place as they enter or leave the dee at the outer edge. The ion beam is taken out with the aid of electrostatic deflector.

The first synchrocyclotron built was at Berkeley, USA. It had magnets with pole piece diameter of 184 inches. In this synchrocyclotron, frequency of the oscillator was varied from 36 MHz to 18 MHz. Thirty-six megahertz was the initial frequency when the protons were moving with non-relativistic velocity and as they gained energy, the frequency was slowly reduced to 18 MHz. At this stage, protons would have gained maximum energy. This change in the frequency was done about 64 times per second. The magnetic field of the magnets is about 2.3 tesla. This accelerator was capable to accelerate protons to 740 MeV.

However, if we compare the output of cyclotron with that of synchrocyclotron, there is a difference. In a cyclotron, the flow of accelerated ions is regarded as continuous, although it actually consists of a series of pulses corresponding to each cycle of the oscillating potential. For a frequency of 20 MHz, there would be 20 million pulses of ions that reach the target per second. In the synchrocyclotron, however, a pulse of ions is carried from the ion source at the centre to the periphery of the dee as the frequency of the oscillating potential is decreased from its initial value (36 MHz) to the final value (18 MHz). The frequency then returns to its original high value (36 MHz) and another pulse of the ions is carried from the ion source to the periphery and so on.

The rate at which ion pulses are produced depends on the repetition rate of the frequency. This rate generally varies from 60 to 1000 per second, which means that at the target 60 to 1000 pulses of ions will be reaching the target per second. Because of this reason, the average value of the ion current is very low in these accelerators.

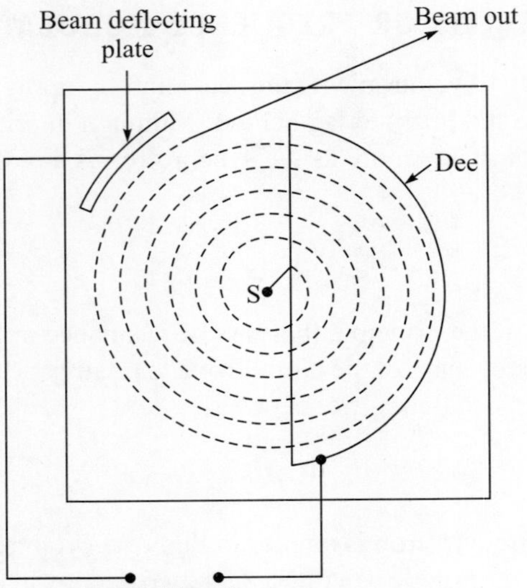

Figure 6.9 A schematic diagram of synchrocyclotron.

6.9.3 Theory

The frequency of revolution or angular frequency ω_c of the ions is

$$\omega_c = \frac{Bq}{m} = \frac{Bqc^2}{mc^2}$$

where mc^2 is the total energy of the ions which includes the kinetic energy T and the rest mass energy m_0c^2. Therefore,

$$\omega_c = \frac{Bqc^2}{T + m_0c^2}$$

Now $$\omega_c = 2\pi f \text{ or } f = \frac{\omega}{2\pi}$$

Here f is taken as the frequency of ac oscillator as it is in phase with the frequency of revolution of ions.

$$f = \frac{Bqc^2}{2\pi(T + m_0c^2)} \qquad (6.22)$$

This frequency will have maximum value when $T \approx 0$, or

$$f_{\max} = \frac{Bqc^2}{2\pi\, m_0 c^2} \tag{6.23}$$

and will have minimum value when $T \sim T_{\max}$, or

$$f_{\min} = \frac{Bqc^2}{2\pi\,(T_{\max} + m_0 c^2)}$$

Solving these two equations for $T_{\max}$, we get

$$T_{\max} = \frac{f_{\max} - f_{\min}}{f_{\max} f_{\min}} \frac{Bqc^2}{2\pi} \tag{6.24}$$

6.9.4 Advantages

1. It is capable of accelerating positively charged particles to very high energies.
2. Normally, low power (~ 15 kW) oscillators are needed in these accelerators as a source of ac potential.
3. With one dee, the electrical and mechanical design become simple.

6.9.5 Limitation

1. The output beam current is very low around fraction of microamperes or even lower than that.

6.10 AZIMUTHALLY VARYING FIELD (A.V.F.) OR SECTOR FOCUSING CYCLOTRONS

It was shown that relativistic effects limited cyclotron energy because of incompatibility of resonance and focusing. Thomas pointed out that this conclusion applied only to azimuthally constant magnetic fields. If one allowed azimuthal variation the limit can be removed. This idea was not utilized by the accelerator builders until the late 1950s.

Thus, in early 1960s, a new type of cyclotron, the sector-focusing cyclotron emerged. Iron sectors were introduced in the pole gap so that an azimuthal variation of the magnetic field was obtained. This azimuthal variation provided a strong vertical focusing on the circulating beam of ions and it was then not necessary to have the azimuthally averaged field to decrease with increasing radius as it had to do in the conventional cyclotron in order to maintain vertical focusing. The frequency of the accelerating voltage can thus be kept constant while maintaining a steady acceleration at each gap traversal; the energy was limited only by the size of the magnet.

6.10.1 Principle

The relativistic increase in mass of high velocity ions can be compensated by increasing the magnetic field B in proportion to the increase in mass m so that the ratio B/m remains constant. Due to this the time period of the ion $t = \pi m/Be$ remains unaffected by increase in mass of the ion.

6.10.2 Construction

This principle can be implemented by designing a magnet in which the magnetic field increases as the distance from the centre increases. But this type of magnets will have an undesirable effect of defocusing the beam as the radius of the orbit of ion beam increases. This design did not find favour. If a charged particle while traversing its path around the dee, encounters a series of alternating regions in which the magnetic field increases then decreases and so on, the same effect is achieved. In this continuously changing field strength, the ion does not move in a circular orbit, but the one, that is distorted.

The basic design of AVF accelerator is same as that of a cyclotron except that the magnetic field is modified making sectors of higher field strength followed by sectors of lower field strength. This is achieved by putting extra soft iron pieces of the shape as shown in Figure 6.10 on both pole pieces. So, the shape of pole pieces look like as shown in Figure 6.11. The orbit of protons in the dees is not circular but depends upon the magnetic field in that sector and looks like as shown in Figure 6.12.

One of the earliest AVF cyclotrons is at Oak Ridge Isochronous Cyclotron Laboratory, USA. and is capable of accelerating protons to 70 MeV. The diameter of pole pieces of the accelerator is about 2 metres and the field strength is about 1.5–1.7 tesla. Another large AVF cyclotron is at TRIUMF, Vancouver in Canada, which is capable of accelerating protons to 520 MeV.

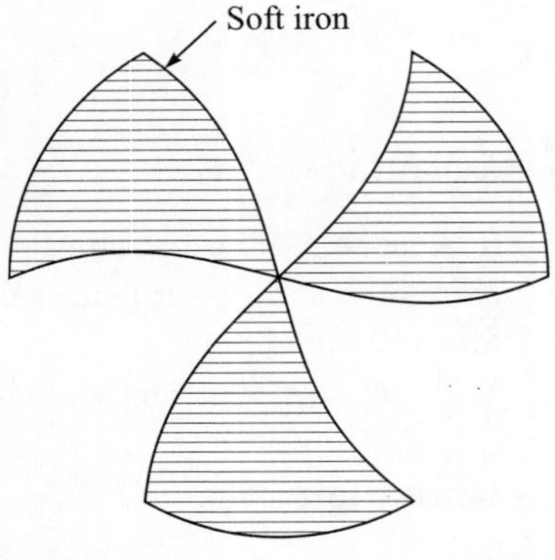

Figure 6.10 Soft iron piece to be placed on the pole pieces of the magnet of A.V.F. cyclotron.

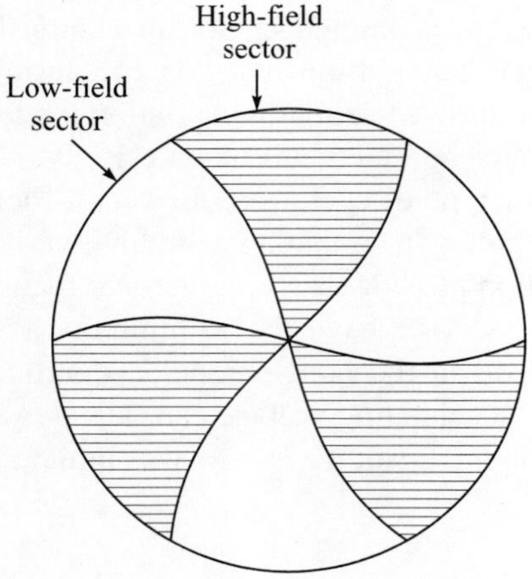

Figure 6.11 Shape of the magnet after placing the soft iron piece.

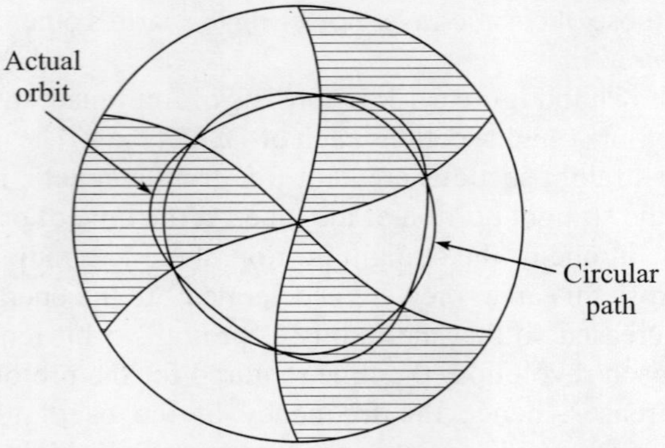

Figure 6.12 High-field and low-field regions in an A.V.F. cyclotron. Particles perform radial oscillations about the circular path.

6.11 SYNCHROTRONS

If we wish to accelerate the particles to still higher energies by cyclotrons or synchrocyclotrons, we have to build machines with larger magnets. Magnet is the costliest part of a cyclotron. Which is not possible due to high-costs (Section 6.7.6). This situation can be overcome by synchrotron accelerators in which both the magnetic field and the resonance frequency vary simultaneously. The main feature that keeps the cost reasonable as the energy is increased is that the particle orbit is very nearly having a constant radius at all the energies. Thus the magnetic field needs to be applied only at the circumference and not throughout the entire circular volume as in a cyclotron. Particles are confined to circular path by annular magnets. Particles follow circular path and are accelerated by the ac electric field as they cross the gap during each orbit. As the energy of the particles increases, they move fast, so the frequency of the ac voltage across the gap must increase to maintain the resonance condition. Along with frequency, the magnetic field must also be increased to keep the radius constant.

Charged particles, generally protons, are injected into synchrotron by giving them some initial kinetic energy and when they attain full energy, they are deflected out of the accelerator with the help of high voltage deflecting plates.

6.11.1 Proton Synchrotron

Principle

The protons are made to move in circular orbits under the action of a magnetic field. As the protons gain energy, the magnetic field is also increased with time so that radius of proton orbits does not change.

Construction

Here we describe the construction of Brookhaven proton synchrotron also known as *Brookhaven cosmotron.* The new accelerator was planned to accelerate protons to previously unheard of energies, comparable to those of cosmic-rays showering the earth's outer atmosphere. That is why it is also called the *Cosmotron*.

Its basic design is shown in Figure 6.13. It consists of four quadrants, each of about 30 feet radius and are joined by four straight sections each of 10 feet long. The magnetic field is applied over four quadrants and straight sections are kept free from magnetic field. Protons of 4 MeV are injected into one of the straight portion of the tube. At the time of proton injection magnetic field is about 300 gauss. In one of the straight portion, there is a radio frequency accelerating cavity. When protons cross this area, they are accelerated. As the energy of protons increases, the magnetic field is increased so that the radius of proton's orbit remains the same. Also as protons gain energy in each revolution, the time required for the protons to make a revolution of constant radius decreases, hence the frequency of the oscillator has to be increased correspondingly. The frequency of oscillator changes from 0.37 MHz to 4.0 MHz. The mean radius of the orbit is about 10 metres. This synchrotron is capable of accelerating protons to 3000 MeV or 3 GeV. Later on, number of more powerful synchrotrons were developed. Few such accelerators are:

- Proton synchrotron at Fermi National Accelerator Laboratory, USA which is accelerating protons from 50 GeV to 500 GeV. This accelerator has an orbital radius of 1000 m.
- There is proposal to built another proton synchrotron which could accelerate protons to 20,000 GeV or 20 TeV. The estimated diameter of this accelerator is about 60 km.

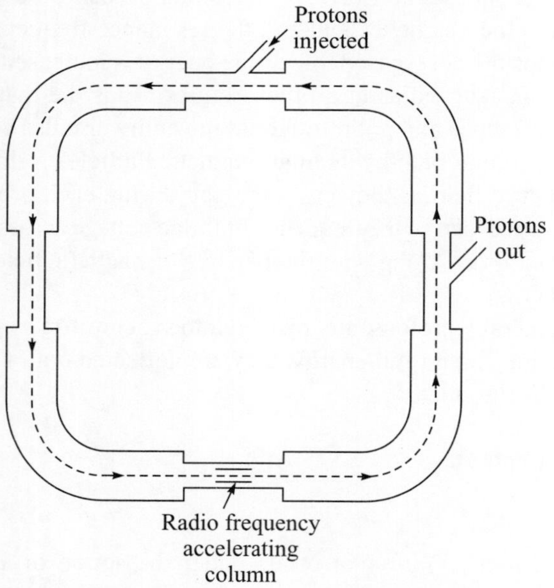

Figure 6.13 Basic structure of proton synchrotron, with four dipole sections to bend the beam and a radio frequency cavity to accelerate the beam.

Storage Rings and Colliding Beam Accelerators

Suppose we accelerate protons and bombard them on hydrogen target to induce a nuclear reaction

$$p + p \rightarrow p + p + X \tag{6.25}$$

where X can be a single particle (say π^0) or can be composite particles (say $p + \bar{p}$), etc. The question arises, what should be the energy of accelerated protons to induce such a reaction? The answer to this question we get from the nuclear reaction where we define threshold energy for a nuclear reaction as

$$E_{\text{th}} = -Q\frac{\text{Total mass of initial and final particles}}{2 \times \text{mass of target particle}}$$

If m_i is the total mass of initial particles and m_f is the total mass of the final products, then

$$Q = (m_i - m_f)c^2$$

Therefore, for reaction (6.25)

$$E_{\text{th}} = m_X c^2 \frac{4m_p + m_x}{2m_p}$$

or

$$E_{\text{th}} = m_X c^2\left(2 + \frac{1}{2}\frac{m_X}{m_p}\right) \tag{6.26}$$

If the particle X is π^0, $m_{\pi^0}c^2 = 135$ MeV, then

$$E_{\text{th}} = 135 \text{ MeV}\left(2 + \frac{1}{2}\frac{135 \text{ MeV}}{938 \text{ MeV}}\right)$$

$$\approx 280 \text{ MeV}$$

Therefore, to produce π^0 of rest mass energy 135 MeV, we require protons of 280 MeV. Out of 280 MeV, only 48% goes in the production of particle and rest of the energy is wasted.

Let the particle X be $p + \bar{p}$, total rest mass energy = 2 × 938 MeV = 1876 MeV. Therefore, the threshold energy according to Eq. (6.26) is

$$E_{\text{th}} = 1876 \text{ MeV}\left(2 + \frac{1}{2}\frac{1876 \text{ MeV}}{938 \text{ MeV}}\right)$$

$$\approx 5630 \text{ MeV}$$

$$= 5.63 \text{ GeV}$$

Therefore, to produce $p + \bar{p}$ pair whose rest mass energy is 1876 MeV, we need protons of about 5.63 GeV. Out of 5.63 GeV, about 1/3 energy goes in the production of new particle and the rest of the energy is wasted.

Finally, if we wish to create a particle whose rest mass energy is 90 GeV (= 90,000 MeV), the threshold energy required to induce this reaction is given by Eq. (6.26)

$$E_{\text{th}} = 90{,}000 \text{ MeV}\left(2 + \frac{1}{2}\frac{90000 \text{ MeV}}{938 \text{ MeV}}\right)$$

$$\approx 45{,}00{,}000 \text{ MeV}$$

$$= 4500 \text{ GeV}$$

Thus, to create a particle of mass 90 GeV, we require protons of 4500 GeV. Only 2% of the energy is used in creating a new particle and the remaining 98% energy is lost. Thus, as the rest mass of the produced particle increases, we require beams of higher and higher energies and most of the energy is lost. Where does this lost energy go? Answer to this question is as follows:

The experiments are being performed in laboratory frame of reference, in which the target (hydrogen) is at rest and projectiles (protons) are moving. In the centre of mass, centre of mass has to be stationary before, during and after the collision. The lost energy is in fact utilized to keep the centre of mass stationary. A resolution to this problem is colliding beam accelerators, in which we bring together two beams of equal energy moving in opposite directions. In fact, we perform the collision in the centre of mass frame. The two beams are moving in opposite directions with equal energy, therefore, the centre of mass is at rest before the collision. Compare this case with the usual experiment in which the target is at rest before the collision. The threshold energy for each beam is only $\frac{m_X c^2}{2}$. Thus, we need only two beams of 45 GeV to produce a 90 GeV particle whereas in the previous case we needed 4500 GeV beam against a fixed target for the same experiment. This enormous relaxation in the energy need comes at the expense of the experimental difficulty. In the two oppositely moving beams the density of the accelerated beam or beam flux is generally quite low of the order of 10^{11} particles/cm^3, compared with a liquid hydrogen target of about 10^{25} hydrogen atoms/cm^3. Thus, the reaction rate with colliding beams would be negligibly small. To counteract this small reaction rate, first particles are stored in storage rings; many particle pulses from an accelerator are kept circulating for time of the order of one day and are properly focused. When the number of particles circulating in opposite directions become desirable in number, they are made to collide by a magnetic or electric field.

One such example of proton–proton colliding beam accelerator is at CERN, Switzerland. In this accelerator, a beam of 28 GeV protons from a proton synchrotron is directed into two storage rings, where the beam orbits are in opposite directions and are made to collide at eight locations around the storage ring. The total 56 GeV (28 GeV + 28 GeV) centre of mass energy is equivalent to beam energy of about 1700 GeV in a fixed target accelerator. Later on, many such facilities have been developed throughout the world.

NUMERICAL PROBLEMS

Section 6.2

Solved Problem

1. In a Cockcroft Walton accelerator, the high voltage applied to the dome is 800 kV. Protons are injected in the high-voltage dome. What is the kinetic energy of protons, when they reach the target, kept at ground potential?

Solution: Energy gained by a particle having q electrons removed in passing through a potential drop of V volts is qV eV.

For proton $q = 1$

Therefore, energy gained by proton = V eV.

Since V = 800 kV = 8,00,000 V

So, energy of the proton = 8,00,000 eV = 800 keV

Unsolved Problems

1. In a Cockcroft Walton accelerator, double charged He ions are injected in the high-voltage dome kept at 600 kV. Calculate the kinetic energy of He ions, when they reach the target, kept at ground potential.

[**Ans.** 1.2 MeV]

2. In a Cockcroft Walton accelerator, O^{+3} ions (3 electrons removed from neutral oxygen atom) are injected in the high-voltage dome kept at 450 kV. Calculate the kinetic energy of O^{+3} ions, when they reach the target, kept at ground potential.

[**Ans.** 1.35 MeV]

Section 6.3

Solved Problem

1. In a Van de Graaff accelerator, ion source injects protons in the accelerating tube in the high-voltage dome. Seven MV is applied to the dome. What is the kinetic energy of protons, when they hit the target?

Solution: Energy gained by a particle having q electrons removed in passing through a potential drop of V volts is qV eV.

For proton $q = 1$

Therefore, energy gained by proton = V (eV).

Since V = 7 MV = 70,00,000 V.

So, energy of the proton = 70,00,000 eV = 7 MeV

2. A 5 MV Van de Graaff generator is equipped to accelerate protons, deuterons and double ionized ^{3}He particles.

(a) What are the maximum energies of the various particles available from this machine?

(b) List at least four reactions by which the isotope ^{15}O can be prepared with this machine.

Solution:

(a) 5 MV Van de Graaff accelerator can accelerate

Protons	–5 MeV
Deuterons	–5 MeV
Double charged ^{3}He	–10 MeV

(b) The possible reactions at these energies are

$$p + {}^{14}N \rightarrow {}^{15}O + \gamma$$
$$d + {}^{14}N \rightarrow {}^{15}O + n$$
$${}^{3}He + {}^{13}C \rightarrow {}^{15}O + n$$
$${}^{4}He + {}^{12}C \rightarrow {}^{15}O + n$$

Unsolved Problems

1. In a Van de Graaff accelerator, Ni^{+3} ions (3 electrons removed from neutral nickel atom) are injected in the accelerating tube. Five MV is applied to the dome. Calculate the kinetic energy of Ni^{+3} ions, when they reach the target, kept at ground potential.

[**Ans.** 15 MeV]

2. In a Van de Graaff accelerator, Ni^{+7} ions (7 electrons removed from neutral nickel atom) are injected in the accelerating tube. Five MV is applied to the dome. Calculate the kinetic energy of Ni^{+3} ions, when they reach the target, kept at ground potential.

[**Ans.** 35 MeV]

Section 6.4

Solved Problem

1. In a tandem accelerator high-voltage terminal is at 15 MV. H^{-1} ions (i.e. on neutral hydrogen atom, one extra electron added) are injected into the accelerating tube. After passing through the carbon stripper foil, what is the kinetic energy of protons, when they reach the target?

Solution: Energy gained by an ion having charge q (i.e. q-electrons removed), when it reached the target after passing through a high voltage of V volts is $(q + 1) \times V$ eV.

Protons when reach the target have $q = 1$.

Therefore, energy of the protons $= (1 + 1)V = 2V$ eV

Here,

$$V = 15 \text{ MV} = 15 \times 10^6 \text{ V}$$

Therefore, kinetic energy of protons

$$= 2 \times 15 \times 10^6 \text{ eV} = 30 \times 10^6 \text{ eV}$$
$$= 30 \text{ MeV}.$$

Unsolved Problems

1. In a tandem accelerator high-voltage terminal is at 17 MV. Ag^{-1} ions (i.e. on neutral lead atom, one extra electron added) are injected into the accelerating tube. After passing through the carbon stripper foil, they reach the target with 15 electrons removed, i.e. as Ag^{+15} ions. What is the kinetic energy of Ag^{+15}, when they reach the target?

[**Ans.** 272 MeV]

2. In a tandem accelerator high-voltage terminal is at 20 MV. Pb^{-1} ions (i.e. on neutral silver atom, one extra electron added) are injected into the accelerating tube. After passing through the carbon stripper foil, they reach the target with 12 electrons removed, i.e. as Pb^{+12} ions. What is the kinetic energy of Pb^{+12}, when they reach the target?

[**Ans.** 260 MeV]

3. A tandem accelerator, whose terminal potential is 5 MV, is capable of accelerating protons, doubly charged ^3He and ^4He particles. What is the maximum energy of each of these particles accelerated with this machine? [**Ans.** 10 MeV, 15 MeV, 15 MeV]

Section 6.5

Solved Problems

1. Electrons are accelerated to 30 GeV in the SLAC linear accelerator. Calculate the difference (in m/s) between the electron's speed and the speed of light.

Solution: We know in relativistic case

$$KE = mc^2 - m_0c^2$$

$$= \frac{m_0c^2}{\sqrt{1-\dfrac{v^2}{c^2}}} - m_0c^2$$

or

$$KE = m_0c^2\left[\frac{1}{\sqrt{1-\dfrac{v^2}{c^2}}} - 1\right]$$

Substituting $m_0c^2 = 0.511$ MeV and $KE = 30{,}000$ MeV, we get

$$30{,}000 = 0.511\left[\frac{1}{\sqrt{1-\dfrac{v^2}{c^2}}} - 1\right]$$

$$\frac{30{,}000}{0.511} = \frac{1}{\sqrt{1-\dfrac{v^2}{c^2}}} - 1$$

$$\frac{1}{\sqrt{1-\dfrac{v^2}{c^2}}} = \frac{30{,}000}{0.511} + 1 = 58709.4$$

which gives

$$\frac{v}{c} = 0.999999999$$

Therefore,

$$v = 0.999999999 \times c$$
$$= 2.999999997 \times 10^8 \text{ m/s}$$

Therefore,

$$c - v = 0.3 \text{ m/s}$$

2. In a drift tube portion of a linear accelerator, protons are accelerated from 0.75 MeV to 100 MeV. AC voltage applied has a frequency of 200 MHz. Find the length of the first and last drift tubes.

Solution: We have Eq. (6.2)

$$v_n = \sqrt{\frac{2nqV_0}{m}}$$

Taking $n = 1$ for the first tube

$$v_1 = \sqrt{\frac{2qV_0}{m}}$$

or

$$V_0 = \frac{v_1^2 \times m}{2q}$$
$$= \frac{1}{2} m v_1^2 \frac{1}{q}$$

We have

$$\frac{1}{2} m v_1^2 = 0.75 \text{ MeV}$$
$$= 0.75 \times 1.6 \times 10^{-13} \text{ J}$$

Therefore,

$$V_0 = 0.75 \times 1.6 \times 10^{-13} \frac{1}{1.6 \times 10^{-19}}$$
$$= 0.75 \times 10^6$$
$$= 750 \text{ kV}$$

The length of the tubes is given by Eq. (6.3)

$$L_n = \frac{1}{2f} \sqrt{\frac{2nqV_0}{m}}$$

For the first tube $n = 1$. Therefore,

$$L_1 = \frac{1}{2f} \sqrt{\frac{2qV_0}{m}}$$
$$= \frac{1}{2 \times 200 \times 10^6} \sqrt{\frac{2 \times 1.6 \times 10^{-19} \times 750 \times 10^3}{1.67 \times 10^{-27}}}$$
$$= 0.03 \text{ m}$$

Therefore, the length of the first drift tube = 0.03 m

Now let us calculate the total number of drift tubes n.

For the nth or the last drift tube, we have

$$\frac{1}{2}mv_n^2 = nqV_0$$

or

$$100 \times 10^6 \times 1.6 \times 10^{-19} = n \times 1.67 \times 10^{-19} \times 750 \times 10^3$$

or

$$n = 127.7 \approx 128 \text{ tubes}$$

Length of the last tube by Eq. (6.3)

$$L_n = \frac{1}{2f}\sqrt{\frac{2nqV_0}{m}}$$

$$= \frac{1}{2 \times 200 \times 10^6}\sqrt{\frac{2 \times 128 \times 1.6 \times 10^{-19} \times 750 \times 10^3}{1.67 \times 10^{-27}}}$$

$$= 0.34 \text{ m}$$

3. Calculate the velocity of electrons having kinetic energy of 1.17 MeV using relativistic considerations.

Solution: We have

$$KE = mc^2 - m_0c^2$$

or

$$KE = \frac{m_0c^2}{\sqrt{1 - \frac{v^2}{c^1}}} - m_0c^2$$

Therefore,

$$KE = m_0c^2\left[\frac{1}{\sqrt{1 - \frac{v^2}{c^2}}} - 1\right]$$

Given $KE = 1.17$ MeV and $m_0c^2 = 0.511$ MeV. Therefore,

$$1.17 = 0.511\left[\frac{1}{\sqrt{1 - \frac{v^2}{c^2}}} - 1\right]$$

which on solving gives

$$v = 0.95c$$

Unsolved Problems

1. In solved problem 1, what would be the energy of a proton moving at the same speed as that of the electron.

   [Ans. 77,700 MeV]

2. Find the average energy gained in any two drift tubes of solved problem 2.

   [Ans. 0.775 MeV]

3. A Los Alamos linear accelerator for accelerating protons to energy of 1 GeV and beam current of 100 μA has been built at Los Alamos. How many protons per second are produced? **[Ans.** 6.25×10^{14}]

4. Two mile long linac at Stanford University was aligned with the aid of a laser. How many wavelengths of red laser light were there in the accelerator? Given the wavelength of red laser is 660 nm.

   [Ans. 4.95×10^{9}]

5. What is the length of the Stanford 2-mile accelerator as viewed by a 25 GeV electron?

   [Ans. 4.09×10^{-5} mile or 0.065 m]

6. What is the velocity of a 10 MeV proton as a fraction of the velocity of light? How long a proton will take to move from the ion source to the target in a uniform 10 MeV accelerating tube of length 3 m? **[Ans.** 0.146, 1.37×10^{-5} s]

7. The relativistic mass of electron is thrice its rest mass. What is its kinetic energy and velocity? **[Ans.** 1.022 MeV, 0.94*c*]

8. Calculate the ratio of relativistic mass to its rest mass of electrons and protons, when the kinetic energy of both is 100 MeV. **[Ans.** 196.69, 1.11]

Section 6.7

Solved Problems

1. In Chandigarh cyclotron, the r.f. potential applied across the dees is 20 kV and a magnetic field of 1.1 tesla. Accelerated protons are extracted from the dees at a radius of 28 cm from the centre of the dees. Find:

   (i) The maximum energy acquired by protons
   (ii) The oscillator frequency
   (iii) The number of revolutions made by proton to achieve the maximum energy.

Solution:

Given:

$$V = 20 \text{ kV} = 20 \times 10^3 \text{ V}$$

$$B = 1.1 \text{ tesla}$$

$$r = 0.28 \text{ m}$$

(i) Maximum energy E_{max} acquired by protons is given by Eq. (6.8)

$$E_{\text{max}} = \frac{B^2 q^2 r^2}{2m}$$

$$= \frac{(1.1)^2(0.28)^2(1.602\times10^{-19})^2}{2\times1.67\times10^{-27}}$$

$$= 7.289 \times 10^{-13}\ \text{J}$$

$$= 4.56\ \text{MeV}$$

(ii) Frequency of the oscillator is given by Eq. (6.6)

$$f = \frac{Bq}{2\pi m}$$

$$= \frac{1.1\times1.602\times10^{-19}}{2\times3.1415926\times1.67\times10^{-27}}$$

$$= 16.77\ \text{MHz}$$

(iii) If N is the number of revolutions required to achieve maximum kinetic energy E_{max}, then

$$E_{max} = NqV$$

or

$$N = \frac{E_{max}}{qV} = \frac{7.289\times10^{-13}}{1.602\times10^{-19}\times20\times10^3}$$

$$= 227\ \text{revolutions}$$

2. Deuterons are to be accelerated with a cyclotron. If its magnet produces a flux density of 2.475 tesla, what must be the frequency of the oscillating potential applied across the dees. Mass of ${}^2_1H$ = 2 amu.

Solution:

Given:

$$\text{Magnetic field} = 2.475\ \text{tesla}$$

$$\text{Mass of deuteron} = 2 \times 1.672 \times 10^{-27}\ \text{kg}$$

$$\text{Charge on deuteron} = 1.602 \times 10^{-19}\ \text{C}$$

$$\text{Frequency } f = \frac{Bq}{2\pi m}$$

Substituting various values

$$f = \frac{2.475\times1.602\times10^{-19}}{2\times3.14159\times2\times1.672\times10^{-27}}$$

$$= 18.85 \times 10^6\ \text{Hz}$$

$$= 18.85\ \text{MHz}$$

3. A cyclotron oscillating frequency of 1 MHz is used to accelerate protons. If the radius of the dee is 60 cm, find the magnetic field in tesla.

Given:

$$\text{Frequency } f = 1 \text{ MHz} = 10^6 \text{ Hz}$$

$$\text{Radius of the dees } r = 60 \text{ cm} = 0.60 \text{ m}$$

$$\text{Charge on proton } q = 1.602 \times 10^{-19} \text{ C}$$

$$\text{Mass of proton } m = 1.672 \times 10^{-27} \text{ kg}$$

Solution: We have Eq. (6.6)

$$f = \frac{Bq}{2\pi m}$$

or

$$B = \frac{2\pi mf}{q}$$

$$= \frac{2 \times 3.14159 \times 1.672 \times 10^{-27} \times 10^6}{1.602 \times 10^{-19}}$$

$$= 0.0655 \text{ tesla}$$

4. A cyclotron in which flux density of 1.4 tesla is used to accelerate protons, what should be the frequency of alternating field applied to dees? Mass of protons is 1.67×10^{-27} kg.

Solution: Given:

$$\text{Magnetic field } B = 1.4 \text{ T}$$

$$\text{Mass of proton } m = 1.672 \times 10^{-27} \text{ kg}$$

$$\text{Charge on proton } q = 1.602 \times 10^{-19} \text{ C}$$

$$f = \frac{Bq}{2\pi m}$$

$$= \frac{1.4 \times 1.602 \times 10^{-19}}{2 \times 3.14159 \times 1.672 \times 10^{-27}}$$

$$= 21.35 \times 10^6 \text{ Hz}$$

$$= 21.35 \text{ MHz}$$

Unsolved Problems

1. For the original Berkeley cyclotron (R = 12.5 cm, B = 1.3 tesla) compute the maximum proton energy (in MeV) and the corresponding frequency of the varying voltage.

[**Ans.** 1.41 MeV, 19.8 MHz]

2. A fixed frequency cyclotron magnet of radius 1 m produces the maximum flux of 1.5 tesla. Calculate the energy of single and doubly charged helium ions.

[**Ans.** 26.95 MeV, 107.78 MeV]

3. Alpha-particles are accelerated in a cyclotron having dees of radius 65 cm. Magnetic field across the pole pieces is 1.8 tesla. What would be the frequency of the oscillating potential applied across the dees? What is the maximum energy of α-particles in this cyclotron? [**Ans.** 3.72 MHz, 65.5 MeV]
4. The beam of cyclotron has a maximum radius of 80 cm. The magnetic field induction is 0.75 W/m^2. Calculate the kinetic energy of (a) protons, (b) deuterons. [**Ans.** (a) 17.2 MeV, (b) 8.62 MeV]
5. The beam of a cyclotron describe a circle of diameter 3.2 m just before emerging out from the dees. The frequency of the applied alternating voltage is 10.0 MHz. Neglecting the relativistic effects, find the flux density of the magnetic field, the energy and speed of deuterons upon emergence. [**Ans.** 0.79 tesla, 76.5 MeV, 1.21×10^8 m/s]
6. Before the Second World War, a cyclotron was fabricated at Berkeley for which the magnetic field was 1.5 tesla. Calculate (i) the frequency in MHz needed to accelerate protons, and (ii) the maximum energy of protons (in MeV). Given the radius of the cyclotron was 15 inches. [**Ans.** (i) 22.9 MHz, (ii) 15.56 MeV]
7. A pulse of 10^8 protons is injected into a cyclotron and is kept circulating in a stable orbit by application of radio frequency of 10 MHz. What is the mean beam current? [**Ans.** 160 μA]
8. A fixed frequency cyclotron magnet of radius 1 m produces maximum magnetic field of 1.5 tesla. Calculate the energy of α-particles accelerated by it. [**Ans.** 108.1 MeV]

Section 6.8

Solved Problem

1. The maximum magnetic field in a betetron is 1 tesla. If the radius of doughnut is 1 m and the frequency of variation of magnetic field is 60 Hz, calculate the energy gained by electron per turn.

Solution: In a betatron, energy gained by electron per turn (Eq. 6.18)

$$= 4e\omega r_0^2 B_0$$
$$= 4 \times e \times 2\pi \times f \times r_0^2 \times B_0$$
$$= 4 \times 1.602 \times 10^{-19} \times 2 \times 3.1415926 \times 60 \times 1^2 \times 1$$
$$= 2.41 \times 10^{-16} \text{ J}$$
$$= 1508 \text{ eV}$$

Unsolved Problems

1. Calculate the maximum energy and number of revolution to attain the maximum energy in the above problem. Electrons are injected in the betatron with an initial energy of 100 keV. [**Ans.** 190 MeV, 1.26×10^5 revolutions]
2. Magnetic field in a betatron is 0.5 tesla, frequency of the field is 50 cycles per second and radius of the doughnut is 0.8 m. Calculate the energy gained by electrons per turn. [**Ans.** 402.6 eV]

Section 6.9

Solved Problems

1. A frequency modulated cyclotron is capable of accelerating protons to 500 MeV. What is the ratio of highest to lowest frequency needed to accomplish this?

Solution: We have a relation between frequency f and magnetic field B Eq. (6.6)

$$f = \frac{Bq}{2m\pi}$$

$$= \frac{Bqc^2}{2\pi mc^2}$$

Now for relativistic case,

Total energy $E = mc^2 = K + m_0c^2$

where K is the kinetic energy of the proton and m_0c^2 is its rest mass energy.

Therefore, according to Eq. (6.22)

$$f = \frac{Bq}{2\pi E} = \frac{Bq}{2\pi(K + m_0c^2)}$$

$f = f_{max}$, when $K = 0$ and $f = f_{min}$ otherwise. Therefore,

$$f_{max} = \frac{Bq}{2\pi\, m_0c^2}$$

and according to Eq. (6.23)

$$f_{min} = \frac{Bq}{2\pi(K + m_0c^2)}$$

$$\frac{f_{min}}{f_{max}} = \frac{m_0c^2}{K + m_0c^2}$$

$$= \frac{938}{500 + 938}$$

$$= 0.65$$

2. Calculate $\frac{w}{B}$ for a completely stripped nitrogen atom with $A = 14$ to move in a stable orbit of a FM cyclotron with a kinetic energy of 1.2 GeV.

Solution: Given:

Kinetic energy $KE = 1.2$ GeV $= 1200$ MeV

$q = 7$, and

Rest mass energy of nitrogen $m_0c^2 = 14 \times 931.47$ MeV $= 13{,}040.58$ MeV

Therefore, total energy $E = 1200 + 13040.58 = 14{,}240.48$ MeV

or

$$E = 14{,}240.58 \times 1.6 \times 10^{-13}\ \text{J} = 2.278 \times 10^{-9}\ \text{J}$$

We also have total energy $E = \dfrac{Bqc^2}{w}$

or

$$\frac{w}{B} = \frac{qc^2}{E}$$

$$= \frac{7 \times 1.6 \times 10^{-19} \times (3 \times 10^8)^2}{2.278 \times 10^{-9}}$$

$$= 4.42 \times 10^7\ \text{m}^2/\text{W}$$

Unsolved Problems

1. Calculate the maximum frequency applied (at the time of injecting α-particles at the centre) to a synchrocyclotron to achieve an energy of 800 MeV. Given B = 1.8 tesla.

[**Ans.** 13.72 MHz]

2. A synchrocyclotron gives a 400 MeV proton beam. What must be the change in frequency? [**Ans.** 29.9%]

3. (i) Calculate the kinetic energy in MeV of a proton that is moving in a circular orbit in a magnetic field of 17,000 gauss in resonance with the applied frequency of 18 MHz.

(ii) Determine the momentum of this proton in MeV/c.

(iii) Determine the linear velocity of the proton.

(iv) Determine the radius of this orbit.

[**Ans.** (i) 412 MeV, (ii) 970 MeV/c, (iii) 0.72c, (iv) 1.91 m]

4. A proton synchrotron with an orbit radius of 9 m operates at the field strength of 1.4 tesla. Calculate the final energy of protons. [**Ans.** 3.78 GeV]

Section 6.10

Solved Problems

1. In a 70 MeV betatron synchrotron, the radius of the stable orbit is 28 cm. Find the value of magnetic field B at this orbit for the given energy.

Solution: Given:

Radius of stable orbit r = 28 cm = 0.28 m

Energy E = 70 MeV = $70 \times 1.6 \times 10^{-13} = 1.12 \times 10^{-11}$ J

Charge on electron e = 1.602×10^{-19} C

We have Eq. (6.21) with the assumption that for 70 MeV electrons $v \approx c$

$$E = Berc$$

or

$$B = \frac{E}{erc}$$

$$= \frac{1.12 \times 10^{-11}}{1.602 \times 10^{-19} \times 0.28 \times 3 \times 10^{8}}$$

or

Magnetic field $B = 0.83$ T

2. What radius is needed in proton synchrotron to attain particles of energy of 12 GeV? Assume that a guide field is 1.9 Wb/m^2 is available.

Solution: Given:

Kinetic energy KE = 12 GeV = 12000 MeV

Magnetic field B = 1.9 Wb/m^2

Charge on proton q = 1.602 × 10^{-19} C

As the kinetic energy of proton is larger than its rest mass energy, so we have to apply relativistic relations. According to the theory of relativity, if m_0c^2 is the rest mass energy and KE is the kinetic energy of a particle, then

Total energy $mc^2 = m_0c^2 + KE$

Total energy mc^2 = 938 + 12,000 MeV

Now

931 MeV = 1 amu

Therefore,

$$12{,}938 \text{ MeV} = \frac{12{,}938}{931} = 13.9 \text{ amu}$$

We have the relation Eq. (6.4)

$$r = \frac{mv}{Bq}$$

Let us calculate the velocity of 12 GeV proton

We have

$$KE = mc^2 + m_0c^2$$

$$KE = \frac{m_0c^2}{\sqrt{1 - \frac{v^2}{c^2}}} - m_0c^2$$

$$= m_0c^2\left[\frac{1}{\sqrt{1 - \frac{v^2}{c^2}}} - 1\right]$$

Substituting for KE and m_0c^2, we get

$$\frac{12{,}000}{938} = \frac{1}{\sqrt{1 - \frac{v^2}{c^2}}} - 1$$

or

$$\frac{1}{\sqrt{1-\frac{v^2}{c^2}}} = \frac{12{,}000}{938} + 1 = \frac{12{,}938}{938} \approx 13.8$$

or

$$\sqrt{1-\frac{v^2}{c^2}} = \frac{1}{13.8}$$

or

$$1-\frac{v^2}{c^2} = \frac{1}{190.44}$$

or

$$\frac{v^2}{c^2} = 0.9947$$

or

$$\frac{v}{c} = 0.9973$$

Therefore,

$$v \approx 0.9973 \times c$$

So, in our calculations we can take $v \approx c$. Substituting various values in the relation $r = \frac{mv}{Bq}$, we get

$$r = \frac{13.9 \times 1.672 \times 10^{-27} \times 3 \times 10^{8}}{1.9 \times 1.602 \times 10^{-19}}$$

$$= 22.91 \text{ m}$$

Unsolved Problem

1. The principle of colliding beam accelerator is similar to the problem discussed below:

(i) A 1500 kg automobile travelling at its maximum speed of 150 km/hr crashes head on into a stationary identical automobile. The speed of both the cars locked together immediately after collision is 90 km/hr. How much energy has gone into the bending and heating of metal?

(ii) Now consider the case when the two automobiles identical to the two given in part (i) move towards each other at their maximum speed of 150 km/hr. After collision, the wreckage is at rest. How much energy has gone into the bending and heating of metal?

[Ans. (i) 3.65×10^5 J, (ii) 2.6×10^6 J]

REVIEW QUESTIONS

Short Answer Type

1. Explain the principle of cyclotron.
2. What is an electron synchrotron?
3. Explain the function of electric and magnetic fields in a cyclotron.
4. What are the limitations of a cyclotron?
5. What is the principle of Cockcroft Walton accelerator?
6. Why are particle accelerators required?
7. What is a betatron? How does it differ from a cyclotron?

Long Answer Type

1. What is a cyclotron? Write the principle, construction, working and theory of cyclotron.
2. What are the limitations of a cyclotron?
3. What is a synchrotron? Give constructional details of a proton synchrotron. Explain its working and theory.
4. Why cannot electrons be accelerated in cyclotron, but can be accelerated in betatron and electron synchrotron? Explain.
5. Discuss the construction, theory and working of a linear accelerator.
6. Explain the theory of betatron and obtain the betatron condition.
7. Why cannot a cyclotron accelerate electrons? How will you accelerate electrons? Discuss in detail the machine used to do so.
8. What is the advantage of cyclic accelerators over linear accelerators?
9. Describe the Cockcroft Walton machine for acceleration of charged particles.
10. What is the principle, construction and working of a betatron? Obtain the betatron condition for acceleration of electrons. Also highlight its advantages.
11. Explain the principle of betatron.
12. What is the difference between a cyclotron and synchrocyclotron?
13. What is meant by particle accelerator? Describe Cockcroft Walton voltage multiplier. Give its principle and usefulness.
14. Discuss the principle, construction, working and theory of a cyclotron. Derive an expression for the maximum kinetic energy achieved by a particle of mass m in terms of applied magnetic field and dee radius. Discuss its limitations.
15. What is a betatron? Derive the betatron condition for successful acceleration of electrons. Describe its principle, construction and function of alternating magnetic field on it.
16. What is the principle of betatron and what is its major advantage?
17. Explain the principle of operation of a synchrotron.
18. Discuss the construction and principle of a betatron.
19. Explain the principle and working of a betatron.
20. Explain the design and principle of a synchrocyclotron.
21. Give an account of principle of working of a cyclotron. Discuss the limitations of energy that can be obtained by this machine and its possible improvement.

Chapter 7

Radiation Detectors

7.1 INTRODUCTION

If we perform any nuclear physics related experiment or apply nuclear science to any problem, nuclear radiation detectors or radiation detectors or simply detectors play vital role in such measurements. Nuclear radiation is a general term and it includes a variety of energetic particles like electrons, protons, α-particles, heavy ions or neutral radiations like neutrons, X-rays or γ-rays, etc. The development of radiation detectors started with the discovery of radioactivity by Henry Becquerel in 1896. He noticed that the radiations emitted by uranium salts blacken photosensitive paper. Almost at the same time Roentgen discovered X-rays. X-rays were also found to blacken photosensitive paper. So, the first radiation detector was a photosensitive paper or X-ray film and was extremely simple. In the beginning of the twentieth century, Rutherford used flashes of light or scintillations produced in ZnS as nuclear radiation detector. These simple detectors used at that time were very primitive. They could simply indicate the presence or absence of radiations.

Nowadays, it is not sufficient only to detect the presence or absence of radiations but one would also like to know the nature of radiations, i.e. whether the radiations are electrons, protons, α-particles, X-rays, γ-rays, etc. On top of that, accurate energy and momentum measurements are often required. In some applications an exact knowledge of the spatial coordinates of the particle trajectories is also of interest.

There are number of ways in which we can classify the detectors. One simple way of classifying various detectors is by what type of signal is provided by the detector. This signal can be an electrical signal or visible light, correspondingly we classify detectors as electrical, which are based on ion collection or optical which are based on the visible light emitted by the detector. Also signal can form the image of the trajectory of charged particle. This classification is shown in Figure 7.1.

Nuclear particles or radiations cannot be directly detected but rather only through their interactions with matter. In case of X-rays and γ-rays, the main interaction processes are photoelectric effect, Compton effect and pair production. These processes have already been discussed in Chapter 5. The electron produced in these processes can be observed through their ionization or excitation in the sensitive volume (the volume in which particle is detected) of the detector. In case of charged particles main mode of interaction with matter is ionization and

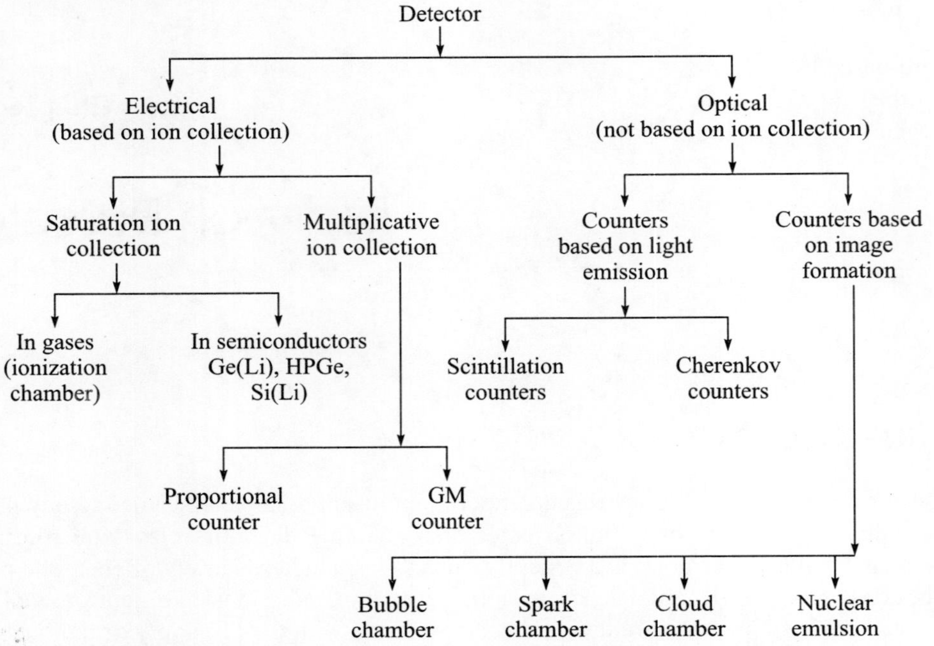

Figure 7.1 Classification of detectors according to signals produced.

excitation of electrons in the matter. In ionization the incident particle transfers an amount of energy equal to the ionization energy of the atom/molecule to permit the ionization process to occur. In this process an ion–electron pair is created. In most of the materials used for radiation detectors, the ionization energy for the least tightly bound electron is between 10 eV and 20 eV.

In excitation process, an electron is elevated to a higher bound state in the atom/molecule without completely removed. Later on, the excited electron may emit visible light and return to its original state.

In this chapter, we shall discuss the following radiation detectors:

1. Gas-filled detectors:
   (i) Ionization chamber.
   (ii) Proportional counter.
   (iii) Geiger–Müller (GM) counter.
2. Scintillation detectors.
3. Solid state detectors.

At the end of this chapter, some specialized detectors such as bubble chamber, spark chamber, cloud chamber, nuclear emulsions, etc. are discussed briefly. First *gas-filled detectors* are discussed.

7.2 GAS-FILLED DETECTORS

Gas-filled detectors are some of the oldest and most widely used detectors for nuclear radiations.

7.2.1 Principle

They are based on the principle of ionization and excitation caused by charged particles while passing through a gas.

A schematic diagram of a gas-filled detector is shown in Figure 7.2.

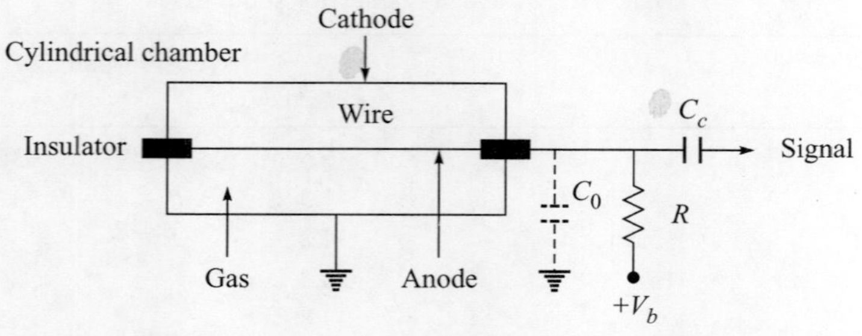

Figure 7.2 Schematic diagram of a gas-filled detector.

7.2.2 Construction and Working

It consists of a cylindrical gas-filled chamber. A thin wire is placed along its axis and is well insulated from the walls of the chamber. Through an external resistance R a positive potential is applied to the central wire, which acts as anode. The cylindrical chamber is grounded and it acts as a cathode. Capacity of the electrodes and stray capacities of the connecting wires, etc. constitutes the total capacity C_0 as shown in Figure 7.2. When a nuclear radiation (like electrons, α-particles, etc.) enters the chamber, it ionizes the gas present in the chamber thus creating number of positive ions and electrons called *ion-pairs*. If there is no electric field present, i.e. $V_b = 0$, the ion-pairs just created recombine forming neutral atoms/molecules. In the presence of the applied electric field ($V_b > 0$), the positive ions move along the radial electric lines of force i.e. towards the cathode or outer walls of the chamber. Similarly, electrons move towards the anode or central wire. Electrons being lighter than positive ions move at a much higher drift velocity ($\sim 10^6$ cm/s). The net effect of this is that a charge Q gets collected on the anode, and this charges the capacitor C_0 to a potential of Q/C_0. This change in the potential drops across R and gives rise to an electric pulse. Thus, when a charged particle passes through the gas present in the detector, the detector gives rise to a pulse, which is processed by external electronic circuit.

7.2.3 Concept of Average Energy Required for Creating Electron–Ion Pair (*W*-value)

We have already stated that a charged particle interacts with gas either through ionization or excitation. In ionization an electron–ion pair is created. The electrons so liberated when they reach the anode give information that a radiation has entered the detector. In excitation no such pair is formed. The energy consumed in excitation is wasted, as during excitation no electron reaches the anode. Therefore, the average energy lost by the incident particle in creating one ion-pair (defined as W-value) is always greater than the ionization energy of that gas. The W-value is a function of the gas present in the detector, type of radiation and its energy. However, experiments show that W-value is not a strong function of any of these parameters. Table 7.1 shows ionization potential I_p and W-value for fast electrons and α-particles for some of the commonly used gases in the radiation detectors.

TABLE 7.1 I_p and W-values for some gases

Gas	I_p (eV)	W (eV/ion-pair)	
		Fast electrons	*α-particles*
H_2	15.6	36.5	36.4
He	24.5	41.3	42.7
N_2	15.5	34.8	36.4
Ar	15.7	26.4	26.3
Air	–	33.8	35.1
CH_4	14.5	27.3	29.1

In most of the cases W lies between 25 eV and 35 eV per ion-pair.

Let us calculate the number of ion-pairs created, when a radiation of particular energy passes through a given gas. Suppose, 1 MeV α-particle is completely stopped within the gas with W-value = 30 eV/ion-pair. Then, the number of ion-pairs created

$$= \frac{10^6 \text{ eV}}{30 \text{ eV/ion-pair}}$$

$$\approx 33{,}000 \text{ ion-pairs}$$

If 3 MeV α-particle is completely stopped within the gas, it creates = 3 × 33,000 ≈ 99,000 ≈ 1,00,000 ion-pairs. This shows that the energy deposited by the incident radiation is directly proportional to the number of ion-pairs created. Hence, there is a possibility of determining the energy of incident radiation, if we can determine the number of ion-pairs created.

The variation of the logarithm of the number of ion-pairs formed or pulse height, which is across the resistor R, with applied voltage V for a gas detector is sketched in Figure 7.3. Two curves are shown in the figure corresponding to two different amount of energies brought in the gas. Lower curve marked as E_1 is for less-energetic particles while the upper curve marked as E_2 is for higher energy particles entering the gas detector. In the diagram symbols (o and Δ) are the measured values and line through the points is only a guide for the eye. Both these curves have been drawn when the pressure inside the chamber is about 0.5 torr (1 torr = 1 mm of Hg) and the spacing between anode and cathode is 5 mm. In this diagram there are five regions marked as I, II, III, IV and V. Details of various regions are given as under.

Region I (~0 V *to* ~30 V)

In this region, the applied voltage is not sufficient to overcome the recombination of ion-pairs formed. As voltage is increased from 0 V to 30 V, more and more electrons start reaching the central anode and hence the pulse height is increasing in this region.

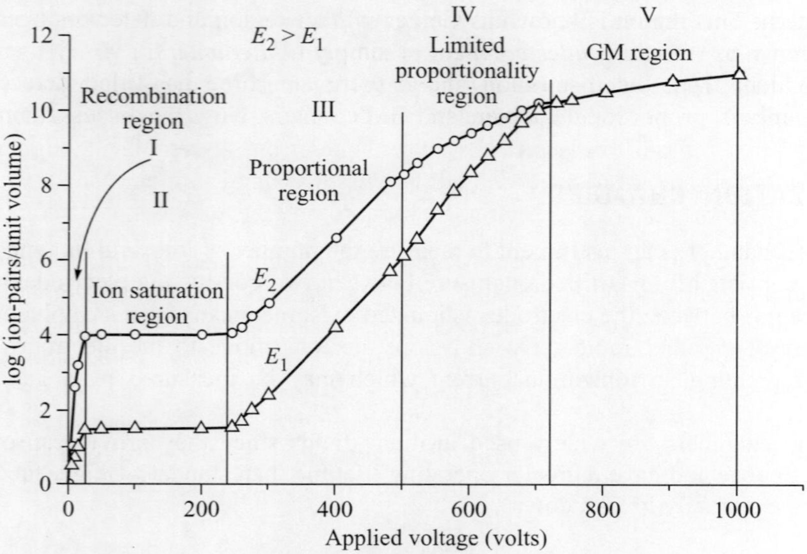

Figure 7.3 Variation of logarithmic pulse height or logarithm of number of ion-pairs formed versus applied voltage. Lower curve marked E_1 is for less energetic particles.

Region II (~30 V *to* ~250 V)

In this region, the curves are almost flat which signifies the collection of all the ion-pairs formed initially. Depending upon the energy of incident radiation about 10^1 to 10^4 ion-pairs are formed in this region due to primary radiation and they all are collected by the respective electrodes resulting in flattening of the curve. This region is known as *ionization region* and detectors operating in this region are called *ionization chambers*. The output pulses produced are of low amplitude, of the order of few mV.

Region III (~200 V *to* ~500 V)

Here, the production and collection of the ion-pairs increase rapidly with the applied voltage (this phenomenon is known as *gas multiplication* and is discussed later on). So, as long as the curves remain approximately parallel, the charge collected is proportional to the amount of the charge produced in the initial event which in turn is proportional to the energy of the incident radiation. This region is known as *proportional region* and the detectors operating in this region are known as *proportional counters*.

Region IV (~500 V *to* ~700 V)

In this region, larger amount of ionization is produced, but with less discrimination between the ion-pairs due to two different energy radiations. This region is known as *region of limited proportionality* and generally no detector operates in this region.

Region V (~800 V *to* ~1000 V)

In this voltage range, there is approximately flat region, also known as plateau and the number of ion-pairs formed per unit volume is ~10^9 to 10^{10}. This number is independent of the amount of initial ionization. This region is known as Geiger–Müller region and detectors operating in this region are known as *Geiger–Müller counters* or simply *GM counters*.

As is evident from the discussion above there are three gas-filled detectors, namely ionization chambers, proportional counters and GM counters. Now, we discuss them one by one.

7.3 IONIZATION CHAMBER

An ionization chamber is an instrument to measure the number of ions within a medium usually gas or air. It consists of a gas-filled enclosure between two conducting electrodes.

When the gas between the electrodes is ionized by some means such as alpha particles, beta particles, X-rays, or other radioactive emission, the ions move to the electrodes of opposite polarity, thus creating an ionization current which may be measured by a galvanometer or electrometer.

Ionization chambers are widely used in many fields since they provide an output that is proportional to dose and have a greater operating lifetime than standard Geiger tubes (in Geiger tubes, the gas eventually breaks down).

7.3.1 Principle

The ionization chamber works on the principle that charged particles passing through the matter remove electrons from the atoms as they move along (process called ionization). If voltage is applied across the material that is ionized, the electrons drift to one side and the leftover positively charged ions drift to the other.

7.3.2 Construction

Schematically an ionization chamber is shown in Figure 7.4. It is in the form of a rectangular box. It consists of two parallel plate electrodes separated by a distance d. These electrode combinations form a capacitor with capacity C. High voltage is applied across these plates through a large bias resistance R. This voltage sets up an electric field across the two plates. These plates are enclosed in a chamber and the chamber is filled with a desired gas. The most common gas in ionization chamber is air. Other gases used in such a detector are He, Ne, isobutene, etc. In some cases, ionization chamber is of cylindrical shape. The cylinder is made up of conducting metals like copper which acts as cathode. A straight thin wire passes in the centre of the cylinder and is isolated from it. It acts as anode.

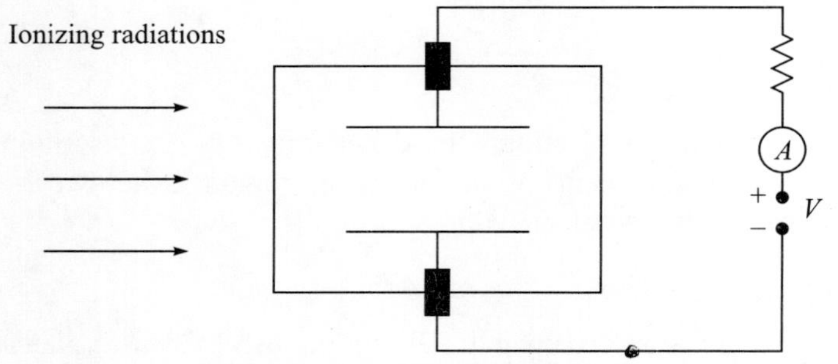

Figure 7.4 Ionization chamber.

7.3.3 Working

When incident radiations enter the ionization chamber, they ionize the gas present in the chamber, producing electrons and positive ions together known as electron-ion pairs. Under the influence of the electric field, electrons and ions move towards their respective electrodes. For a voltage gradient of 100 V cm^{-1}, ions at a speed of about 1 m/s in air at STP take about 0.02 s to cross a gap of ~2 cm. Electrons being lighter move about 1000 times faster and are collected at anode quickly. The slow movement of positive ions means that the ionization chamber is inoperative till all the +ve ions get neutralized at the cathode.

The electrons collected at anode constitute a current in the circuit. The number of electron-ion pairs produced in the gas and hence the current in the circuit is calculated as under.

Consider that a 3.5 MeV α-particle is moving in an ionization chamber, filled with air. Air has W-value ~35 eV. This α-particle produces

$$\sim \frac{3.5 \times 10^6}{35} \approx 10^5 \text{ ion-pairs.}$$

Let us suppose that the source is emitting about 10^4 α-particles per second. These particles will produce a current of

$$1.6 \times 10^{-19} \times 10^5 \times 10^4 \approx 1.6 \times 10^{-10} \text{ A}$$

This current, though small, can be measured with most of the current meters.

7.3.4 Main Uses

Ion chambers have a good uniform response to radiation over a wide range of energies and are the preferred means of measuring high levels of gamma radiation. They are widely used in the nuclear power industry, research labs, radiography, other medical radiation fields, and in environmental monitoring.

They are also used as radiation survey instruments, for calibrating or measuring the activity of radioactive sources, measuring radioactive gases, etc.

7.3.5 Main Drawback

These detectors are not suitable to measure individual particles entering the chamber at a higher rate.

7.4 PROPORTIONAL COUNTERS

A proportional counter is a device to count particles of ionizing radiation and measure their energy. It works on the same principle as the Geiger–Müller counter, but uses a lower operating voltage. A proportional counter differs from an ionization chamber in that the operating voltage is sufficiently high that the drifting electrons gain enough energy over a mean free path to create further ion-pairs when they collide with other neutral atoms of the gas.

By measuring the total charge (time integral of the electric current) between the electrodes, the particle's kinetic energy can be calculated.

7.4.1 Principle

Proportional counters use the principle that nuclear radiations while passing through a medium ionize the medium.

7.4.2 Construction

The proportional counter is another gas-filled detector that was introduced in the late 1940s. These detectors work in the proportional region III (Figure 7.3). The proportional counters are available in a variety of shapes and sizes. The most commonly used proportional counters are cylindrical in shape. The cylinder is made up of metal, which acts as a cathode. In some cases, this cylinder is made up of glass, the inner walls of which are coated with a conducting metal like copper or silver. In the centre of the cylindrical cathode, a straight conducting wire, which is insulated from the outer cylinder, acts as anode. The schematic view of the proportional counter is shown in Figure 7.5. Usually a mixture of gases is used containing gases like neon, argon, krypton, etc. which favor high amplification and a complex gas like carbon dioxide, methane, pentane etc., which helps in stopping the phenomenon of secondary ionization. Common proportional counters used in the laboratories contain a mixture of about 90% argon and 10% methane. The total gas pressure in the tube is generally less than one atmosphere. A thin glass, mica or polymer window (~10 μm) isolates the gases present in the proportional counter from the atmosphere.

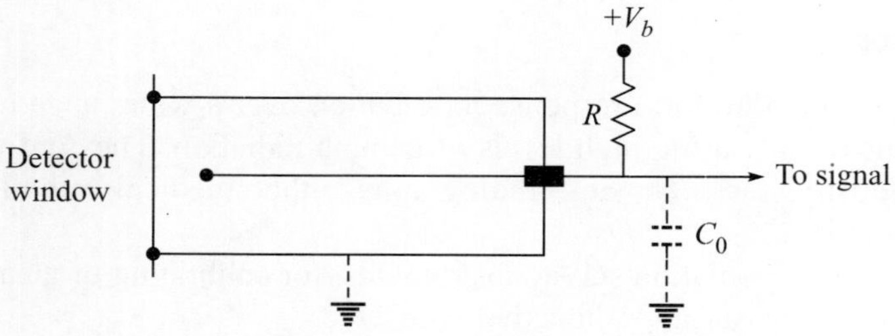

Figure 7.5 Schematic diagram of a proportional counter.

7.4.3 Working

As discussed earlier, the output current signal in the ionization chambers is very low. One common method of amplifying the signal is to increase the electric field till electrons gain sufficient energy between two successive collisions with gas atoms to cause further ionization.

A high voltage (V) applied to anode is so adjusted that the tube operates in the proportional region (Figure 7.3). For cylindrical geometry, the strength of electric field at a radial distance r from the central wire is given by

$$E(r) = \frac{V}{r \log\left(\frac{b}{a}\right)}$$

where a and b are the radii of the central wire and the inner radius of the cylinder, respectively. The electric field in the neighbourhood of anode wire is extremely high. For example, if $a = 0.005$ cm, $b = 1.0$ cm, and $V = 2000$ V, then

$$E(r) = 5.18 \times 10^6 \text{ V/m near the wire.}$$

In proportional counters, cylindrical geometry is preferred. For parallel plate geometry when spacing between the plates is 1 cm, then to get this field gradient, voltage of about 51,800 V is required for this field gradient. This is practically unworkable.

In such a high electric field gradient, a phenomenon called *gas multiplication* sets in. This can be explained as under.

Gas Amplification

Because of the increasing voltage, the electrons librated by the primary ionizing event get sufficient kinetic energy. These energetic electrons while moving towards the anode collide with neutral gas atoms, causing ionization of the neutral atoms. The electrons so produced may cause further ionization. This multiplication effect is often called a *Townsend avalanche* or *Townsend cascade*. The total number of ion-pairs produced by a single primary ion-pair is called the *gas multiplication factor*. This factor is unity in ionization chamber and is 10^2 to 10^4 in proportional counters. For example, if 10^4 ion-pairs are originally created by the incident particle and are multiplied by a factor of 1000, then instead of 10^4 pairs now the detector has 10^7 ion-pairs.

At such a high field, the electrons acquire sufficiently high velocities. For example, if the field gradient is 10^4 V m^{-1}, the velocity of the electrons is approximately 4×10^4 m/s. With such high velocities, electrons generated near the cathode take less than 1 μs to reach the anode. Thus, the proportional counter works at a much faster rate compared to the ionization chamber.

A proportional counter has two main features. First because of its built-in amplification, it can conveniently detect low energy radiations and second because of its proportionality, one can produce an energy spectrum of the incoming radiations.

Figure 7.6 shows a typical low energy X-ray spectrum obtained with a proportional counter. This shows two peaks, first at 22.1 keV due to K_α line of Ag and second at 25.0 keV due to K_β line of Ag. One important parameter used in such detectors is Energy Resolution or Full Width at Half Maximum (FWHM). From the experimental spectrum, it is measured as under. Let us consider 22.1 keV peak. The maximum counts at this peak are 2000. The width of the peak at exactly half of 2000, i.e. 1000 counts, is about 1300 eV. This number 1300 eV is known as energy resolution or FWHM of this detector at 22.1 keV. The physical meaning of energy resolution is that if there are two peaks with energy difference greater than the resolution, the peaks are seen as two separate peaks; otherwise only a single peak is visible. For example, in the present detector where the energy resolution is 1300 eV, if we draw the spectrum with two X-rays at 22.1 the and 22.2 keV, the detector shows only a single peak.

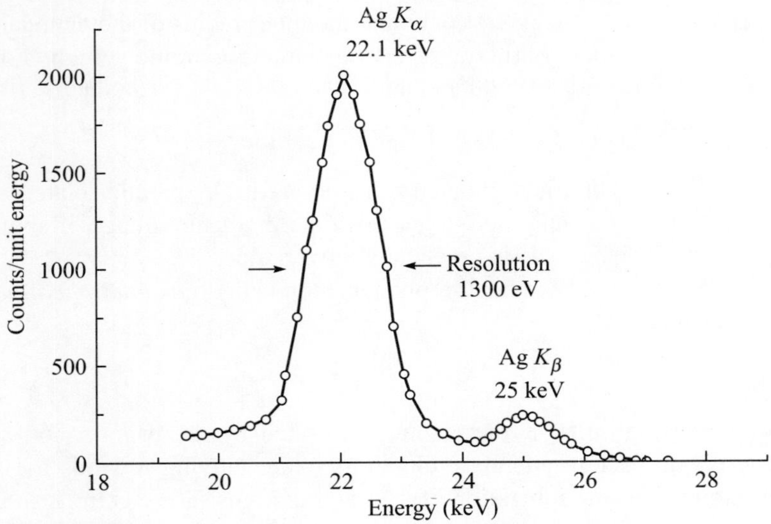

Figure 7.6 X-ray spectrum of Ag having two X-rays at 22.1 keV and 25.2 keV recorded with proportional counter.

7.4.4 Uses

The main use of proportional counters is detecting charged particles and low energy X-rays and recording their energy spectrum. $^{10}BF_3$-filled proportional counters are used for detecting slow neutrons. Slow neutrons produce $^{10}B(n, \alpha)^7$ Li reaction. α-particles ejected in the reaction are detected in the proportional counter.

7.4.5 Main Disadvantage

The main disadvantage of the proportional counter is the need for an expensive highly stabilized power supply to maintain the proportionality characteristics.

7.5 GEIGER–MÜLLER (GM) COUNTERS

If we continue to increase the voltage across the electrodes (Figure 7.3), then we pass from the proportional region through the region of limited proportionality to the GM region. In this case, the electron cascade spreads from the immediate neighbourhood of the initial track, until it encompasses the whole length of the anode. The resulting pulse is thus very much larger than that obtained from a proportional counter, but is obviously independent of the initial particle energy. This means that all pulses from a GM tube are of the same amplitude regardless of the number of original ion-pairs that started the process or is independent of the energy brought into the tube by the incident particle. Therefore, a GM counter cannot be used for measuring energy of the radiations.

This counter was named after Hans Geiger and W. Müller, who invented it in the 1920s. It is sometimes called a Geiger counter or a GM counter or GM tube and is the most commonly used portable radiation instrument. Although useful, cheap and robust, a counter using a GM tube can only detect the presence and intensity of radiation. Geiger counters are used to detect ionizing radiation, usually alpha and beta radiation and other types of radiation like low energy X-rays, etc.

7.5.1 Principle

It works on the principle that nuclear radiations while passing through the gas contained in GM counter ionize it.

7.5.2 Construction

The schematic diagram of a GM counter is identical to that of a proportional counter and is shown in Figure 7.5. In majority of the cases, The GM tube is cylindrical in shape. This cylinder is either made up of a conducting metal like copper, which acts as a cathode or it is made up of glass, in which a metallic cylinder is supported, which acts as a cathode. However, in some cases when the tube is of glass, inner sides of the tube are coated with a conducting metal like copper or silver. This coating acts as a cathode. Anode is generally made up of tungsten wire which stretches in the centre of the tube and is isolated from the cylindrical tube. The thickness of this wire is generally 0.02 to 0.05 mm.

Depending upon the requirement, the GM tubes are available in a variety of sizes. The diameters of the tubes vary from 0.3 cm to about 15 cm while their lengths vary from 1 cm to 100 cm.

Commercially available GM tubes are filled with 90% argon (as the filling gas) and 10% ethyl alcohol (as the quenching gas) or 0.1% either chlorine or bromine (as the quenching gas) and the rest is neon or argon gas. A thin glass, mica or polymer window (~10 μm) isolates the gases present in the GM counter from the atmosphere.

7.5.3 Working

The charged particles entering the GM tube ionize the gas present in the tube. This event is known as *primary ionization*. The electrons librated in the primary ionization are accelerated towards the central anode wire. Because of high potential present on the wire, the electrons may gain sufficient energy to cause further ionization of the neutral gas molecules (known as secondary ionization). This leads to a chain of ionizing events, which is already discussed and is known as Townsend avalanche.

Besides these processes, there is a possibility of excitation of molecules and atoms. Such excited molecules and atoms, while de-exciting, may emit ultra violet (UV) or visible photons. These photons again lead to the production of electrons due to ionization of gas atoms or molecules or due to photoelectric interaction with walls of the counter. Each such librated electron would again cause a Townsend avalanche. Such a series of avalanches would lead to discharge in the tube called Geiger discharge. In such a state, there is a formation of dense envelop of electron–ion pairs distributed around the anode. The voltage applied to anode collects all the electrons pertaining to single event leading to Geiger discharge.

Quenching

Practically the process is not as simple as discussed above. During the Geiger discharge, there is a dense envelop of ions and electrons around the anode. The electrons would drift towards the anode and positive ions would drift towards the cathode. The positive ions drifting towards the cathode having ionization potential greater than the work function of the cathode material lead to exchange of electron from the cathode and become neutral. The excess energy may be dissipated with the emission of either a photon or an electron from the cathode the latter process takes place only if excess energy is greater than the work function of the cathode material. This again initiates another Geiger discharge. The result of this is that the tube would always be in continuous Geiger discharge and hence is not able to measure any radiation.

To overcome this problem, the concept of quenching is introduced. There are two types of quenching:

1. Organic quenching
2. Halogen quenching

Organic quenching: This involves addition of small quantity of organic gas having a complex molecular structure. This prevents the continuous Geiger discharge by charge transfer collision principle. The positive ions on their path collide with organic molecules to get neutralized. This makes only ions of organic gas to reach the cathode and get neutralized. Any excess energy released would lead to dissociation of organic molecules. Thus, multiple Geiger discharge could be avoided.

A typical filling of organic quenched GM tubes is 90% argon and 10% ethyl alcohol. When the organic gas gets depleted to a sufficient extent, there is the occurrence of multiple discharges frequently and thus the plateau length gets decreased with slope increased.

The organic quenched GM tubes are characterized by short life time and are not suitable for operation in very high fields which leads to large count rate. To overcome this, the technique of halogen quenching was introduced.

Halogen quenching: This involves the addition of small quantity of halogen gas such as chlorine or bromine. A typical filling is about 0.1% of chlorine to argon or neon. The quenching action is the same as that of the organic quenching process. The diatomic halogen gas molecules also get dissociated in quenching but get recombined to replenish the gas molecules and thus the counter life gets extended.

Characteristics of GM Tubes

The important parameters which decide the quality of functioning of GM tubes are as follows:

1. Dead time
2. Recovery time
3. Geiger plateau length and plateau slope

Dead Time: As discussed above, the positive ions take considerable time to reach the cathode compared to electrons. The reason is that the mobility of electrons is about 1000 times greater than that of positive ions. Due to low drift velocity of positive ions, there is a formation of cloud of positive ions. The electric field of these positive ions forms a sheath of positive ions (space charge) close to the anode and reduces the intense electric field sufficiently. Thus, approaching electrons do not gain sufficient energy to start new avalanches. The detector is then inoperative (dead) for the time required for the ion sheath to migrate outwards far enough for the field gradient to recover above the avalanche threshold. The dead time of a Geiger tube is defined as the time between the initial pulse and the time at which a second Geiger discharge (regardless of its size) can be developed. In most Geiger tubes, this time is of the order of 50–100 μs.

Dead time is the time immediately following the introduction of a particle during which the counter tube is insensitive for detection of another particle.

Recovery time: After the dead time is over, some of the electrons and +ve ions recombine and form neutral atoms. During this time, when a new ionizing radiation enters the tube, pulses of lower amplitude are formed, which cannot be counted by the counter.

The total time taken for the tube to recover to its fully sensitive state to give the next pulse is called *recovery time*.

The sum of the *dead time* and *recovery time* is called *Resolving time*.

The dead time, recovery time and resolving time are shown in Figure 7.7. During the dead time, no new pulses are formed. However, during the recovery time, pulses of lower amplitude are formed, which are not counted by the counting system. It is only after the recovery time that the pulses of larger amplitude are formed, capable of getting counted.

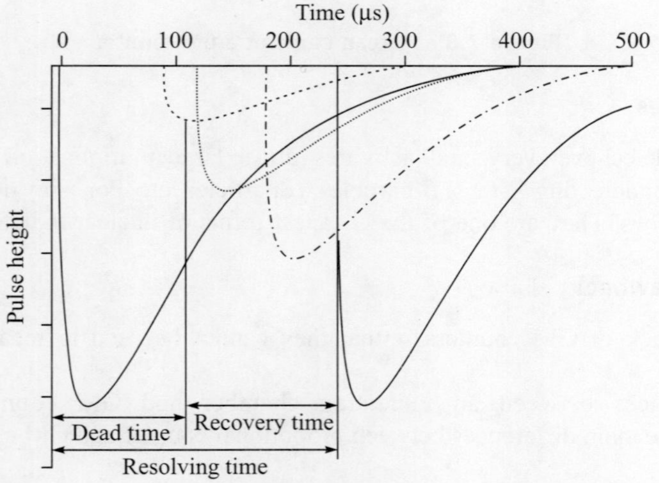

Figure 7.7 Relation between dead time, recovery time and resolving time of a GM counter.

Geiger plateau: In Figure 7.8, a plot of count rate versus the anode voltage is shown. In this figure, the source of radiations is kept at a fixed distance from the counter. The potential difference at which the counting just starts is called threshold potential. The typical threshold potential depends upon the gas and its pressure in the counter. It generally varies between 400 and 900 volts. After a very sharp rise, the counting rate remains almost constant with increasing voltage. *This flat region, the plateau, where the number of counts recorded per second remains independent of applied voltage is called Geiger region.* The length of the plateau is about 100 to 300 volts. When the plateau region becomes shorter and steeper, it means that the

GM tube is nearing the end of its useful life. If a Geiger–Müllar counter is operated in this region, it eliminates the need of a costly and highly regulated power supply.

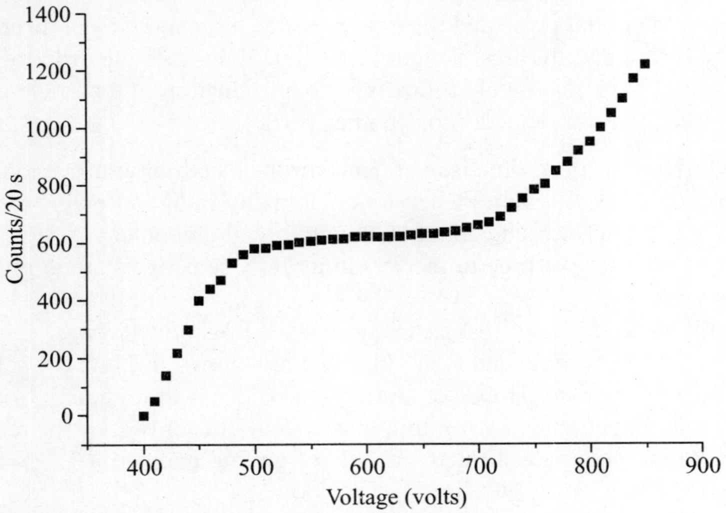

Figure 7.8 Plateau curve of a GM counter.

7.5.4 Main Uses

GM counters can detect even very small activities (due to large multiplication factor) and they are extensively used for detecting X-rays, β-particles, α-particles, etc. For γ-ray detection, we require bigger size GM tubes. They are one of the cheapest forms of nuclear radiation detectors.

7.5.5 Main Drawback

The major drawback of GM counters is that they cannot be used to measure the energy of radiations.

Main differences between an ionization chamber and GM counter are shown in Table 7.2, while the main differences between propotional counter and GM counter are shown in Table 7.3.

TABLE 7.2 Differences between ionization chamber and GM counter

S. No.	*Ionization chamber*	*GM counter*
1.	It operates in the ionization region (region II, Figure 7.3).	It operates in the GM region (region V, Figure 7.3).
2.	It operates at relatively low voltages (~ 30 V to ~ 250 V).	It operates at much higher voltages (~ 800 V to ~ 1000 V).
3.	The output pulse height is low, so an amplifier is needed.	The output pulse height is large, so no amplifier is needed.
4.	Power supply used to feed voltage to ionization chamber must be highly regulated.	Power supply used to feed voltage to a GM counter need not be regulated.

TABLE 7.3 Differences between proportional counter and GM counter

S. No.	*Proportional counter*	*GM counter*
1.	It works in the proportional region (region III, Figure 7.3).	It works in the GM region (region V, Figure 7.3).
2.	It generally operates at lower voltages (~200 V to ~500 V).	It generally operates at higher voltages (~ 800 V to ~1000 V).
3.	Output pulse height depends upon the energy of the incident particle.	Output pulse height do not depends upon the energy of incident particle.
4.	Energy of the incident radiation is measured by it.	Energy of the incident radiation cannot be measured by it. However, intensity/flux of incident radiation can be measured by it.
5.	The output pulse height is low, so an amplifier is needed.	The output pulse height is large, so no amplifier is needed.
6.	Power supply used to feed voltage to proportional counter must be highly regulated.	Power supply used to feed voltage to a GM counter need not be regulated.

7.6 SCINTILLATION DETECTORS

We are familiar with ZnS screen used by Rutherford in his famous α scattering experiment. α-particles, when fall on ZnS, produce light flashes or scintillations. These flashes can be seen with a microscope. ZnS was the first scintillation detector put to actual use. Another most common and familiar example of the use of scintillation mechanism is the television screen, which emits visible photons under electron bombardment in a cathode ray tube.

It was found later on that along with ZnS, there are a large number of materials which exhibit the property of scintillations. Some of them are sodium iodide, cesium iodide, anthracene, naphthalene, etc. Some gases particularly noble gases like xenon, etc. also exhibit this property. Scintillation detectors became the most powerful and widely used radiation detectors with the development of photo multiplier (PM) tubes. These PM tubes are extremely sensitive detectors of light. Thus, there is no need of using a microscope to view the scintillations.

A scintillation spectrometer consists of three components:

1. Scintillation detector/crystal
2. Photomultiplier tube
3. Electronic circuitry

Below we discuss these components one by one.

7.6.1 Scintillation Detector/Crystal

Scintillators are available in many forms. Some commonly used scintillators are

1. **Inorganic crystals**
   - NaI(Tl): This is the most commonly used scintillator for γ-ray work. The efficiency for the conversion of the energy deposited into light output is one of the highest among all known scintillators. It is available in a variety of shapes and sizes. One drawback with this scintillator is that it is hygroscopic and therefore it has to be sealed in airtight containers. Another drawback of this scintillator is that its density is low (~ 3.67 g/cm^3).

- CsI(Tl) and CsI(Na): These crystals are non-hygroscopic and have light output compared to that of NaI(Tl). The only difficulty with these scintillators is that growing big single crystals is quite difficult.
- CaF_2(Eu): This again is a non-hygroscopic crystal and it is useful to detect low energy (<100 keV) X- or γ-rays.
- ZnS: It was the first scintillator used in nuclear physics. However, making big transparent single crystals is quite difficult.
- $Bi_4Ge_3O_{12}$ (BGO): It is non-hygroscopic, mechanically and chemically more durable compared to NaI(Tl). Its density is much larger (7.13 g/cm^3) than that of NaI(Tl) scintillator. This would enable us to use a smaller size crystal compared to NaI(Tl) crystal having the same efficiency. However, it has low light output and its energy resolution is inferior compared to that of NaI(Tl). They are more expensive than NaI(Tl) detectors.

2. **Organic scintillators:** Common organic scintillators are anthracence, stilbenek, etc. These scintillators have very poor efficiency compared to NaI(Tl) detector. However, in these scintillators, the decay time of light produced due to ionizing radiations is much shorter compared to that in NaI(Tl) detector. These detectors can be operated at higher count rate. These detectors are practically more suitable to β counting.
3. **Liquid scintillators:** In these detectors, scintillator material is dissolved in certain suitable liquid. In some applications, radioactive source is thoroughly mixed with liquid scintillator. These detectors are mostly used to count very low activities specially that of β-particles.

Many scientists have studied the mechanism of light emission by scintillators in detail. Without going into the detailed mechanism of light emission, certain terms commonly used while working with scintillation detectors are discussed.

The process by which visible light is emitted by the scintillator within 10^{-8} s of falling of radiations is known as *prompt fluorescence* or simply *fluorescence*. There is another mechanism of visible light emission known as *phosphorescence*, in which the light of longer wavelength is emitted and this process occurs in a time which is much longer (few seconds to few minutes). There is a third mechanism, which is known as *delayed fluorescence* in which the light emitted is of the same wavelength as in fluorescence but its emission time is longer (hours to days).

Another term commonly used in scintillators is the *scintillation efficiency*. It is defined as the fraction of the energy deposited by the incident radiation that is converted into light.

A good scintillator should have high scintillation efficiency and should convert a larger fraction of the emitted light by the process of prompt fluorescence. Single crystal of sodium iodide with traces of thallium written as NaI(Tl) is one such crystal, which is extensively used for detecting and measuring the energy of γ-rays. Tl in scintillators is also known as *wave length shifter*. The emission spectrum of pure sodium iodide has maxima at a wavelength of 303 nm. Most of the photomultiplier tubes are not sensitive to this wavelength. Addition of a small quantity of Tl (0.1 to 0.5% mole fraction of Tl) shifts this wavelength to 410 nm. Most of the photomultiplier tubes are sensitive to this wavelength.

Incident-charged particles falling on the scintillator lose all the energy in the scintillator and light photons are produced. However, if the incident radiations are X- or γ-rays, they interact with the scintillator in mainly three different ways (Chapter 5).

1. Photoelectric effect
2. Compton effect
3. Pair production

In all these three processes, electrons are produced and ultimately total energy of X- or γ-rays is converted into the kinetic energy of electrons. These electrons give up their kinetic energy in ionization and excitation in the scintillator material, resulting in production of scintillations in the scintillator.

Scintillators are available in a variety of shapes and sizes. Scintillators commonly used in laboratories have dimensions as 1″ × 1″ or 1½″ × 1½″. 1″ × 1″ means it is a cylinder of base diameter of 1″ and height of 1″, etc.

All the light produced in the scintillator should reach the photomultiplier tube. For this purpose, the scintillator is covered with light reflector MgO from all sides except one and this uncovered side is optically coupled to PM tube. Optical coupling means that the whole of the light photons produced in the scintillator reaches the photo cathode of the PM tube without any loss.

7.6.2 Photomultiplier Tubes

Photomultiplier tubes (PMTs) are extremely sensitive detectors of light in the ultraviolet, visible and near infrared. These detectors multiply the signal produced by incident light by as much as 10^6 to 10^8. The combination of high gain, low noise and large area of collection means that these devices find applications in nuclear and particle physics.

Photomultipliers are constructed from a glass vacuum tube which houses a photocathode, several dynodes and an anode as shown in Figure 7.9. Incident photons strike the photocathode of the PM tube, and electrons are produced as a consequence of the photoelectric effect. The photocathode is generally coated with either antimony-rubidium-cesium or antimony-potassium-cesium. These electrons are directed by the focusing electrode towards the electron multiplier, where electrons are multiplied by the process of secondary emission.

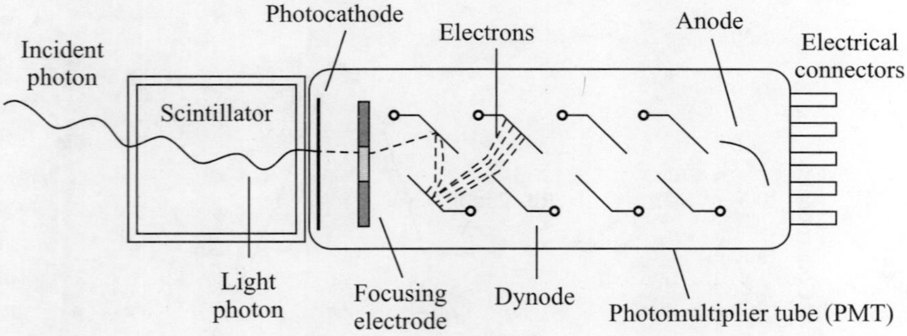

Figure 7.9 Photomultiplier tube.

The electron multiplier consists of a number of electrodes, called *dynodes*. Each dynode is held at a more positive voltage than the previous one (50 to 100 V). The electrons are ejected by the photocathode. As they move towards the first dynode, they are accelerated by the electric field and arrive with much greater energy. On striking the first dynode, more low energy electrons are emitted and these, in turn, are accelerated towards the second dynode. The geometry of the dynode chain is such that a cascade occurs with an ever-increasing number of electrons being produced at each stage. Finally the anode is reached where the accumulation of charge results in a sharp current pulse indicating the arrival of a photon at the photocathode.

7.6.3 Electronic Circuitry

A block diagram of a complete NaI(Tl) γ-ray spectrometer is shown in Figure 7.10. In this figure, pre-amp is a pre-amplifier. The purpose of the pre-amplifier is to match impedance between photomultiplier and amplifier. Amplifier amplifies the signal arriving at its input. The amplified output of the amplifier is in direct proportion to the energy deposited by γ-rays in scintillation detector. For example, if energy absorbed in the NaI(Tl) detector is 100 keV, the amplifier gives output pulse of 1 V (say) and if the energy deposited in the NaI(Tl) is 600 keV, the output pulse would be 6 V and so on. The SCA stands for single channel analyzer. It gives the number of pulses in a certain voltage interval, for example, the number of pulses between voltage interval 500 mV and 600 mV, etc., depending upon the setting of SCA. These pulses are fed to counter/timer, which counts the number of pulses for a fixed time. A typical γ-ray spectrum of ^{137}Cs emitting one γ-ray of 662 keV is shown in Figure 7.11. The resolution of the scintillation spectrometer at 662 keV is 75 keV. Similarly the γ-ray spectrum of ^{60}Co emitting two γ-rays of 1172 and 1332 keV is shown in Figure 7.12. The two peaks are clearly visible in the spectrum.

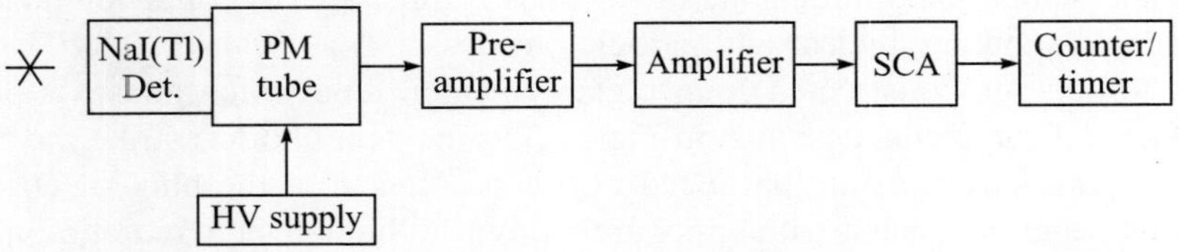

Figure 7.10 Block diagram of a scintillation spectrometer.

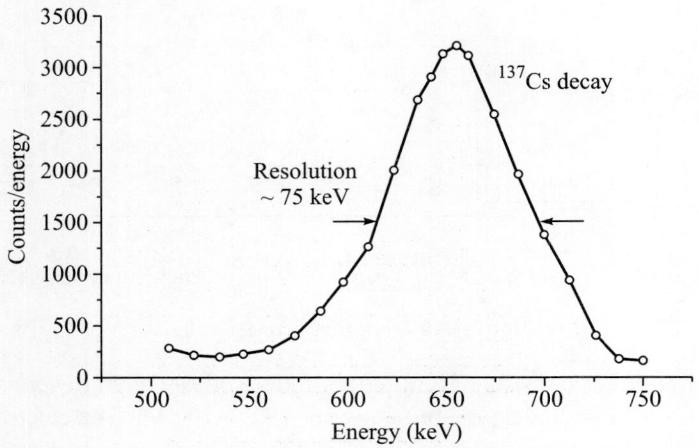

Figure 7.11 Photo peak (662 keV) of γ-ray spectrum of ^{137}Cs as recorded by a scintillation spectrometer. The photo peak shows an energy resolution of ~75 keV.

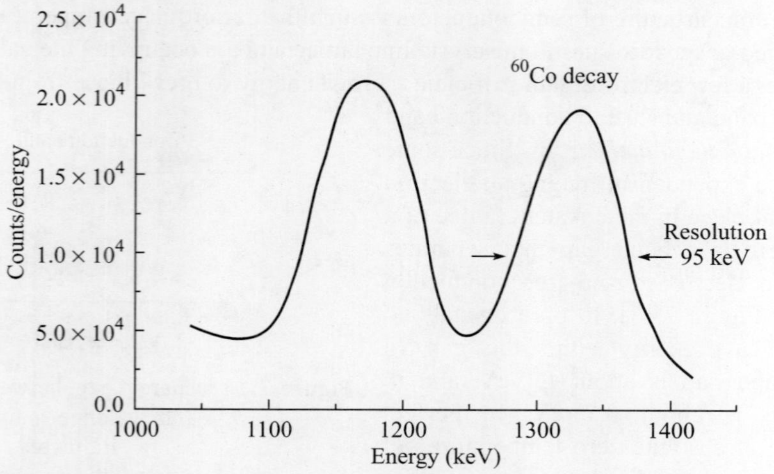

Figure 7.12 Photo peaks of γ-ray spectrum of ^{60}Co. It has two γ-rays of 1172 keV and 1332 keV.

7.6.4 Uses of Scintillation Detectors

- Inorganic scintillation detectors are used to measure the energy of X- and γ-rays.
- Organic scintillation detectors are used to detect β and α-particles.
- They generally have high efficiency. Efficiency is nearly 100% in case of NaI(Tl) detectors.
- They are rugged and can with mechanical jerks.
- Scintillation detectors have very short dead time ($\sim 10^{-9}$ s) permitting very high counting rates.

7.6.5 Limitations of Scintillation Detectors

- Energy resolution of these detectors is poor. For example, in case of NaI(Tl) detector, the energy resolution is 8–10% at 662 keV.
- The NaI(Tl) detector is hygroscopic. If it absorbs moisture, it is completely damaged.

7.7 SEMICONDUCTOR RADIATION DETECTORS

The field of nuclear physics has been completely revolutionalized with the development of semiconductor radiation detectors. A semiconductor radiation detector is a solid-state version of a gaseous ionization chamber. It is simply a reverse biased p-n junction diode. In these diodes, ionizing radiations produce electron-hole pairs in the depletion region. This process is just like creation of electron-ion pairs in ionization chamber. These pairs are collected by the electric field applied and thus the detector gives an electric pulse. The amplitude of this pulse is proportional to the energy of ionizing radiation. Since semiconductor detector is a reverse biased, therefore, the signal is much larger than the noise due to leakage current.

The electronic structure of semiconductors is such that, at ordinary temperatures, nearly all electrons are tied to specific sites in the crystalline lattice and are occupying the valence band. At any given time, a few electrons gain sufficient thermal energy to break loose from localized sites in the valence band and shift to conduction band and are called *conduction electrons*. Since some energy must be expended in freeing an electron from its normal place in the covalent lattice of a crystal as there is a band gap that separates bound valence electrons from free conduction electrons (see Figure 7.13). In pure crystals no electrons can have energy within this gap. In silicon the band gap is about 1.1 eV, and in germanium it is about 0.7 eV. In perfect materials held at absolute zero temperature, all electrons are theoretically bound to specific lattice sites, so that the valence band is completely filled and the conduction band is empty. The thermal energy available at ordinary temperatures allows some electrons to be freed from specific sites and be elevated across the band gap to the conduction band. Therefore, for each conduction electron that exists, an electron is missing from a normally occupied valence site. This electron vacancy is called a *hole*, and in many ways it behaves as though it is a point positive charge. If an electron jumps from a nearby bond to fill the vacancy, the hole can be thought of as moving in the opposite direction. Both electrons in the conduction band and holes in the valence band can be made to drift in a preferred direction under the influence of an electric field.

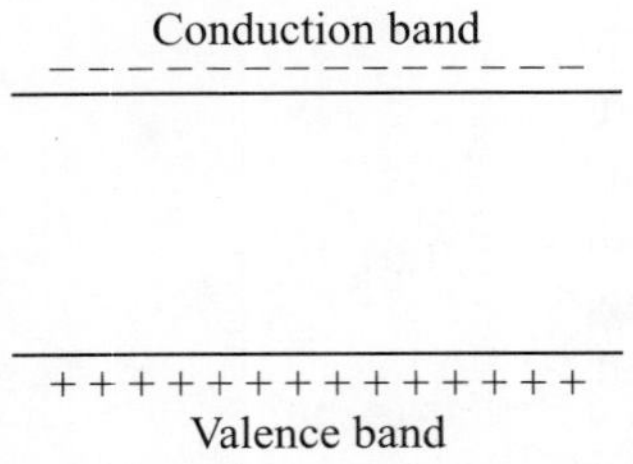

Figure 7.13 Energy gap between conduction and valence band for silicon. – sign indicates electrons and + sign indicates holes.

In semiconductor detectors, an electric field is present throughout the active volume, i.e. the volume of the detector or total volume of the depletion region, where incident radiations are detected. The subsequent drift of the electrons and holes toward electrodes on the surface of the semiconductor material generates a current pulse, in much the same manner as the motion of ion-pairs in a gas-filled ion chamber.

The minimum energy required for the creation of an electron-hole pair is the band-gap energy of about 1 eV. Experimental measurements show that, as in the production of an ion-pair in a gas, about three times the minimum energy is required on the average to form an electron-hole pair. Thus, a 1-MeV charged particle losing all its energy in a semiconductor creates about 3,00,000 electron-hole pairs. This number is about 10 times larger than the number of ion-pairs that would be formed by the same particle in a gas. As a consequence, the charge packet for equivalent energy loss by the incident particle is 10 times larger, thus improving the signal-to-noise ratio as compared with a pulse-type ion chamber. More significant is the improvement in energy resolution. The statistical fluctuations in the number of charge carriers per pulse (that often limit energy resolution) become a smaller fraction as the total number of carriers increases. Thus, semiconductor detectors offer the best energy resolution provided by common detectors, and values of a few tenths of a per cent are not uncommon.

Another benefit is derived from the fact that the detection medium is a solid rather than a gas. In solids, the range of heavy charged particles such as alphas is only tens or hundreds of micrometers, as opposed to a few centimetres in atmospheric pressure gases. Therefore, the full energy of the particle can be absorbed in a relatively thin detector. More importantly, it is practical to fully absorb fast electrons such as beta particles. As opposed to ranges of metres in gases, fast electrons travel only a few millimetres in solids, and semiconductor detectors can be fabricated that are thicker than this range. Therefore, spectroscopic methods can be employed to measure the energies of fast electrons.

Semiconductor detectors offer many advantages over scintillation or gas-filled detectors. Some of the advantages are:

1. Excellent energy resolution.
2. Linear response over wide-energy range of incident radiation.
3. These detectors are generally compact and are small in size.
4. These detectors are fast detectors. The mobility of electrons is about 1500 cm^2/V s and the mobility of holes is about 500 cm^2/V s, which is much faster compared to mobility ions in gas counters.
5. These detectors can be fabricated in a wide range of sensitive depth and geometry.

Initially, these detectors were developed for detection of heavy charged particles. Later on, with the advancement in semiconductor technology, detectors were developed for detection of electrons, X- and γ-rays. Let us discuss some commonly used semiconductor detectors in detail.

7.7.1 Diffused Junction Detector

In a *p*-type silicon crystal (doped with boron) with a resistivity of about 500–10,000 Ω cm, *n*-type impurity usually phosphorus is diffused. A *p-n* junction is formed at a depth of 0.1–2.0 μm from the surface. Generally, gold is evaporated on both sides of the crystal for good electrical connection. As shown in Figure 7.14, when this *p-n* junction is reverse biased, a depletion region is formed which is free from majority charge carriers. This depletion region is identical to a gas in the ionization chamber. It offers ideal conditions for the detection of ionizing radiation. Ionizing radiations entering the depletion region create electron–hole pairs as shown in Figure 7.15. Electrons and holes drift towards respective electrode due to applied field and a pulse is obtained in the output circuit.

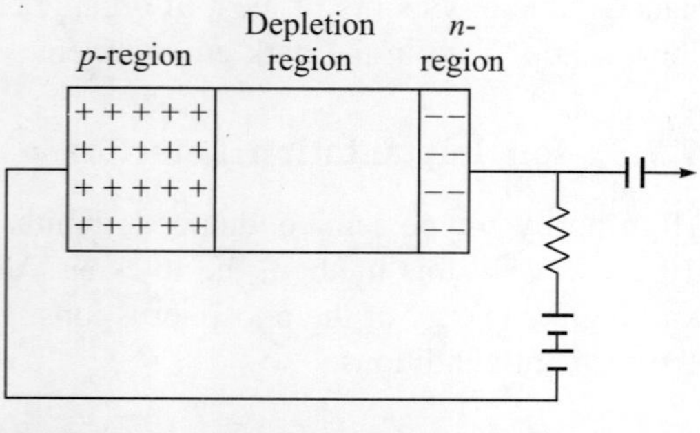

Figure 7.14 Reverse biased diffused *p-n* junction.

These detectors are rugged and are less prone to radiation damage. The main disadvantage of these detectors is that incident radiation has to pass through relatively thick *n*-type layer and the radiation may lose considerable amount of energy in the detector window.

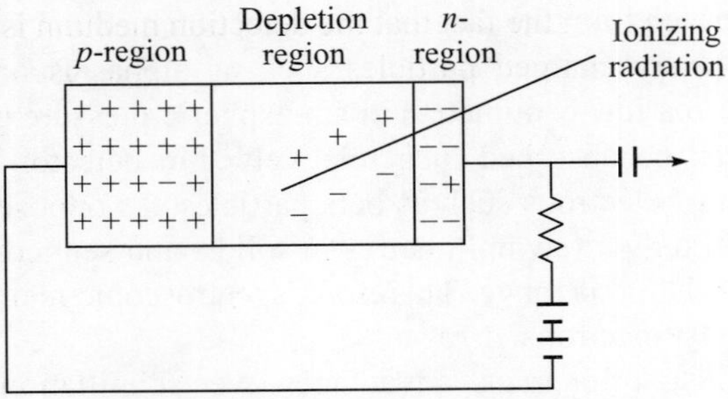

Figure 7.15 Ionizing radiation entering depletion region.

7.7.2 Silicon Surface Barrier Detectors

Silicon surface barrier detectors have extremely thin *p*-type window and are fabricated by using a high purity *n*-type silicon crystal. One face of a high purity *n*-type silicon is etched with acid and is exposed to air. An extremely thin layer of oxide is formed on this face which acts as *p*-type layer (also called *dead layer*) and is generally <0.1 micron thin. On this side a very thin layer (~20–40 $\mu g/cm^2$ or $1–2 \times 10^{-6}$ cm) of gold is evaporated and on the back side 200–400 $\mu g/cm^2$ or $70–150 \times 10^{-6}$ cm aluminium is evaporated for making electrical connections. They are small in size and have long-term stability. They are widely used in charged particle spectroscopy.

These detectors are less-rugged and need care while handling them. They are susceptible to damage due to vapours of most of organic chemicals. They are sensitive to light and, therefore, they are operated in the dark environment.

7.7.3 Ion Implantation Detectors

High purity *p*-type semiconductor is bombarded by say phosphorus ions accelerated to about 10–20 keV. In this method, the thickness of *n*-type dead layer can be precisely controlled by varying the energy of the phosphorus ions. These detectors are more stable and less sensitive to the ambient conditions.

7.7.4 Silicon Lithium Si(Li) Detectors

Till 1968, the techniques available could not reduce the impurity concentration in silicon to less than about 10^{12} atoms/cc. With these impurity concentrations, it was not possible to fabricate a good semiconductor detector. This situation was overcome by using special techniques. One such commonly used technique is to begin with *p*-type silicon and diffuse into its surface Li atoms, which tend to act as donor atoms and this creates a thin *n*-type region. This *p-n* junction is now put under reverse bias of about 500 V and at a slightly higher temperature (~120°C). This causes the Li to drift into the *p*-type region, making a quite thick (3–5 mm) depletion region. Such a detector is known as Si(Li) (pronounced as sili) detector. These detectors must be kept at liquid nitrogen temperature (–196°C or 77K), otherwise the Li will migrate out of its lattice sites, destroying the effectiveness of the detector. Keeping the detector at liquid nitrogen temperature also reduces the background electrical noise in the detector.

Si(Li) detectors are mostly used for measuring energy of X-rays. They offer excellent energy resolution in the X-ray region. Typical energy resolution of these detectors is of the order of 160–180 eV for 5.9 keV X-rays from ^{55}Fe. A typical X-ray spectrum of ^{55}Fe recorded with Si(Li) detector is shown in Figure 7.16. The energy resolution of this detector is 180 eV at 5.9 keV.

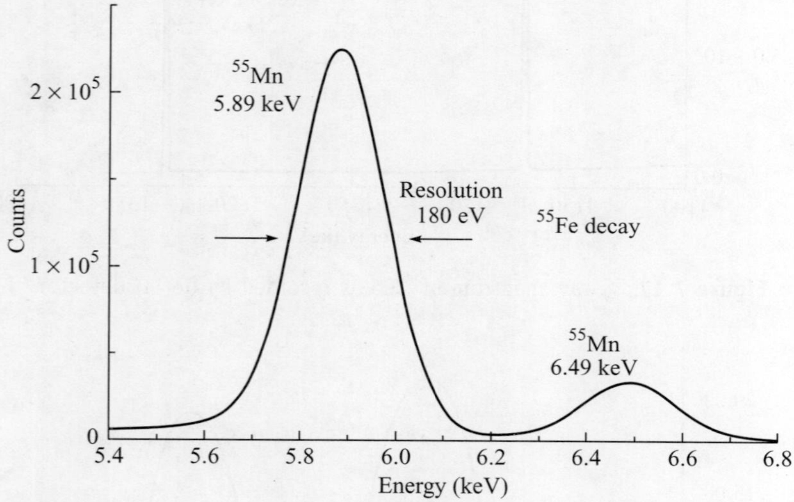

Figure 7.16 X-ray spectrum of ^{55}Fe as recorded by Si(Li) detector.

7.7.5 Germanium Lithium Ge(Li) Detectors

The photoelectric cross-section varies as Z^5, therefore, germanium is a better detector for high energy gamma rays compared to silicon detector. The method of fabrication of Ge(Li) detector is similar to that of Si(Li) detector. In Ge(Li) detectors large active volume (i.e. the total volume of the depletion region) has been obtained like Si(Li) detectors. Ge(Li) detectors with active volume of 5 cc to 100 cc are commonly available.

Like Si(Li) detectors, Ge(Li) detectors are always maintained at the liquid nitrogen temperature, while in use or when they are stored. If by mistake, Ge(Li) detector comes to room temperature, then due to migration of Li, the detector is permanently damaged. The energy resolution of Ge(Li) detector is ~1.8 keV for 1332 keV ^{60}Co γ-ray. The typical spectrum of ^{60}Co is shown in Figure 7.17. In order to visualize the comparative energy resolution of Ge(Li) and NaI(Tl) detectors, γ-ray spectrum from both these detectors under similar conditions is shown in Figure 7.18. The energy resolution of Ge(Li) detector is about 2 keV, whereas for Nai(Tl) the energy resolution is 70 keV.

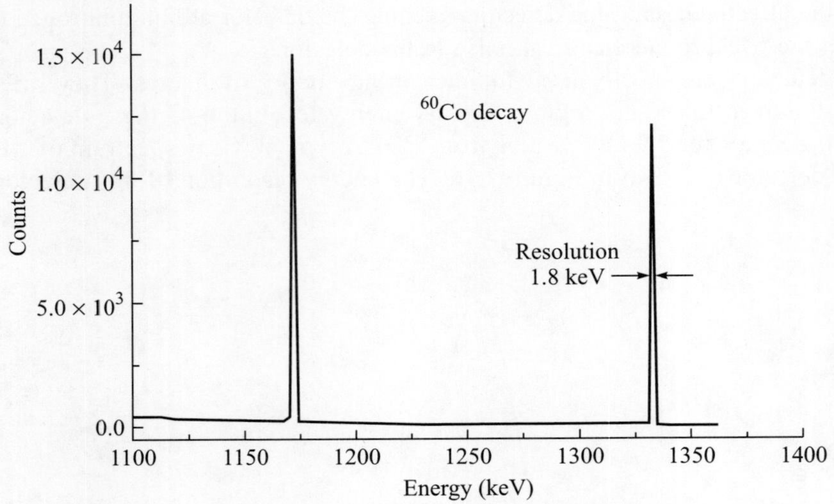

Figure 7.17 γ-ray spectrum of ^{60}Co as recorded by Ge(Li) detector.

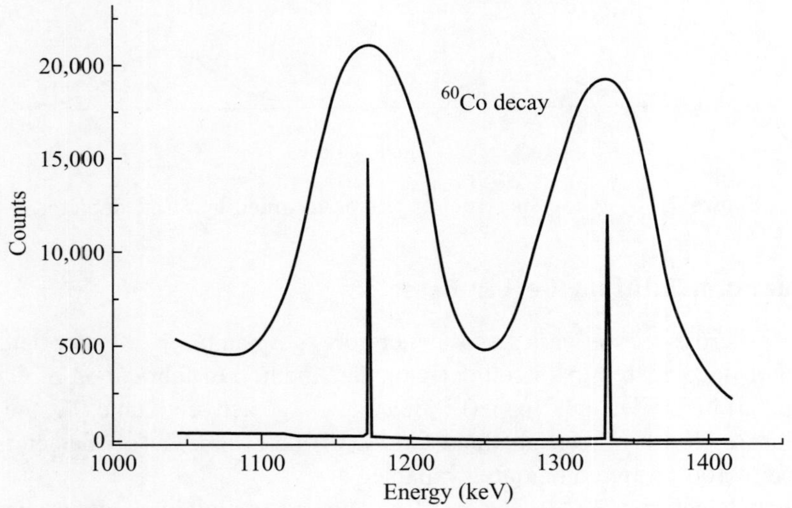

Figure 7.18 Comparative γ-ray spectrum of ^{60}Co as recorded by Ge(Li) and scintillation [NaI(Tl)] detectors.

7.7.6 High Purity Germanium Detectors

With the advancement of semiconductor technology, it became possible to get germanium of very high purity, i.e. about 1 atom of impurity for about 10^{13} to 10^{14} atoms of germanium. Because of this high purity, it has become possible to fabricate thick detectors without lithium compensation. These germanium detectors are commonly referred as *High Purity Germanium (HPGe) detectors*. The biggest advantage of this is that the HPGe detectors can be maintained or stored at room temperature. There is no lithium in these detectors. However, HPGe detectors like Ge(Li) or Si(Li) detectors must be operated at liquid nitrogen temperature. Keeping the detector at liquid nitrogen temperature reduces the background electrical noise generated in the detector due to thermal energy. The energy resolution of this detector is also is 1.8 keV at 1332 keV. Like Ge(Li) detectors, these detectors are also commonly available with active volume of 5 to 100 CC.

7.8 CLOUD CHAMBER

It is one of the detectors, which provides visual trajectory of a charged particle like electron, proton, α-particles, etc. Cloud chamber also known as *Wilson Chamber* was built by C.T.R. Wilson in 1911.

7.8.1 Principle

It is based on the principle that when dust-free air saturated with vapours of a liquid (like water, alcohol, ether, etc.) is allowed to expand adiabatically, supersaturation occurs. If at this stage an ionizing particle enters the chamber and creates ion-pairs, tiny droplets of liquid condense on these ions and form a visible track along the path of the ionizing radiation. These visible tracks can be photographed. In some cases, cloud chamber is subjected to a strong magnetic or electric field. Such a field causes the charged particles to travel in curved path. The curvature of the curved path gives information about the mass and charge of the ionizing particle.

7.8.2 Construction

A simplified form of a cloud chamber is shown in Figure 7.19. It consists of a transparent cylindrical chamber CC in which a piston P is fitted at the bottom. On the top of this cylindrical chamber is an optically-flat glass plate G and a camera from the top views inside the chamber. This chamber is illuminated from a side with the help of a strong light source.

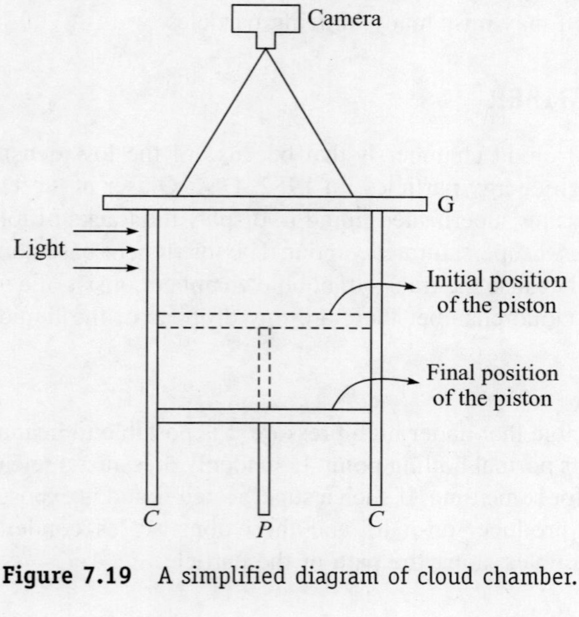

Figure 7.19 A simplified diagram of cloud chamber.

7.8.3 Working

Air saturated with given liquid is filled in the space between the movable piston P and glass plate G. The pressure inside the chamber is kept high. The pressure in the chamber is lowered by moving the piston down suddenly due to which the temperature of the saturated liquid falls and vapours become super-saturated. If at this moment, a charged particle passes through the chamber, it will produce ion-pairs. The supersaturated vapours condense on the ions and a trail of droplets along the path of the charged particle is seen. These tracks are known as cloud tracks. These tracks have distinctive shapes (for example, an α-particle track is broad and straight while an electron's trajectory is thinner and shows a zig-zag trajectory). If the chamber is illuminated with light, a camera can take a photograph of the track, which appears as a white line on a dark background. When a vertical magnetic field is applied, positively and negatively charged particles curve in opposite directions.

7.8.4 Advantages

1. When subjected to electric or magnetic field, cloud chamber is used to find charge on the ionizing particles and their momentum.
2. With cloud chamber, the range of high energy particles can easily be determined.
3. By seeing the broadness of a cloud track, we can immediately get an idea whether the track is due to heavy particle (like α-particle) or light particle (like electron).

7.8.5 Limitations

1. If the energy of the ionizing particle is high, it may not completely stop in the cloud chamber and may come out of the chamber. So, we will not get full information about the particle.
2. The recovery time of the cloud chamber is relatively very long, 10–60 s after the expansion, so it may miss many ionizing particles.

7.9 BUBBLE CHAMBER

The basic drawback of cloud chamber is that because of the low density of the gas, it is not possible to observe high energy particles. In 1952, D.A. Glaser at the University of Michigan, conceived the idea of using superheated liquid to display the tracks of ionizing particles, just as a cloud chamber utilizes a supersaturated vapour. The instrument based on this concept is known as a *Bubble Chamber*, because the tracks in bubble chamber consist of a series of closely spaced bubbles, whereas in a cloud chamber there are tiny droplets of the liquid.

7.9.1 Principle

It is based on the principle that under high pressure it is possible to heat a liquid without bubble formation well above its normal boiling point. If suddenly pressure is released, the liquid remains in a superheated state for some time. If such a superheated liquid is exposed to ionizing particles, the ionizing particles produce ion-pairs and these ions act as condensation centres for the formation of vapour bubbles along the path of the particle.

7.9.2 Construction

The schematic diagram of the bubble chamber is shown in Figure 7.20. The main body of the chamber is made up of stainless steel with thick glass ports at the top for a viewing camera. A box of thick-walled glass is filled with liquid hydrogen and is connected to the expansion pressure system. In order to maintain the chamber at constant temperature, it is surrounded by liquid nitrogen. High energy particles are allowed to enter the chamber from a side window W.

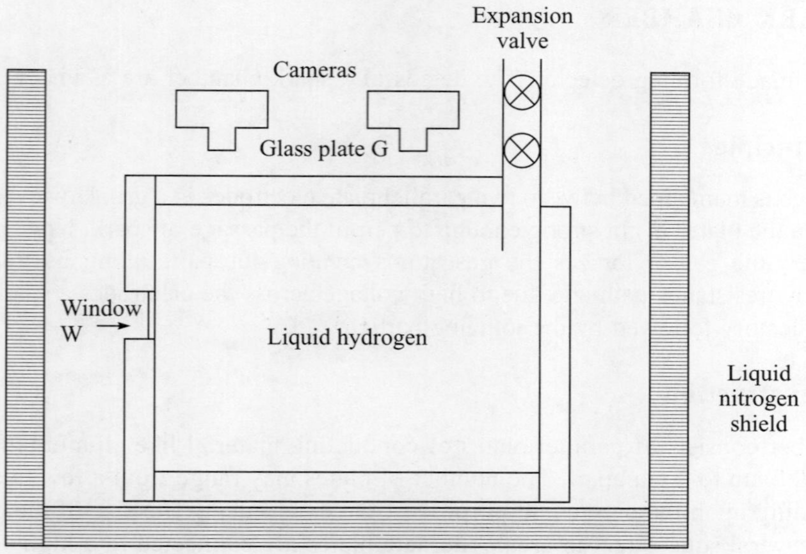

Figure 7.20 A schematic diagram of bubble chamber.

7.9.3 Working

Initially, the liquid hydrogen is kept under high pressure, but when a charged particle is passing through it, the pressure is released so that the liquid is in superheated state. The liquid vapours get condensed in the form of bubbles on the ions formed by ionizing particle and photographs of the tracks formed are obtained by the cameras. Generally, bubble chamber is subjected to a strong magnetic field in order to distinguish the sign of the charge on the ionizing particles and to measure their momenta from the radius of curvature of the bubble tracks. Though most commonly used liquid in bubble chamber is liquid hydrogen, other liquids such as deuterium, helium, xenon, propane, pentane, etc. are also used in some cases.

7.9.4 Advantages

1. Due to high density of the liquid, even high energy cosmic rays can also be recorded in bubble chamber.
2. The bubble chamber is sensitive to both high- and low-ionizing particles.
3. As bubbles grow rapidly, the tracks formed in bubble chamber are clean and undistorted.

7.9.5 Limitations

1. The time during which bubble chamber is sensitive is only few milliseconds, the entry of ionizing particles and photographing the tracks formed must take place during this short time.
2. Bubble chambers are the costliest detectors.

7.10 SPARK CHAMBER

It is also an image-forming detector. The details of a spark chamber are as under.

7.10.1 Principle

A high voltage is maintained between two parallel plate electrodes in a gas. However, the electric field between the plates is not strong enough to permit the passage of spark. Now, if an ionizing particle enters the gas, it ionizes the gas atoms creating ion-pairs along its trajectory. This provides a low-resistance path and due to high voltage across the electrodes a spark takes place along the trajectory followed by the ionizing particle.

7.10.2 Construction

Spark chamber consists of parallel plates of conducting material like aluminium, etc. spaced from about 0.2 cm to 3 cm apart. The number of plates may range from a few to a hundred or more depending upon the nature and purpose of the experiment. The size of the plates can be as large as several square feet in area. Alternate plates are connected to a high voltage direct pulse generator so that electric field of the order of 10,000 to 15,000 volts/cm can be applied between adjacent pair of plates. The spark chamber is generally filled with neon or argon or a mixture of 90% neon and 10% helium. The schematic diagram of spark chamber is shown in Figure 7.21.

7.10.3 Working

When a high energy particle enters the spark chamber, the scintillation detector outside the chamber sends a pulse to the high-voltage pulse generator, which in turn sends a high-voltage pulse to the spark chamber plates. When a high energy charged particle passes through the chamber, it produces large number of ion-pairs along its path. A visible spark jumps along the path of the ionizing particle between two plates. Camera records the visible track of the particle.

7.10.4 Advantages

1. It is capable of fast operations.
2. It provides good definition of direction of the ionizing particles.
3. It is relatively cheaper compared to bubble and cloud chambers.

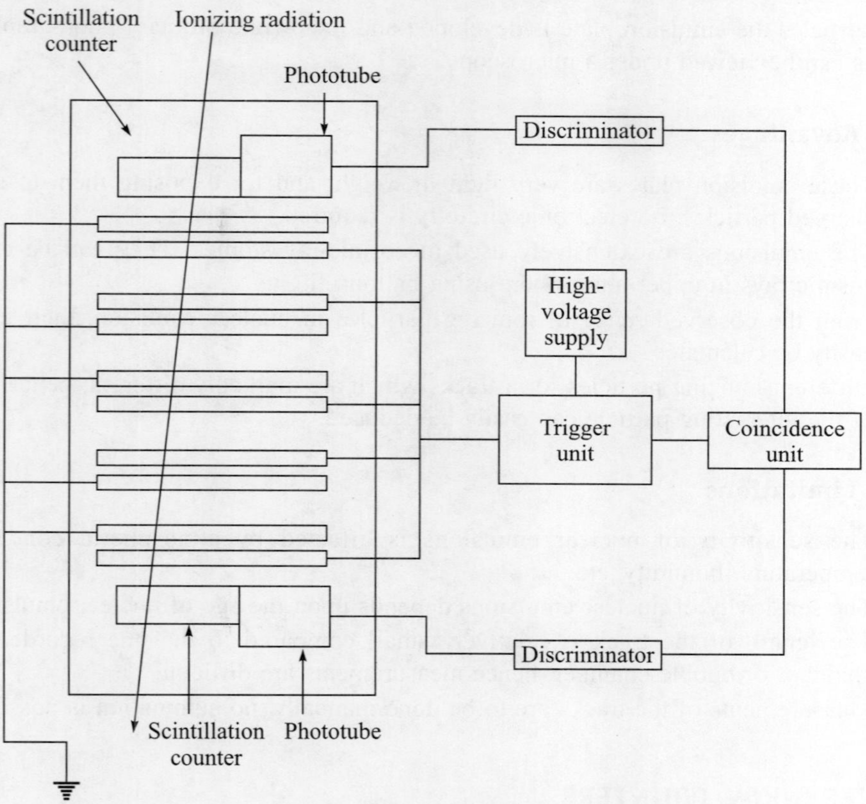

Figure 7.21 Spark chamber.

7.10.5 Limitation

1. Spark scatter 15 to 20 thousandths of an inch, and the path uncertainity increases as the path of the particle becomes parallel to the plates.

7.11 NUCLEAR EMULSIONS

Henry Becquerel first used photographic plates for detection of radioactivity in 1886. When an ionizing radiation passes through an emulsion, it interacts with the silver halide grains in the gelatin present on photographic plate. When this plate is developed, the affected silver halide grains change into black grains of metallic silver. It has been found experimentally that optical photographic emulsions are not suitable for qualitative work with nuclear radiations. The sensitivity of ordinary photographic plates is low and the tracks formed do not show clear range because the developed crystal grains are large and widely spaced. In nuclear emulsions, the grain size of silver halide is kept much smaller (0.1–0.6 micron) compared to optical emulsions (1–3.5 micron). Similarly, the thickness of nuclear emulsions is kept large (50–2000 micron) compared to that of optical emulsions (2–4 micron). After exposing the nuclear emulsions to charged particles the emulsion plate is developed and fixed (like ordinary photographic plate). The tracks can be viewed under a microscope.

7.11.1 Advantages

1. These emulsion plates are very light in weight and for exposing them to a beam of charged particles, no electronic circuitry is required.
2. The emulsions are extensively used in cosmic-ray studies. They can be exposed to cosmic rays in upper atmosphere using balloon flights.
3. From the observed range of ionizing particles in nuclear emulsion, their energy can easily be calculated.
4. Different ionizing particles form tracks which are markedly different, hence the nature of the interacting particle can easily be deduced.

7.11.2 Limitations

1. The sensitivity of nuclear emulsions is affected by atmospheric conditions like temperature, humidity, etc.
2. The sensitivity of nuclear emulsions depends upon the age of nuclear emulsion.
3. The length of the track is relatively small compared to the one recorded in cloud chamber or bubble chamber, hence measurements are difficult.
4. Measurements of the tracks are to be done manually, no automation is possible.

7.12 CERENKOV COUNTERS

In 1934, Russian physicist P.A. Cerenkov demonstrated that transparent dielectric substances such as water, glass, mica, etc. when exposed to γ-rays, emit a weak bluish-white light in the visible region. Later on, it was found that when a charged particle travels through a medium with velocity greater than the velocity of light in that medium (velocity of light in glass of refractive index (n)1.5 is 2×10^8 m/s), it emits visible radiations called *Cerenkov radiations*. The nature of this radiation is different from that of fluorescence as well as Bremsstrahlung. Unlike Bremsstrahlung radiations, it does not depend on the atomic number of the medium in which the charged particle is moving or the mass of the charged particle. The emission of Cerenkov radiation is explained as under.

Along its path, an electromagnetic pulse associated with fast moving charged particle causes polarization of the atoms of the medium by displacing the bound electrons. This time variation of the polarization produced by the field causes the atoms to radiate electromagnetic waves. If the particle is moving slowly, the radiations reaching a distant point interferes destructively, and the resultant intensity of the light is zero. If the velocity of the particle, however, is greater than c/n, a cone of light is emitted. As the mass of the electron is small, the velocity of electron even at lower kinetic energy is greater than c/n, therefore, the energy loss due to Cerenkov radiations is more important for electrons. However, the energy loss due to Cerenkov radiations in case of electrons is smaller than due to ionization and Bremsstrahlung. The Cerenkov radiations come out in a cone of half angle θ with the particle at apex as shown in Figure 7.22. This angle θ is related to the velocity v of the particle as under

$$\sin \theta = \frac{\left(\frac{c}{n}\right)t}{vt} = \frac{c}{nv}$$

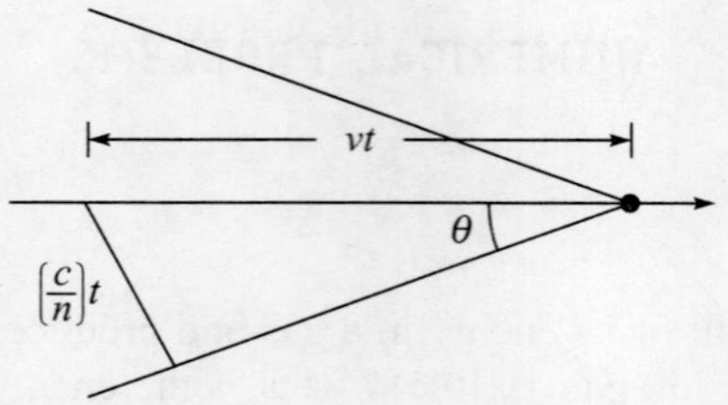

Figure 7.22 Cerenkov radiations wavefront in a medium.

The visible light emitted can easily be detected and angle θ measured. This makes it possible to measure the speed and hence the kinetic energy of the particle. Cerenkov radiations are seen as a bluish glow when intense beam of electrons is involved.

The apparatus used for Cerenkov detector is shown in Figure 7.23.

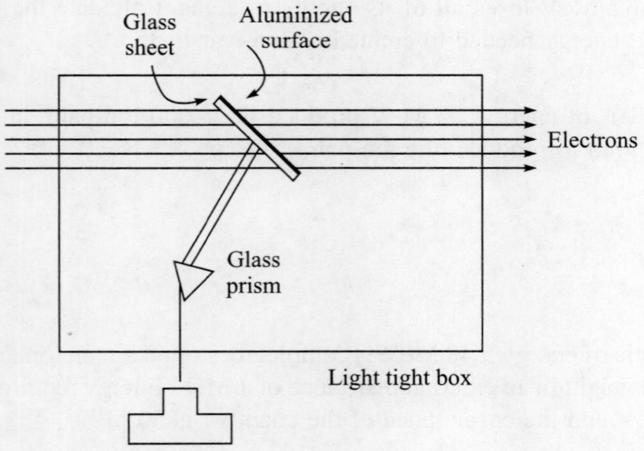

Figure 7.23 Cerenkov detector.

7.12.1 Advantages

1. The duration of light pulse at any point is vanishingly small ($\leq 10^{-10}$ s), so these detectors are fast detectors.
2. High efficiency and high counting rate.
3. Directional emission of light with an angle dependent on velocity of the incident particle.
4. Intensity of light emitted by the incident particles depends upon the velocity of the particle.

7.12.2 Limitations

1. For slow moving particles, these detectors are ineffective.
2. These counters are highly directional because the Cerenkov radiations are emitted only in the forward cone.

NUMERICAL PROBLEMS

Section 7.2

Solved Problem

1. An α-particle loses all of its energy in a gas and produces 150,000 ion-pairs. If energy needed to create one ion-pair is 30 eV, what is the energy of an α-particle?

Solution: Energy required to create one ion-pair = 30 eV

Energy required to create 1,50,000 ion-pairs = 30 × 1,50,000

= 45,00,000 eV = 4.5 MeV

Therefore, the energy of an α-particle = 4.5 MeV

Unsolved Problems

1. A proton of 2 MeV loses all of its energy in a gas. Calculate the number of ion-pairs produced, if energy needed to create one ion-pair is 35 eV.

[**Ans.** ≈ 57,140 ion-pairs]

2. A ^{3}He nucleus of energy 7.2 MeV produces 2,25,000 ion-pairs in a gas. Calculate the energy required to produce one ion-pair.

[**Ans.** 32 eV]

Section 7.3

Solved Problems

1. An α-particle of energy 5.48 MeV is completely stopped in an ionization chamber. What is the pulse height in an external resistance of 1 MΩ? Energy required to produce an ion-pair is 35 eV and the capacitance of the chamber is 50 pF.

Solution:
Given:

$$E = 5.48 \text{ MeV} = 5.48 \times 10^6 \text{ eV}$$

$$C = 50 \times 10^{-12} \text{ F}$$

$$R = 1 \text{ M}\Omega = 10^6 \ \Omega$$

$$\text{Number of ion-pairs produced } n = \frac{5.48 \times 10^6}{35}$$

$$= 1.566 \times 10^5 \text{ ion-pairs.}$$

$$\text{Pulse height } V = \frac{ne}{C}$$

$$= \frac{1.566 \times 10^5 \times 1.6 \times 10^{-19}}{50 \times 10^{-12}} = 0.501 \times 10^{-3} \text{ V}$$

$$= 0.501 \text{ mV}$$

$$\text{Current} = \frac{V}{R}$$

$$= \frac{0.501 \times 10^{-3}}{10^6} = 0.501 \times 10^{-9} \text{ A}$$

$$= 0.501 \text{ nA}$$

2. A proton is stopped in an ionization chamber producing 10^5 ion-pairs. Energy required to produce an ion-pair is 35 eV. (i) What is the kinetic energy of proton entering the ionization chamber? (ii) What is the amount of charge collected on each plate?

Solution: (i) Number of ion-pairs produced = 10^5

Energy required to create one ion-pair = 35 eV

Kinetic energy of protons = $10^5 \times 35$ eV = 3.5×10^6 eV

= 3.5 MeV

(ii) Amount of charge collected on each plate = Number of ion pairs × Charge on each plate

= $10^5 \times 1.6 \times 10^{-19}$ C = 1.6×10^{-14} C

Unsolved Problems

1. ^{235}U undergoes fission producing fission fragments having total kinetic energy of 180 MeV inside an ionization chamber. Calculate the resultant pulse height, if capacitance of ionization chamber is 100 pF. Energy required to create one ion-pair is 35 eV.

[**Ans.** 8.2 mV]

2. A proton is completely stopped in an ionization chamber and produces 3.1×10^5 ion-pairs. What is the kinetic energy of proton? Also calculate the pulse voltage and current flowing in an external circuit having load resistance of 0.5 MΩ. Given capacitance of the ionization chamber is 25 pF and W for the gas is 32 eV.

[**Ans.** 9.92 MeV, 1.984 mV, 3.97×10^{-9} A]

3. A 15 MeV proton when completely stopped in an ionization chamber produces 10^6 ion-pairs. Calculate the pulse height, if total capacitance of the circuit is 100 pF.

[**Ans.** 1.6 mV]

4. An ionization chamber produces 4.5×10^8 ions per discharge. If it responds to an incident radiation 10 times per second, what is the average ion current?

[**Ans.** 7.2×10^{-10} A]

Section 7.4

Solved Problems

1. A 20 MeV α-particle enters a proportional counter long enough to stop the α-particle completely. If the gas multiplication factor is 1000, how many coulombs flow in the counter, when α-particle is absorbed. If the current flows for 1 ms through an external resistance of 100 kΩ, estimate the height of voltage pulse. Given energy required to create one ion-pair = 30 eV.

Solution:

$$\text{Kinetic energy of an } \alpha\text{-particle} = 20 \text{ MeV}$$

$$\text{Energy required to create one ion-pair} = 30 \text{ eV}$$

$$\text{No. of ion-pairs created} = \frac{20 \times 10^6}{30}$$

$$= 6.667 \times 10^5$$

$$\text{Multiplication factor} = 1000$$

$$\text{Number of ion-pairs after multiplication} = 6.667 \times 10^5 \times 1000$$

$$= 6.667 \times 10^8$$

$$\text{Charge carried by these ions} = 6.667 \times 10^8 \times 1.6 \times 10^{-19}\text{C}$$

$$= 1.0667 \times 10^{-10}\text{C}$$

$$\text{Current through } 100 \text{ k}\Omega \text{ resistance} = \frac{\text{Charge}}{\text{Time}}$$

$$= \frac{1.0667 \times 10^{-10}}{10^{-3}}$$

$$= 1.0667 \times 10^{-7}\text{A}$$

$$\text{Amplitude of voltage pulse} = I \times R$$

$$= 1.0667 \times 10^{-7} \times 1{,}00{,}000$$

$$= 1.0667 \times 10^{-2} \text{ V}$$

$$= 10.67 \text{ mV}$$

2. The radius of the central wire of a proportional counter is 0.1 mm and the radius of the cylindrical tube is 2 cm. Calculate the electric field developed at the surface of the wire, when the potential difference of 1500 volts is applied between the two.

Solution: We have

$$E(r) = \frac{V}{r \ln \frac{b}{a}}$$

Here,

$$b = 2 \text{ cm} = 0.02 \text{ m}$$

$$a = 0.1 \text{ mm} = 0.0001 \text{ m and}$$

$$V = 1500 \text{ V.}$$

At the surface of the wire $r = 0.0001$ m
Therefore,

$$E(r) = \frac{1500}{0.0001 \times \ln \dfrac{0.02}{0.0001}}$$

$$= 2.83 \times 10^6 \text{ V/m}$$

Therefore, the electric field developed around the wire = 2.83×10^6 V/m.

Unsolved Problems

1. A proportional counter is to be operated with electric field of 10^6 V/m. What is the applied voltage, if the radii of the wire is 0.01 cm and that of the tube is 1.0 cm?
[**Ans.** 465 V]

2. Calculate the number of ion-pairs formed in a proportional counter, if a 60 MeV ^{28}Si ion is completely stopped in it. Given gas multiplication factor is 1200 and energy required to create one ion-pair is 32 eV.
[**Ans.** $\sim 2.25 \times 10^9$]

3. A ^{16}O ion of 70 MeV is completely stopped in a proportional counter. If the gas multiplication factor is 1500, how much charge will be collected at the plates? If this pulse of charge flow for about 1 ms in an external resistance of $10^5\ \Omega$, estimate the height of the current flowing in the resistor and voltage pulse developed across it. Given energy required to create one ion-pair is 35 eV.
[**Ans.** 4.8×10^{-10} C, 0.48 μA, 0.048 V]

4. Calculate the relative size of the pulses produced by 1 MeV ^{1}H, 2 MeV ^{2}H and 3 Mev ^{3}H.
[**Ans.** 1:2:3]

5. Calculate relative sizes of the pulses produced in a proportional counter by 10 MeV protons, 20 MeV deuterons and 40 MeV α-particles. [**Ans.** 1:2:4]

Section 7.5

Solved Problems

1. Calculate the electric field at the surface of the wire of a GM counter. The radius of the wire is 0.1 mm and the inner radius of the outer cylinder is 2 cm. The potential applied between the two electrodes is 2000 volts.

Solution: Electric field is given by

$$E = \frac{V}{r \ln \dfrac{b}{a}}$$

where

b = radius of the cylinder = 2 cm

a = radius of the wire = 0.1 mm = 0.01 cm

V = 2000 V

$$E = \frac{2000}{0.01 \log \frac{2}{0.01}}$$

$$= 37{,}750 \text{ V/cm}$$

Therefore, electric field at the surface of the wire is 37,750 V/cm.

2. Estimate the useful life of a GM counter operating daily for 3 hours at 2000 counts per minute. The guaranteed counts for GM counter are 10^9 counts.

Solution: Suppose the total useful life = x years.
GM works only up to time when it records 10^9 counts
In one day GM counter records total counts while operated for 3 hours per day = 2000 × 3 × 60
Let 10^9 counts are recorded in x years or in $x \times 365$ days

$$\text{In one day counter records} = 2000 \times 3 \times 60 \text{ counts}$$

$$\text{In } x \times 365 \text{ days counter records} = 2000 \times 3 \times 60 \times x \times 365 \text{ counts}$$

This number must be equal to 10^9

or $$x \times 3 \times 365 \times 60 \times 2000 = 10^9$$

or $$x = \frac{10^9}{3 \times 365 \times 60 \times 2000}$$

$$= 7.62 \text{ years}$$

Therefore, the useful life of the GM counter = 7.62 years.

3. A 10 MeV α-particle loses all its energy in a GM counter, one electron ion-pair is produced for each 30 eV of energy lost in the gas of the GM counter. The GM counter has a multiplication of 5000 and the total capacitance between the two electrodes is 50 pF. Calculate the amplitude of voltage pulse developed.

Solution: $$\text{Number of ion-pairs produced} = \frac{10 \times 10^6}{30}$$

$$= 3.333 \times 10^5$$

$$\text{Total number of ion-pairs after multiplication} = 3.333 \times 10^5 \times 5000$$

$$= 1.667 \times 10^9$$

$$\text{Total charge present on each of the ion } Q = 1.667 \times 10^9 \times 1.6 \times 10^{-19} \text{ C}$$

$$= 2.667 \times 10^{-10} \text{ C}$$

$$\text{Amplitude of the voltage pulse developed} = \frac{Q}{C}$$

$$= \frac{2.667 \times 10^{-10}}{50 \times 10^{-12}}$$

$$= 5.3 \text{ V}$$

4. A GM counter with dead time of 2×10^{-4} s registers 30,000 counts per minute. Find the true counting rate of the incoming beam of particles in terms of the particles received per second.

Solution:

$$\text{The true counting rate} = \frac{\text{Observed counting rate}}{1 - \text{Observed counting rate} \times \text{Dead time}}$$

$$\text{Observed counting rate} = \frac{30{,}000}{60}$$

$$= 500 \text{ counts/s}$$

$$\text{The true counting rate} = \frac{500}{1 - 500 \times 2 \times 10^{-4}}$$

$$= 556 \text{ counts/s}$$

Therefore, the true counting rate is 556 counts/s.

Unsolved Problems

1. A GM tube operates at 1000 volts and the central wire has a radius of 0.12 mm. The inner radius of the outer cylindrical electrode is 2 cm and the tube has a guaranteed lifetime of 10^9 counts. Calculate the maximum radial field and the life of the counter assuming that the counter works on an average for 4 hours per day at 3000 counts per minute.
[**Ans.** 16,290 V/cm, 3.805 years]

2. A GM counter records 16,000 counts per minute, while the true counting rate is 17,400 counts per minute. What is the dead time of the GM counter?
[**Ans.** 297 μs]

3. A GM counter has dead time of 250 μs. It records 20,000 counts per minute. What is the actual count rate? [**Ans.** 21,820]

4. A GM counter has gas amplification of 2.5×10^8 and is operating a circuit with a capacitance of 200 pF. What is the amplitude of the pulse generated by it? (Assume that one ion-pair is being produced.) [**Ans.** 0.2 V]

5. Calculate the relative size of the pulses in a GM counter produced by 1 MeV protons and 4 MeV α-particles. [**Ans.** 1:1]

6. A Geiger counter consists of a 50 mm diameter grounded tube with a wire of 25 μm diameter at +700 V in the centre. What is the electric field at the wire?
[**Ans.** 74 kV/cm]

7. A Geiger counter consists of a 50 mm diameter grounded tube with a wire of 25 μm diameter at +700 V in the centre. What is the electric field at the tube wall?
[**Ans.** 3.72 MV/cm]

Section 7.6

Solved Problems

1. ^{22}Na emits two gamma rays of energy 511 keV and 1275 keV. These γ-rays are detected by a NaI(Tl) spectrometer. These two γ-rays fall at 1202 and 3000 channels. Full width

at half maximum (FWHM) of these two γ-rays is 97 and 82 channels respectively. Find the energy resolution at these energies.

Solution: Here, we introduce a concept of channel in the γ-ray spectrum. It is as under:

We know that the output pulse amplitude of the amplifier is directly proportional to the energy of γ-ray absorbed by the detector. Suppose, the maximum amplitude of an amplifier pulse is 8 V (which is a common case). This 8 V is divided into say 2048 (2K) or 4096 (4K) or 8192 (8 K), etc. equal bins or equal amplitude. Each bin is called a channel. Therefore, channel number is directly proportional to the energy of the γ-ray absorbed by the detector. Therefore, for 511 keV γ-ray:

$$\text{\% resolution (in terms of channels)} = \frac{\text{FWHM in channels}}{\text{Peak position in channels}} \times 100$$

$$= \frac{97}{1202} \times 100 \approx 8.1\%$$

$$\text{\% resolution} = 8.1\%$$

$$\text{Resolution (keV)} = \frac{\text{FWHM in channels}}{\text{Peak position in channels}} \times 511 \text{ keV}$$

$$= \frac{97}{1202} \times 511 \approx 41 \text{ keV}$$

$$\text{Resolution (keV)} = 41 \text{ keV}$$

For 1275 keV γ-ray:

$$\text{\% resolution (in terms of channels)} = \frac{\text{FWHM in channels}}{\text{Peak position in channels}} \times 100$$

$$= \frac{82}{1202} \times 100 \approx 6.8\%$$

$$\text{\% resolution} = 6.8\%$$

$$\text{Resolution (keV)} = \frac{\text{FWHM in channels}}{\text{Peak position in channels}} \times 1275 \text{ keV}$$

$$= \frac{82}{1202} \times 1275 \approx 87.0 \text{ keV}$$

$$\text{Resolution (keV)} = 87.0 \text{ keV}$$

2. A NaI(Tl) scintillation crystal is coupled to a 10 stage photomultiplier. This system is used to detect 1 MeV γ-ray. The output circuit of the photomultiplier has a capacitance of 200 pF. The average electron multiplication at each stage is 4. Assume light collection efficiency is unity, photocathode efficiency is 0.1 and 250 eV energy is required to create one photon. What is the amplitude of the output voltage pulse?

Solution:

$$250 \text{ eV produces} = 1 \text{ photon}$$

$$1 \times 10^6 \text{ eV will produce} = \frac{1}{250} \times 1 \times 10^6 = 4000 \text{ photons}$$

Light collection efficiency = 1

Photocathode efficiency = 0.1

Therefore, number of photoelectrons emitted = $4000 \times 1 \times 0.1 = 400$

Each dynode produces 4 secondary photoelectrons. Therefore, after passing through 10 stages number of photoelectrons = $400 \times 4^{10} = 4.2 \times 10^8$ photoelectrons

$$\text{Capacitance} = 200 \text{ pF} = 200 \times 10^{-12} \text{ F}$$

Therefore,
$$\text{output pulse amplitude} = \frac{4.2 \times 10^8 \times 1.602 \times 10^{-19}}{200 \times 10^{-12}} = 0.34 \text{ V}$$

Amplitude of the output pulse = 0.34 V

3. In a scintillation detector, the 662 keV photopeak of ^{137}Cs source is observed at 6.0 V and the full width at half the maximum of the photopeak is 0.72 V. Find the % resolution of the scintillation detector. Also find its resolution in keV.

Solution:
$$\% \text{ resolution} = \frac{\text{Full width at half the maximum in volts}}{\text{Peak position in volts}} \times 100$$

$$= \frac{0.72}{6.0} \times 100 = 12.0\%$$

$$\text{Resolution in keV} = \frac{\text{Resolution in \%}}{100} \times \text{Energy of photopeak in keV}$$

$$= \frac{12}{100} \times 662 = 79.4 \text{ keV}$$

Unsolved Problems

1. A NaI(Tl) detector is coupled to a photomultiplier tube having certain stages for electron multiplication. This detection system is used to detect γ-rays of 1115 keV. The output circuit of photomultiplier has a capacitance of 250 pF. The average electron multiplication at each stage is 3. Assume light collection efficiency is unity, photocathode efficiency is 0.12 and 300 eV energy is needed to create one photon and the amplitude of output pulse is 0.25 V. Find the number of multiplication stages in the photomultiplier.

[**Ans.** 10]

2. Estimate the relative size of the pulses produced at the output of a photomultiplier tube in a scintillation counter, when 1 MeV and 3 MeV γ-rays are absorbed in it.

[**Ans.** 1 : 3]

3. Two photopeaks due to two γ-rays of energies 1172 keV and 1332 keV of ^{60}Co falls on channels 3516 and 3996, when they are absorbed in a NaI(Tl) detector. What is the energy of the photopeak falling at channel 3345?

[*Hint:* Use the relation: Energy = $a \times$ channel + const.) [**Ans.** 1115 keV]

4. Two photopeaks due to two γ-rays of energies 1172 keV and 1332 keV of ^{60}Co falls on channels 2871 and 3264, when they are absorbed in a NaI(Tl) detector. What is the energy of the photopeak falling at channel 3125?
(*Hint:* Use the relation: Energy = $a \times$ channel + const.) **[Ans.** 1275 keV]

Section 7.7

Solved Problems

1. Calculate the thickness of the depletion layer of a silicon detector of active area 1.6 cm^2. The capacitance of the detector is 100 μF. The dielectric constant of silicon is 12. What is the amplitude of the voltage pulse developed across this detector, when ^{16}O ion of 12 MeV is completely stopped in it? Energy required to create one electron-hole pair is 3.2 eV.

Solution: Relative permittivity $\varepsilon_r = 12$

Permittivity of free space $\varepsilon_0 = 8.85 \times 10^{-12}$

Absolute dielectric constant $\varepsilon = \varepsilon_r \varepsilon_0 = 12 \times 8.85 \times 10^{-12}$

$C = 100 \times 10^{-12}$ F

Area $A = 1.6 \text{ cm}^2 = 1.6 \times 10^{-4} \text{ m}^2$

We have

$$d = \frac{\varepsilon A}{C}$$

Therefore,

$$d = \frac{12 \times 8.85 \times 10^{-12} \times 1.6 \times 10^{-4}}{100 \times 10^{-12}}$$

$$= 170 \times 10^{-6} = 170 \text{ micron}$$

Thickness of depletion layer = 170 micron

$$\text{Number of ion-pairs produced} = \frac{12 \times 10^6}{3.2} = 3.75 \times 10^6 \text{ ion-pairs}$$

$$\text{Charge on these ion-pairs} = 3.75 \times 10^6 \times 1.602 \times 10^{-19} = 6.0075 \times 10^{-13}\text{C}$$

Therefore,

$$\text{voltage developed} = \frac{6.0075 \times 10^{-13}}{100 \times 10^{-12}} = 0.0060075 \text{ V}$$

Amplitude of voltage pulse = 6.0075 mV

2. A silicon surface barrier detector of area 2.0 cm^2 depletion depth 100 micron and dielectric constant 12. Calculate its capacitance. An α-particle of energy 5.48 MeV enters this detector and is completely stopped. What is the amplitude of the voltage pulse developed across this capacitance? Energy required to create one electron-hole pair is 3.0 eV.

Solution: Capacitance is given by

$$C = \frac{\varepsilon A}{d}$$

$$\varepsilon = 12 \times 8.85 \times 10^{-12}$$

$$A = 2.0 \times 10^{-4}\ \text{m}^2$$

$$d = 100 \times 10^{-6}\ \text{m}$$

$$C = \frac{12 \times 8.85 \times 10^{-12} \times 2 \times 10^{-4}}{100 \times 10^{-6}} = 2.124 \times 10^{-10}\ \text{F}$$

$$\text{Number of ion-pairs produced} = \frac{5.48 \times 10^6}{3.0}$$

$$= 1.827 \times 10^6$$

$$\text{Total charge present on the ions } Q = 1.827 \times 10^6 \times 1.6 \times 10^{-19}\ \text{C}$$

$$= 2.923 \times 10^{-13}\ \text{C}$$

Amplitude of the voltage pulse developed across the detector capacitance

$$V = \frac{Q}{C}$$

$$= \frac{2.923 \times 10^{-13}}{2.124 \times 10^{-10}}$$

$$= 1.376 \times 10^{-3}\ \text{V}$$

$$= 1.376\ \text{mV}$$

Unsolved Problems

1. Calculate the area of a silicon detector with depletion layer of thickness 80 micron. The dielectric constant of silicon is 12 and the capacitance of the detector is 150 μF. Also calculate the amplitude of the voltage pulse developed across this detector, when 15 MeV ^{28}Si ion is completely stopped in it. Energy required to create one electron-hole pair is 3.2 eV.

[**Ans.** 1.13 cm^2, 5 mV]

2. The 662 keV gamma ray of ^{137}Cs falls on channel 2648, when it is recorded by a HPGe detector. What is the energy of an unknown photopeak, which falls on channel 2044?

[**Ans.** 511 keV]

3. Two photopeaks due to two γ-rays of energies 1172 keV and 1332 keV of ^{60}Co fall on channels 3516 and 3996, when they are absorbed in a HPGe detector. What is the energy of the photopeak falling at channel 3345?

(*Hint:* Use the relation: Energy = $a \times$ channel + const.) [**Ans.** 1115 keV]

4. Two photopeaks due to two γ-rays of energies 1172 keV and 1332 keV of ^{60}Co fall on channels 2871 and 3264, when they are absorbed in a HPGe detector. What is the energy of the photopeak falling at channel 3132?

(*Hint:* Use the relation: Energy = $a \times$ channel + const.) [**Ans.** 1275 keV]

5. ^{22}Na emits two γ-rays of energies 511 keV and 1275 keV. The spectrum of these γ-rays is recorded using HPGe detector. The photopeaks due to these two γ-rays fall on 1457 and 3636 channels. Full width at half maximum (FWHM) of these two γ-rays is 4.0 and 5.1 channels respectively. Find the energy resolution of the HPGe detector at these two energies.

[**Ans.** 1.4 keV, 1.8 keV]

6. 5.48 MeV α-particles emitted by a ^{241}Am source are detected by a silicon charged particle detector. The photopeak due to α-particles falls at channel 3460.5. Full width at half maximum for this peak is 13.2 channels. What is the energy resolution for this detector in keV?

[**Ans.** 20.9 keV]

7. ^{55}Fe radioactive source emits X-ray photons of energy 5.9 keV. A Si(Li) detector is used to detect these X-rays. This X-ray photopeak falls at 3756.00 channel. Full width at half maximum is 114.60 channel. What is the energy resolution of the detection system at this energy?

[**Ans.** 180 eV]

REVIEW QUESTIONS

Short Answer Type

1. What are the limitations of a GM counter?
2. What are the dead time and recovery time of a GM counter?
3. What is the principle and significance of a photomultiplier tube in a scintillation counter?
4. Give the basic principle of conversion of light into an electrical pulse in a scintillation counter.
5. Explain the principle of ionization chamber.
6. What are the advantages of a semiconductor detector?
7. Give the properties of a good quality photomultiplier tube.
8. How are Ge(Li) detectors more suitable over Si(Li) detectors for electromagnetic radiations?
9. Explain the phenomenon of quenching in GM counter.
10. On what factors, does the efficiency of a detector depend?
11. Why are solid-state detectors preferred over scintillation detectors?

Long Answer Type

1. What are gas-filled, ionization-based nuclear detectors? Discuss the curve between pulse height and applied voltage for a gas-filled counter serving as
   (i) an ionization chamber.
   (ii) a proportional counter.
   (iii) a Geiger counter.
2. Describe the construction, principle and working of an ionization chamber. Explain the difference between the ionization chamber and the GM counter.

3. Explain the principle and operation of a scintillation counter.
4. Explain the principle and operation of a nuclear emulsion charged particle detector.
5. Discuss the principle, construction and working of a proportional counter. Give its applications and explain how it can be used for neutron detection.
6. Discuss completely the construction, working and theory of a GM counter. Give its drawbacks.
7. What is the principle of a bubble chamber? Discuss its construction. What are its merits?
8. Describe the principle and working of an ionization chamber and compare it with a proportional counter.
9. What is nuclear emulsion technique? How is it used?
10. What is nuclear emulsion technique? Explain in detail how it is useful for measurement of nuclear radiation.
11. What is a semiconductor detector? Explain its working.
12. Briefly explain the principle, construction and working of a GM counter. What are its dead time and recovery time?
13. Discuss the principle, construction and working of a semiconductor detector. Give its main advantages over other detectors.
14. Write the principle, working and construction of a proportional counter. How is it different from a GM counter?
15. Name three nuclear radiation detectors based on image formation. Discuss briefly one of them.
16. Describe the principle of GM counter. What are its limitations?
17. Write short notes on:
    (i) Cerenkov counters. (ii) Cloud chamber.
    (iii) Bubble chamber.
18. Describe the principle of GM counter and explain the various portions of the characteristic curve.
19. Describe the principle of a scintillation counter and compare its function with a GM counter.

Chapter **8**

Particle Physics

8.1 INTRODUCTION

The discovery of elementary particles, which forms all the matter in the universe, has a long history. Scientists have often made claims to identify the basic building blocks of matter[1]. However, penetrating deeper into the matter has revealed smaller and smaller particle constituents. The past three decades have achieved a remarkable progress in the physics of elementary particles at theoretical as well as experimental levels. In the beginning of the last century, it was thought that the atoms form the basic units that constitute matter. However, soon it was found to contain now well-known particles like electrons and protons. The study of black-body radiation by Planck led to the realization that electromagnetic waves were quantized; the quanta are now known as the photons. Apart from photon, electron the first elementary particle was discovered in 1897 in the classical work of J.J. Thomson. Later, it was found by Rutherford in 1911 that the atom consists of a positively-charged nucleus and negatively-charged electrons revolving around the nucleus. Around 1932, it was established by the work of Chadwick on the nuclear reactions that the nucleus consists of protons and neutrons. So it was thought that electron, proton and neutron are the fundamental particles. Later, host of other particles, like π, K-mesons, Λ- and Σ-baryons, were discovered that were also regarded as elementary particles.

Usually, an elementary particle was thought to be point-like particle having no substructure. However, structure can be probed only up to a given scale, which is decided by the available energy of the probe. So at higher energy scales, one can probe deeper into the structure of a particle, which was once considered to be elementary. For instance, the protons and the neutrons are found to have an inner structure revealing smaller elementary particles called *quarks*. Thus, in Particle Physics also known as High Energy Physics, quarks and leptons should be treated as elementary at the present energy scales. However, in this chapter, we take the traditional perspective to define elementary particle and their properties. Later we shall introduce salient

1. The search for ultimate constituents of matter is indeed very old. All the ancient civilizations, Indian, Greek, Chinese, etc., have tackled this fundamental and frontier question in various ways. For instance, Kanada (800 B.C.) proposed the concept of Paramanus (literally meaning ultimate constituents) theory of matter and causality responsible for observed transformations in them in his Vaisheshshika philosophy.

features of the quark model also. Now a question arises, from where, we get these elementary particles? In the following, we briefly discuss these issues.

8.2 PRODUCTION OF ELEMENTARY PARTICLES

Fundamental particles are not permanent entities. They can be created and destroyed. Different elementary particles are produced by different methods. For instance, electrons are easiest to produce. Simply heat a piece of metal sufficiently and electrons come boiling off. To obtain protons, hydrogen gas can be ionized. Besides these, there are three main sources to observe and create elementary particles:

1. Cosmic rays.
2. Particle accelerators.
3. Nuclear reactors.

8.2.1 Cosmic Rays

Cosmic rays are high-energy particles, which consist mostly of protons, most of them arriving from some unknown source in deep space (which is still a mystery) and are continuously bombarding the earth's atmosphere. When these high-energy particles hit atoms of gases present in the upper atmosphere, they produce showers of variety of secondary particles, which rain down on us all the times. The discovery of several elementary particles has been associated with investigations of the cosmic rays. For instance, examination of the electronic showers observed in cloud chambers led to the discovery of the positron. The main advantage of the elementary particles produced in cosmic rays is that they have enormous energies (and of course free). These energies are far-greater than what could be produced in the laboratories even today. The major disadvantages are that flux of cosmic rays is extremely small and their energies are completely uncontrolled. For more details on cosmic rays, refer to Chapter 9.

8.2.2 Particle Accelerators

As discussed in Chapter 6, particle accelerators accelerate charged particles, like electrons and protons, to very high energy and bombard them on some suitable targets, like hydrogen. These charged particles may also collide head-on with each other. Depending upon the energy of the accelerated charged particles, the resulting reaction debris contain particles like pions, kaons, muons, and even antiparticles like antiproton, etc. With accelerators it is possible to generate most intense beams of the elementary particles, and by proper choice of magnets and absorbers, beam of desired elementary particles can be extracted. The major disadvantage of the accelerator produced elementary particles is the limit on the energy available which depends on the design of the accelerator. This also limits the number of particles produced. Further, high-energy accelerators are extremely costly.

8.2.3 Nuclear Reactors

When heavy nuclei like ^{235}U undergo fission in the nuclear reactors, they emit variety of elementary particles like neutrons (n), neutrinos (ν), β-particles (e^+, e^-) and γ-rays. Nuclear reactors thus are intense source of these elementary particles.

8.3 TYPES OF INTERACTIONS

In order to understand the behaviour of the elementary particles, i.e. their formation and decay, one needs to understand different types of fundamental interactions[2] among them. These interactions which fall into the following four categories:

1. Gravitational interactions.
2. Electromagnetic interactions.
3. Strong interactions.
4. Weak interactions.

In fact, all the known processes occurring in nature (starting from subnuclear to extra-galactic levels) can be understood as manifestations of these four interactions. Note that in particle physics, the words force and interaction are used for the same purpose. Now we give brief introduction of these interactions.

8.3.1 Gravitational Interactions

This was the first interaction known to humans. First law of gravitation was formulated in 1666 by Newton. The gravitational force is mainly responsible for keeping heavenly bodies like moons, planets, stars, etc. in their orbits. This force depends upon the masses of the two objects and the distance between them. It is always attractive and its range is infinite. In order to compare the relative strength of different interactions, we define a dimensionless coupling constant as

$$g_m = G_N \frac{m^2}{\hbar c}$$

where G_N is gravitational constant = 6.66×10^{-11} Nm^2/kg^2, m is the mass of the objects between which gravitational interaction is operating, $\hbar = \dfrac{h}{2\pi}$, h is the Planck's constant and c is the speed of light. The dimensionless coupling constant between two nucleons can be calculated by taking $m \approx 1.67 \times 10^{-27}$ kg

$$g_m = \frac{2 \times \pi \times 6.66 \times 10^{-11} \times (1.67 \times 10^{-27})^2}{6.626 \times 10^{-34} \times 3 \times 10^8} \approx 6 \times 10^{-39}$$

This interaction is universal, i.e. every particle, whether massive or massless, is subject to the gravitational force. The carrier of gravitational interactions is conjectured to be a graviton. The

2. One also hears about many other interactions like chemical, electrical, mechanical, atomic forces, but all such interactions belong to the above-mentioned interactions.

mass of graviton is zero and thus it travels with velocity which is equal to that of the light. This interaction is extremely small for elementary particles, and hence is usually neglected.

8.3.2 Electromagnetic Interactions

The theory of electromagnetism was formulated on the great experimental works, started with appearance of Coulomb law of electrostatics in 1776. Later due to the works of Faraday and Maxwell, a proper field theory of electromagnetic forces was finalized around 1860. The electromagnetic interaction provides a force between two charges, which is repulsive, if the two charges are similar (i.e. both positive or both negative) and attractive, if the charges are opposite (i.e. one is positive and other is negative). All the chemical and biological reactions (*including life*) are basically controlled by the electromagnetic interactions. This interaction is also responsible for formation of electron–positron pair from γ-rays, and vice versa. This interaction also plays a role in nuclear structure through the mutual repulsion between protons. It causes several electromagnetic decays like decay of neutral pion into two photons. The dimensionless coupling constant for electromagnetic interactions is given by

$$\alpha = \frac{e^2}{4\pi\varepsilon_0 \times \hbar c}$$

$$= \frac{2\pi \times (1.602 \times 10^{-19})^2}{4\pi \times 8.854 \times 10^{-12} \times 6.626 \times 10^{-34} \times 3 \times 10^8}$$

$$= \frac{1}{137}$$

The characteristic interaction time of electromagnetic interactions is $\approx 10^{-16}$ s. The carrier of this interaction is a photon with rest mass zero and the range of this interaction is infinite.

Until 1930s, only the above two forces were known. Around 1934, it was realized that there are other kinds of natural forces, which were called strong and weak forces.

8.3.3 Strong Interactions

Strong interactions hold several nucleons together in a nucleus against the electromagnetic repulsion of the protons. Strong force is charge independent, i.e. it is the same between p–p, n–n and p–n. The range of the strong interactions is about 10^{-15} m and the interaction time is about 10^{-23} s. This is roughly the time taken by light to travel across a proton. In 1935, Yukawa explained this force at nuclear level through the exchange of particles called *pions.* However, the dimensionless pion-nucleon coupling constant representing the strength of strong interactions is about $\frac{g^2_{NN\pi}}{\hbar c} = 14$, indicating that the strength of electromagnetic interactions is about a one thousandth part (10^{-3}) of the strong interactions. Strong interactions are responsible for α-decay, nuclear reactions, production of hadrons, etc. In fact, strong interactions occur at the quark level through exchange of so-called *gluons* carrying spin 1 and massless, and nuclear forces are expected to be residual forces arising out of the quark level interactions, similar to the van der Waals forces in electromagnetism.

8.3.4 Weak Interactions

Unlike other interactions, weak interactions are unique in that they do not form any bound system, and only show through decay of the elementary particles. For instance, weak interaction is always responsible for all radioactive decays in which β-particles (e^- or e^+) are involved. The range of weak interactions is very small $\approx 10^{-18}$ m, which is about 0.1% of the diameter of a proton. The dimensionless coupling constant for weak interactions is $g_w \approx 10^{-13}$ and the interaction time $\approx 10^{-10}$ s. In 1934, Fermi developed the first theory of weak interactions known as the *Fermi theory of beta decay* with coupling $G_F = 1.166 \times 10^{-5}$ GeV^{-2}. Later, the weak interactions were thought to be manifestation of the exchange of certain spin one particles. Final theory of weak interactions, known as the Glashow, Weinberg and Salam theory in 1967, involves intermediate vector *bosons*, $W^{\pm}$, Z^0, as the carrier of weak interactions. They have masses greater than 80 GeV and spin = 1. A summary of various interactions is shown in Table 8.1.

TABLE 8.1 Basic properties of fundamental interactions

Interaction	*Strength*	*Interaction time* (s)	*Range* (m)	*Quanta of force*	*Spin*
Strong	1	10^{-23}	10^{-15}	π-mesons, gluons	0 1
Electromagnetic	$\sim 10^{-3}$	10^{-16}	Infinite	Photon	1
Weak	10^{-13}	10^{-10}	10^{-18}	*W*, *Z* bosons	1
Gravitational	10^{-39}	–	Infinite	Graviton[3]	2

8.4 CLASSIFICATION OF ELEMENTARY PARTICLES

In cosmic rays and high-energy accelerator experiments, a large variety of elementary particles were discovered. The total number of elementary particles around 1957 was about 30 and by 1965, this number rose to about 200 and now it has become over 500. As the number kept on growing further, so need was felt to classify these particles into different groups. All fundamental forces can act on the elementary particles at the same time. Since they have differences in their strengths, interaction time and range, etc., these parameters can be used to classify the elementary particles. The most common classification is as shown in Figure 8.1, in which the elementary particles are divided into two main categories:

- Quanta of forces.
- Matter particles.

Matter particles are classified into the following two categories:

- Leptons.
- Hadrons.

The hadron family further consists of baryons and mesons. This classification has some similarities with atomic structure. All the leptons are electron-like, quanta are photon-like, whereas the baryons are nucleon-like, and mesons are pion-like in nature.

3. Graviton is a hypothetical particle, whose existence is yet to be confirmed experimentally.

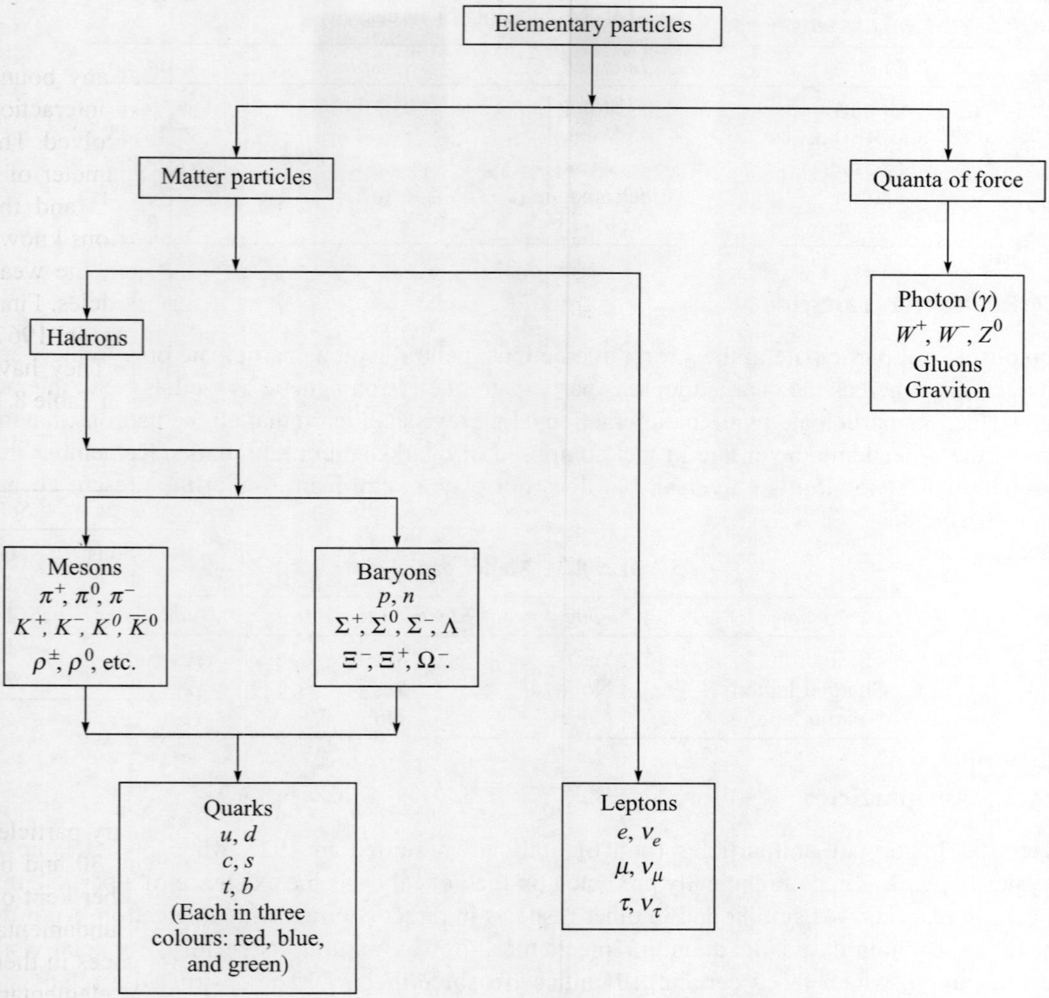

Figure 8.1 Classification of elementary particles.

8.4.1 Quanta of Forces

Corresponding to four different interactions, strong, weak, electromagnetic and gravitational, there are certain quanta of the forces which act as the mediator of these interactions. Their properties in brief are given in Table 8.2. *W*- and *Z*-bosons, having respective decay widths of 2.1 GeV and 2.5 GeV, decay to lepton pairs or hadrons. Decay width (Γ) of any unstable particle is related to its lifetime (τ) as

$$\Gamma = \hbar/\tau$$

Photon and gluons are electrically neutral and massless. These are treated as stable particles. There are eight kinds of gluons differing in their colour quantum number.

TABLE 8.2 Quanta of forces

Quanta	*Interaction*	*Spin*	*Mass*
Gluons	Strong	1	0
W^+ and W^- bosons	Weak	1	80.4 GeV
Z boson	Weak	1	91.2 GeV
Photon	Electromagnetic	1	0
Graviton	Gravity	2	0

8.4.2 Matter Particles

Leptons do not participate in the strong interactions; neutral leptons participate only in the weak interactions, whereas the charged leptons participate in electromagnetic as well as weak interactions. Hadrons participate in all the interactions. Later we shall learn that all the hadrons, having size of the order femtometer, are in fact composed of quarks and/or antiquarks. Remember that gravitational interaction is universal, but does not play a significant role at the present energy scales (Table 8.3).

TABLE 8.3 Matter particles

Particles	*Strong*	*Electromagnetic*	*Weak*
Hadrons	Yes	Yes	Yes
Charged leptons	No	Yes	Yes
Neutral leptons	No	No	Yes

8.4.3 Antiparticles

Scientific history of antiparticles (and of antimatter) started in 1928 when Nobel laureate physicist P.A.M. Dirac dramatically predicted on theoretical basis the existence of a particle that in certain respects was similar and in other respects in direct contrast with an electron. After the successes of nonrelativistic quantum mechanics, it was natural to perform its relativistic generalization, which faced certain difficulties. To solve them, Dirac constructed a relativistic equation, now known as the *Dirac equation*, which has four solutions. First two solutions were used to describe electron (e^-), the last two solutions, having negative energies, were interpreted as antiparticle to the electron. Imagine the amazement when in 1932 Carl Anderson observed this antiparticle in a cloud chamber. Later, it was christened positron which has the same mass as that of the electron but opposite charge (e^+). The discovery of the positron was an important event for two reasons. Firstly, it marked the beginning of a series of discoveries of new elementary particles in the cosmic radiation made in the next two decades. Secondly, it provided the first experimental proof of relativistic quantum theory. There has been no looking back on the story of antiparticles since then, as the idea of antiparticle is not limited to electrons only. In later experiments, pions and muons were found to come with both particle and antiparticle species. In 1957, *CPT* (C = charge conjugation, P = parity, T = time reversal) theorem was formulated (see Section 8.7.10) which enunciates that "*for every particle that exists in nature there is a corresponding antiparticle*". The theorem also predicts that the antiparticle has the same mass as its particle, the same lifetime (if the particle is not stable), the opposite charge and the opposite of all other internal quantum numbers. Certain particles act as their own antiparticle, like photon and π^0.

8.4.4 Observation and Production of Antimatter

Now a question arises. How to create antiparticles? A pair of a particle and its antiparticle can be created from energy in the form of electromagnetic radiation. This phenomenon is known as *pair creation*. In fact, in the Anderson experiment, high-energy cosmic rays hit the lead piece in the cloud chamber and created electron–positron pairs, and the two particles curved in opposite directions in the applied magnetic field. Reverse of this process, known as *pair annihilation* is also possible. A particle and its antiparticle upon coming in contact with each other annihilate yielding energy in the form of radiation. The phenomena of pair production and particle–antiparticle annihilation prove conclusively the equivalence[4] and interconvertibility of mass and energy.

Today, antiparticles of all the basic constituents of matter, such as leptons and quarks, and quanta of forces have been observed experimentally. Also antiproton ($\overline{p}$), antineutron ($\overline{n}$) and a variety of others in this category (hadrons) also have been observed. In 1955, an accelerator Bevatron was built at Berkeley which gave just enough energy to create antiprotons. This was achieved by accelerating protons (p) to sufficient energy (6.2 GeV), colliding them with nuclei to generate the following process:

$$p + p \rightarrow 3p + \overline{p}$$

This was followed by the discovery of antineutron ($\overline{n}$) in 1956 and anti-Λ hyperon in 1958. Then there are instances in radioactive substances where an antiparticle, like positron or antineutrino, is emitted by the nucleus which is made up of matter. In recent years researchers have attempted to produce atoms (or better term is antiatoms) made up of antiparticles only. Simplest antiatom[5] should be *antihydrogen* formed by a positron orbiting around an antiproton. Experiments, using antiproton decelerator and plasma bottles to collect antiprotons and positrons, are in progress which aim to synthesize antihydrogen at sub-kelvin temperatures.

8.5 MASS SPECTRA AND DECAYS OF ELEMENTARY PARTICLES

Except for a few particles, like electrons, neutrinos, and protons, almost all the particles are unstable, and decay to other lighter particles through one or more of these interactions. Every particle is denoted by a special symbol, and its electric charge is usually shown as superscript. Besides the charge, the particles carry unique properties, like mass, magnetic moments, and other quantum numbers. Unstable particles may have several decay modes.

---

4. Incidentally, this equivalence was predicted by Albert Einstein much earlier (1905) in his theory of relativity. He demonstrated that energy, (E) momentum $(\vec{p})$ and mass (m) of a free particle are related as $E^2 = p^2c^2 + m^2c^4$. Thus, a particle at rest carries a certain amount of energy ($E = mc^2$) called rest mass energy.
5. Another system of interest could be atomic protonium, a bound state of proton and antiproton. It is similar to a hydrogen atom, except that its electron is replaced by antiproton. However, this system is highly unstable, due to rapid $p\overline{p}$ annihilation.

8.5.1 Leptons

Electron, the first lepton, was discovered in 1897. In 1937 Anderson and others found a particle with a mass of about 105 MeV, which was named muon[6]. The experiments on the energy spread of the electrons from nuclear β-decay, made in the 1920s, indicated an apparent failure of the energy and momentum conservation laws. Later work showed that the conservation of angular momentum was also violated. In 1933, W. Pauli suggested that these conservation laws could be saved by postulating the existence of a spin $1/2\,\hbar$ particle of zero (or nearly zero) mass and no electrical charge, which was later called neutrino. In 1962, it was recognized that there are two kinds of neutrinos. Until 1975, four leptons[7] were known: electron, muon, electron-neutrino and muon-neutrino. In 1975, a fifth lepton named tau was discovered by Perl and his collaborators. This lepton has a mass about 3500 times the mass of the electron, that means even greater than that of the proton. Tau-lepton also has its corresponding neutrino partner, called tau-neutrino. So now the total number of leptons is six, three charged and three neutral: e^-, μ^-, τ^-, ν_e, ν_μ and ν_τ; and of course their antiparticles. Leptons do not have any internal structure at the present energy scales, and all of them have spin 1/2. Range of their masses lies from zero to 1.78 GeV. Basic properties of the leptons and antileptons are given in Tables 8.4 and 8.5. Some of their special features are discussed below.

TABLE 8.4 Summary of lepton properties

Particle	*Charge of particle*	*Mass* (MeV)	*Spin* ($\hbar$)	*Lifetime*	*Decay modes*
e^-	−1	0.511	$\frac{1}{2}$	Stable	Nil
ν_e	0	<2 eV	$\frac{1}{2}$	Stable	Nil
μ^-	−1	105.7	$\frac{1}{2}$	2.2×10^{-6} s	$\mu^- \to e^- + \nu_\mu + \overline{\nu}_e$
ν_μ	0	<0.19	$\frac{1}{2}$	Stable	Nil
τ^-	−1	1777	$\frac{1}{2}$	0.29×10^{-12} s	$\tau^- \to \mu^- + \overline{\nu}_\mu + \nu_\tau$ $\tau^- \to K^- + \nu_\tau$
ν_τ	0	<18.2	$\frac{1}{2}$	Stable	Nil

6. In the beginning this was also called μ-mesons, but now the word meson has been reserved for different type of particles, i.e. hadrons.
7. Note the nomenclature, lepton, meson and baryon, has only historical relevance. For instance, originally, leptons (meaning *light: Greek*) were expected to have small masses. Similarly, mesons (meaning *intermediate*) were though to have intermediate masses, smaller than that of the baryons (meaning *heavy*). However, now the tau-lepton and many meson (D and B) states, heavier than the proton and even other baryons, have been observed in the high-energy accelerator experiments.

TABLE 8.5 Summary of antilepton properties

(Mass, spin and lifetime of antileptons are identically the same as that of the corresponding lepton partner)

Antiparticle	*Charge of antiparticle*	*Decay modes*
e^+	+1	Nil
$\bar{\nu}_e$	0	Nil
μ^+	+1	$\mu^+ \to e^+ + \bar{\nu}_\mu + \nu_e$
$\bar{\nu}_\mu$	0	Nil
τ^+	+1	$\tau^+ \to \mu^+ + \nu_\mu + \bar{\nu}_\tau$ $\tau^+ \to K^+ + \bar{\nu}_\tau$
$\bar{\nu}_\tau$	0	Nil

Electron (e^-)

(i) This particle is negatively charged. e^--electron and its antiparticle (e^+) is known as the positron.

(ii) Electrons e^- are generally produced in two ways:

(a) In metals there are lot of free e^-, which can be ejected from the surface of metal simply by heating the metal piece to high temperature. They can also be produced by ionizing the gases and extracting e^-, using suitable electric or magnetic fields.

(b) In radioactivity, e^- are emitted when a neutron decays to a proton

$$n \to p + e^- + \bar{\nu}_e$$

The source of e^+ is also radioactive decay in which a proton (inside nucleus) decays to a neutron

$$p \to n + e^+ + \nu_e$$

or by electron–positron pair production by gamma rays

$$\gamma \to e^+ + e^-$$

(iii) Mass of both electron and positron is 0.511 MeV.

(iv) Magnetic moment of electron is one Bohr magneton, $\mu_e = e\hbar/2m_e c$.

(v) e^- is stable particle. However, when an e^+ meets e^-, both of them annihilate emitting photons as

$$e^+ + e^- \to \gamma + \gamma$$

Muon (μ^-)

(i) After positron, muon was the next particle observed in the cosmic ray experiments in 1936. Like electron, this particle occurs as negatively charged μ^- and its antiparticle is μ^+.

(ii) μ-mesons are generally produced in the pions decays:

$$\pi^- \to \mu^- + \bar{\nu}_\mu$$
$$\pi^+ \to \mu^+ + \nu_\mu$$

(iii) Mass of both muons $\mu^{\pm}$ is 105.7 MeV.
(iv) Magnetic moment of muon is $\mu_{\mu} = e\hbar/2m_{\mu}c$.
(v) Muons $\mu^{\pm}$ decay as under:

$$\mu^- \rightarrow e^- + \overline{\nu}_e + \nu_{\mu}$$
$$\mu^+ \rightarrow e^+ + \nu_e + \overline{\nu}_{\mu}$$

(vi) Lifetime of $\mu^{\pm}$ is 2.2×10^{-6} s.

Tau-Lepton (τ^-)

(i) This particle occurs as negatively charged τ^- and its antiparticle is τ^+. These particles are originally observed in the electron–positron collision at very high energies, as

$$e^+ + e^- \rightarrow \tau^+ + \tau^-$$

followed by decays of the tau-leptons.
(ii) Mass of both $\tau^{\pm}$ is 1.777 GeV.
(iii) Like muon, $\tau^{\pm}$ can also decay through pure leptonic modes

$$\tau^- \rightarrow \mu^- + \overline{\nu}_{\mu} + \nu_{\tau}$$
$$\tau^+ \rightarrow \mu^+ + \nu_{\mu} + \overline{\nu}_{\tau}$$

however, being massive they have several decay modes. Their dominant modes emit one or more hadrons. For instance,

$$\tau^- \rightarrow K^- + \nu_{\tau}$$

(iv) Lifetime of $\tau^{\pm}$ is much smaller than that of the muon, i.e., 0.29×10^{-12} s.

Neutrinos

Like charged leptons, neutrinos have antipartners called *antineutrinos*, which have been observed experimentally. All the neutrinos (and antineutrinos) are electrically neutral, massless[8] and have spin ½ $\hbar$. They participate in the weak interactions only.

Electron-Neutrino (ν_e)

(i) ν_e and $\overline{\nu}_e$ are produced when in a nucleus a proton or a neutron decays as:

$$p \rightarrow n + e^+ + \nu_e$$
$$n \rightarrow p + e^- + \overline{\nu}_e$$

It may be noted that proton can convert to neutrons only when inside a nucleus.
(ii) Following neutrino-nucleon reactions take place in nature:

$$\nu_e + n \rightarrow p + e^-$$
$$\overline{\nu}_e + p \rightarrow n + e^+$$

However, the cross-section for these reactions is extremely small $\sim 1.2 \times 10^{-43}$ cm^2.

8. Neutrinos were originally thought to be massless; however, some recent experiments on solar-neutrinos and neutrino oscillations observed in super kamiokande detector in Japan in 1998 endow some tiny special masses with them.

(iii) Experimental upper limit on its mass is 2 eV based on tritium beta decay.

In order to detect this small cross-section, very large amount of target material and high flux of neutrinos were required. Finally, the existence of neutrino (in fact antineutrino) was established in nuclear reactor experiment by Reines and Cowan in 1959. (See Sections 3.6.6 and 3.6.7.)

Muon-Neutrino (ν_μ)

(i) Since neutrinos are also emitted along with muons in the decays of pions and kaons, a question of considerable significance was asked: Are the neutrinos associated with muons necessarily the same as those emitted in nuclear β-decay? In 1962, experimental evidence showed that they are different, and thus muon-neutrino was identified, when it was realized that the reaction

$$\nu_\mu + n \to \mu^- + p,$$

performed with neutrinos beam obtained from the pion or kaon decays, always takes place but never the following reaction:

$$\nu_\mu + n \to e^- + p$$

In this experiment, muon-neutrinos were taken from the pion decay. Muon-neutrinos are thus different from the electron-neutrinos.

(ii) ν_μ and $\overline{\nu}_\mu$ are produced in the decay of π-mesons as under:

$$\pi^+ \to \mu^+ + \nu_\mu$$
$$\pi^- \to \mu^- + \overline{\nu}_\mu$$

(iii) Experimental upper limit on their mass is 0.19 MeV.

Tau-Neutrino (ν_τ)

(i) ν_τ and $\overline{\nu}_\tau$ are produced in the tau decays.
(ii) Experimental upper limit on their mass is 18.2 MeV.

8.5.2 Hadrons

The hadron family contains baryons and mesons:

(i) Baryons are particles having mass in the 938 MeV to few GeV regions and spin half-integral multiple of $\hbar$. Since they follow Fermi–Dirac statistics, these are known as the fermions.
(ii) Mesons are particles with spin angular momentum either zero or integral multiple of $\hbar$. They follow Bose–Einstein statistics and are known as the bosons. Some of the properties of the mesons are given in the following paragraph.

The hadrons appear in certain grouping, which are designated by their spin (J) and parity (P). That means, all the particles in one grouping carry the same spin and same parity. The list of the hadrons started growing with H. Yukawa's prediction of three pions (π^+, π^0, π^-) in 1934 to explain the nuclear force between protons and neutrons. These were called *mesons* (meaning *middle: Greek*) as their masses lie between that of the electron and the proton. Pions and many such particles were discovered in the pioneering works on cosmic rays, which were known to leave traces in a *cloud chamber*. With the construction of high-energy particle accelerators, hundreds of new and unstable particles, carrying even high spins (up to 6 $\hbar$ for mesons, and 13/2 $\hbar$ for baryons), were created in the middle of the 20th century. Collectively these are known as *hadrons* (meaning *strong: Greek*). Most of the observed hadrons having spin higher than 1/2 are found to decay within around 10^{-23} s, and are called *resonances*, whereas the lowest spin states decay through weak interaction or electromagnetic interaction (lifetimes ~ 10^{-16} s). Only proton so far has been found to be stable.

8.5.3 Mesons

These are the particles, whose masses lie between ~ 130 MeV to few GeV. Mesons with spin zero to 6 $\hbar$ have been observed. So all of them are bosons. Most of them decay through strong interactions, and are called *meson resonances*. Lightest of them, having spin zero and negative parity ($J^P = 0^-$), π-mesons (or pions), K-mesons (or kaons) and η-mesons belong to this group. Obviously they have zero magnetic moments. In Table 8.6, mass, lifetime and dominant decay modes of $J^P = 0^-$ mesons are given. We now discuss the some more properties of these mesons.

TABLE 8.6 Masses, lifetime and dominant decay modes of mesons ($J^P = 0^-$)

Particle	*Antiparticle*	*Mass* (MeV)	*Lifetime* (s)	*Dominant decay modes*
π^-	π^+	139.6	2.6×10^{-8}	$\pi^- \to \mu^- + \overline{\nu}_\mu$
π^0	π^0	135.0	0.84×10^{-16}	$\pi^0 \to 2\gamma$
K^+	K^-	493.7	1.24×10^{-8}	$K^+ \to \mu^+\nu_\mu/2\pi$
K^0	$\overline{K}_0$	497.6	$K_S^0 = 0.90 \times 10^{-10}$	$K^0 \to 2\pi/3\pi/\pi e\nu_e$
			$K_L^0 = 5.1 \times 10^{-8}$	
η	η	547.5	0.5×10^{-18}	$\eta^0 \to 3\pi/2\gamma$

π-Mesons (Pions)

In 1935, H. Yukawa proposed that the strong force between nucleons in a nucleus must be mediated by particles, called *pions* (π), of around 140 MeV mass. Pions exist as positively (π^+), negatively (π^-) and neutral (π^0) states. In 1947, a University of Bristol group was the first to provide the evidence of charged pions by observing their weak decays in the nuclear emulsion photographs. It was also found that the muons arising in these decays possessed a unique energy, thus indicating that the pions decayed into two particles as

$$\pi^+ \to \mu^+ + \nu_\mu \to e^+ + \nu_e + \overline{\nu}_\mu + \nu_\mu$$

$$\pi^- \to \mu^- + \overline{\nu}_\mu \to e^- + \overline{\nu}_e + \nu_\mu + \overline{\nu}_\mu$$

Evidence for the neutral pion was obtained in 1950 through the appearance of electron–positron pairs in the nuclear emulsions, since π^0 decays electromagnetically as

$$\pi^0 \to \gamma + \gamma$$

and the photons are materialized as electron–positron pairs.

(i) Pions exist as positively (π^+), negatively (π^-) and neutral (π^0) states.
(ii) Pions have also been produced by the following reactions:

$$p + p \rightarrow \pi^+ + d$$
$$p + n \rightarrow \pi^- + p + p$$
$$\gamma + p \rightarrow \pi^0 + p$$

(iii) Mass of $\pi^\pm$-meson is 139.9 MeV and that of π^0 is 135.0 MeV.
(iv) Charged pions have several possible weak decay modes, which are:

$$\pi^+ \rightarrow \mu^+ + \nu_\mu$$
$$\pi^- \rightarrow \mu^- + \bar{\nu}_\mu$$
$$\pi^\pm \rightarrow \pi^0 + e^\pm + \nu_e$$
$$\pi^- \rightarrow e^- + \bar{\nu}_e$$

The first two are the dominant decay modes. π^0 decays through several channels electromagnetically, however, its dominant modes are:

$$\pi^0 \rightarrow \gamma + \gamma$$
$$\pi^0 \rightarrow e^+ + e^- + \gamma$$

(v) Lifetime of charged pions is 2.60×10^{-8} s, and that of neutral pion is 0.84×10^{-16} s.

K-Mesons (Kaons)

(i) Kaons exist as positively (K^+), negatively (K^-) charged and their respective neutral partners (K^0 and $\bar{K}^0$). The first evidence of neutral and charged kaons was obtained in 1947 in the cosmic radiation studies with cloud chamber and nuclear emulsion by observing their decays emitting two charged pions ($\pi^+ + \pi^-$) or three charged pions ($\pi^+ + \pi^+ + \pi^-$).
(ii) Kaons having mass around 495 MeV are produced in the following reactions:

$$\pi^- + p \rightarrow K^0 + \Lambda^0$$
$$\pi^- + p \rightarrow n + K^+ + K^-$$
$$\pi^+ + p \rightarrow K^+ + \bar{K}^0 + p$$

(iii) Charged kaons decay dominantly through the following leptonic modes:

$$K^+ \rightarrow \mu^+ + \nu_\mu$$

the first evidence for which was obtained by Paris Group using double cloud chambers, or nonleptonically as

$$K^+ \rightarrow \pi^+ + \pi^0$$

(iv) Lifetime of $K^\pm = 1.24 \times 10^{-8}$ s.

η–Meson

(i) η-meson occurs only in uncharged state.
(ii) η-meson can be produced by the following reactions:

$$\pi^+ + {}^2_1\text{H} \rightarrow p + p + \eta$$
$$K^- + p \rightarrow \Lambda + \eta$$

(iii) η-meson decays through the following modes:

$$\eta \rightarrow \gamma + \gamma$$
$$\eta \rightarrow \pi^0 + \pi^0 + \pi^0$$
$$\eta \rightarrow \pi^+ + \pi^- + \pi^0$$
$$\eta \rightarrow \pi^+ + \pi^- + \gamma$$

(iv) Decay width Γ of η is 1.30 keV, giving its lifetime $\tau = 0.5 \times 10^{-18}$ s, since

$$t = \frac{\hbar}{\Gamma}$$

8.5.4 Baryons

Following the discovery of several mesons in cosmic rays and laboratory experiments, considerable research was carried out on the heavy unstable particles during the early 1950s, which revealed a rich variety of particles which were classified according to their masses, lifetimes and decays.

In Table 8.7, mass, magnetic moment and lifetime of $J^P = \frac{1}{2}^+$ baryons are given. The spin $1/2\hbar$ baryons are classified into two categories:

1. Nucleons
2. Hyperons

TABLE 8.7 Mass and lifetime of particles and antiparticles of spin 1/2 baryons

Baryon	*Antibaryon*	*Mass* (MeV)	*Baryon magnetic moment* (*nuclear magneton*)	*Lifetime* (s)
p	$\overline{p}$	938.3	2.79	Stable
n	$\overline{n}$	939.6	−1.91	885.7
Λ	$\overline{\Lambda}$	1115.7	−0.61	2.6×10^{-10}
Σ^+	$\overline{\Sigma}^-$	1189.4	+2.46	0.80×10^{-10}
Σ^0	$\overline{\Sigma}^0$	1192.6	–	7.4×10^{-20}
Σ^-	$\overline{\Sigma}^+$	1197.4	−1.16	1.48×10^{-10}
Ξ^0	$\overline{\Xi}^0$	1314.8	−1.25	2.90×10^{-10}
Ξ^-	$\overline{\Xi}^+$	1321.3	−0.65	1.64×10^{-10}

1. **Nucleons (p, n):** Protons, neutrons and their antiparticles belong to this category. The proton (meaning *first: Greek*) owes its name to the fact that the simplest nucleus, that of the first element—hydrogen, is just the proton itself. Both these particles have spin $1/2\,\hbar$ and hence they are fermions. Under the influence of nuclear force, both proton and neutron are stable inside the stable nuclei. However, a free neutron is unstable and decays as under:

$$n \rightarrow p + e^- + \overline{\nu}$$

with a half-life of about 886 s, i.e. 15 minutes. However, it acts as stable particle inside the nucleus. Unlike free neutron, a free proton is considered to be a stable particle. Proton and neutron have antibaryon partners called *antiproton* and *antineutron* respectively.

2. **Hyperons:** The term hyperons refer to those unstable particles which carry additional quantum numbers like strangeness to distinguish them from the nucleons. The well-known hyperons are: Λ (lambda), Σ (sigma), Ξ (xi) carrying spin ½ and Ω (omega) hyperons, which has spin 3/2.

(i) *Λ-Hyperon*

It was first detected in the cosmic rays in 1947. Later on it was also produced in laboratory by shooting 1.5 GeV π^--mesons in a cloud chamber containing hydrogen gas. In this reaction Λ was produced along with another particle K^0

$$\pi^- + p \rightarrow \Lambda + K^0$$

The charge of Λ is zero, and its antiparticle is $\overline{\Lambda}$. Mass of Λ-hyperon was found to be 1115.7 MeV and its half-life is 2.63×10^{-10} s. Main decay modes of Λ-hyperon are:

$$\Lambda \rightarrow p + \pi^-$$
$$\Lambda \rightarrow n + \pi^0$$

Like beta decay, Λ can also emit a lepton pair or even a photon as

$$\Lambda \rightarrow p + e^- + \overline{\nu}_e$$
$$\Lambda \rightarrow n + \gamma$$

though with a less probability. The last decay involves both weak and electromagnetic interactions.

(ii) *Σ-Hyperons*

There are three Σ-hyperons Σ^+, Σ^0 and Σ^-; their antiparticles are $\overline{\Sigma}^+$, $\overline{\Sigma}^0$ and $\overline{\Sigma}^-$. The Σ^+-hyperon was first identified in a cloud chamber in 1953. Next year, the negative counterpart, Σ^-, was observed in a diffusion chamber exposed to negative pion beam of the Brookhaven cosmotron. Lifetime of Σ^+ is 0.80×10^{-10} s and mass is 1189.4 MeV. The main decay modes of Σ^+ are the following weak decays:

$$\Sigma^+ \rightarrow p + \pi^0$$
$$\Sigma^+ \rightarrow n + \pi^+$$

However, it can also emit a lepton pair

$$\Sigma^+ \rightarrow \Lambda + e^+ + \nu_e$$

and even photon in the following decay,

$$\Sigma^+ \rightarrow p + \gamma$$

arising through a mixture of weak and electromagnetic interactions.
Lifetime of Σ^- is 1.48×10^{-10} s and mass is 1197.4 MeV. The main decay mode of Σ^- is

$$\Sigma^- \rightarrow n + \pi^-$$

and with less probability it decays to

$$\Sigma^- \rightarrow n + e^- + \overline{\nu}_e$$

The neutral member of this family, Σ^0, was observed later in accelerator experiments. Mass of Σ^0 is 1192.6 MeV, its main decay mode is

$$\Sigma^0 \rightarrow \Lambda + \gamma$$

with half-life of 7.4×10^{-20} s.

(iii) *Ξ-Hyperons*

Ξ-hyperon also known as cascade hyperon exists in two forms Ξ^- and Ξ^0; their corresponding antiparticles are $\overline{\Xi}^+$ and $\overline{\Xi}^0$.

Ξ^- has half-life 1.64×10^{-10} s and mass 1321.3 MeV. It decays dominantly as

$$\Xi^- \rightarrow \Lambda + \pi^-$$

However, it can also decay through the following decay modes:

$$\Xi^- \rightarrow \Lambda + e^- + \overline{\nu}_e$$
$$\Xi^- \rightarrow \Lambda + \mu^- + \overline{\nu}_\mu$$
$$\Xi^- \rightarrow \Sigma^- + \gamma$$

It is to be noted that no positively charged counterpart of Ξ^- has ever been observed. The Ξ^- hyperon was postulated to be a charge doublet by Gell-Mann with neutral hyperon Ξ^0 as its partner. During the period 1954–56, the emphasis shifted from the studies of cosmic rays to experiments on the particle accelerators, in which several above mentioned (e.g. K, Λ, Σ and Ξ) and new hadrons were observed. In such experiments, Ξ^0 has half-life 2.90×10^{-10} s and mass 1314.8 MeV. Its dominant decay is

$$\Xi^0 \rightarrow \Lambda + \pi^0$$

though it also decays through the following modes:

$$\Xi^0 \rightarrow \Sigma^+ + \mu^- + \overline{\nu}_\mu$$
$$\Xi^0 \rightarrow \Sigma^+ + e^- + \overline{\nu}_e$$
$$\Xi^0 \rightarrow \Lambda + \gamma$$
$$\Xi^0 \rightarrow \Sigma^0 + \gamma$$

Here also photon emitting decays occur through combined weak and electromagnetic forces.

(iv) *Ω-Hyperon*

Ω^- was discovered in the following reaction:

$$K^- + p \rightarrow \Omega^- + K^+ + K^0$$

It has negative charge, and its main decay modes are:

$$\Omega^- \rightarrow \Xi^0 + \pi^-$$
$$\Omega^- \rightarrow \Xi^- + \pi^0$$
$$\Omega^- \rightarrow \Lambda + K^-$$

Its half-life is 0.8×0^{-10} s and mass is 1672.4 MeV. Its magnetic moment is −2.02 nuclear magneton.

Baryon Resonance States

A large number of baryon states carrying spin $1/2\hbar$ to $11/2\hbar$ have been observed. Out of these baryons, eight $J^P = 1/2^+$ and ten $J^P = 3/2^+$ are called *low-lying baryons*. Like mesons, most of the baryons with spin $>1/2\hbar$ decay quickly through strong interactions, living around 10^{-23} s, roughly the time taken by light to travel across a proton. These are called *baryon resonances*. Out of the ten $J^P = 3/2^+$ baryons, only Ω^- being stable against the strong interactions, decays through the weak interactions.

8.6 QUANTUM NUMBERS

In physics a conservation law states that a particular property of an isolated physical system does not change as the system evolves in time. In classical physics, we know that a process can take place if it is allowed kinematically. We are familiar with some of such properties like energy, momentum, angular momentum, and charge, which are conserved in a classical reaction. These properties also remain conserved in a quantum process or reaction. Besides these, there are certain new properties called *quantum numbers*, which are relevant to microscopic physics only. We do not know all the quantum numbers at the start, their presence is in fact deduced from experimental observations of various decays and reactions involving elementary particles. Therefore, before discussing the conservation laws in particle physics, we discuss those quantum numbers, which are relevant to nuclear and particle physics.

8.6.1 Spin (S)

All the elementary particles can be classified as either bosons or fermions depending upon whether their spins are integral multiple of $\hbar$ or half-odd integral multiple of $\hbar$ respectively. All quanta of the fundamental interactions and mesons are bosons, whereas the leptons, baryons and quarks are fermions. There exist $2S + 1$ components of the spin along third axis, which are denoted by S_3. Thus, for a spin $\frac{1}{2}\hbar$ particle, say electron, $2S + 1 = 2$, so the possible values of $S_3 = \frac{1}{2}$ (for spin up) and $S_3 = -\frac{1}{2}$ (for spin down).

For composite states, spin adds vectorially to the orbital angular momentum to give rise to total angular momentum,

$$\vec{J} = \vec{L} + \vec{S}$$

that means J can have discrete values starting from $|L - S|$ to $(L + S)$ differing by unity. For example, $L = 2$ and $S = \frac{1}{2}$, add to give

$$J = 3/2,\ 5/2$$

whereas the third component will have $(2J + 1)$ values.

For example, if $J = \frac{5}{2}$, the values of the third component will be

$$\frac{-5}{2}, \frac{-3}{2}, \frac{-1}{2}, \frac{1}{2}, \frac{3}{2}, \frac{5}{2}$$

8.6.2 Charge (*Q*)

All the elementary particles carry either positive, negative or zero electric charge. Measuring in units of proton charge, they carry +1, –1 or zero (rarely +2) value of the electric charge. Charges of the various elementary particles have already been given in respective tables.

8.6.3 Parity (*P*)

Parity causes reflection of the wave function, i.e. changes x co-ordinate to $-x$, y to $-y$ and z to $-z$, or

$$P\psi(\vec{r}) = \psi(-\vec{r})$$

If operator P is applied second time, the original wave function is obtained, or

$$P^2\psi(\vec{r}) = P\psi(-\vec{r}) = \psi(\vec{r})$$

Therefore,

$$P^2 = 1 \quad \text{or} \quad P = \pm 1$$

If $P = +1$, then wave function is said to have even parity. And if $P = -1$, then wave function is said to have odd parity. Thus, depending upon the behaviour of particle wave functions,

$$\psi(-x, -y, -z) = \pm\psi(x, y, z)$$

they carry either positive or negative parity quantum number. Mesons (bosons) can carry positive or negative parity. Baryons are assigned positive parity, whereas their antiparticles (antibaryons) carry negative parity. Similarly, antileptons carry opposite parity to leptons. Photon and gluons have negative parity. Since weak interactions can violate parity, no fixed parity can be assigned to *W*- and *Z*-bosons.

Parity is a multiplicative quantum number, i.e. parity of a composite state is given by product of parities of its constitutents.

8.6.4 Lepton Number (*L*)

All the leptons are assigned a lepton number +1 and their respective antiparticles are assigned a lepton number –1. Therefore, lepton number of e^-, μ^-, τ^-, ν_e, ν_μ and ν_τ is +1, and for their antiparticles, it is –1. In fact, there are three lepton numbers: electron-type, muon-type and tau-type. Thus, the electron and its neutrino ν_e have electron-lepton number $L_e = 1$ and positron and antineutrino $\bar{\nu}_e$ have $L_e = -1$. All other particles have $L_e = 0$. Muon (μ^-) and its neutrino ν_μ have muon-lepton number $L_\mu = 1$ and μ^+ and antineutrino $\bar{\nu}_\mu$ have $L_\mu = -1$. All other particles have $L_\mu = 0$, and similarly for the τ-lepton and its neutrino. Leptons can thus be split into three families, namely (e^-, ν_e), (μ^-, ν_μ) and (τ^-, ν_τ), each having its own family lepton number. For all other particles, like hadrons and quanta of forces, lepton number is taken to be zero.

8.6.5 Baryon Number (*B*)

All baryons are assigned a baryon number +1 and for the corresponding antiparticles, the baryon number is –1. Thus, baryon number of *p*, *n*, Λ^0, Σ^+, Σ^-, Σ^0, Ξ^-, Ξ^0 and Ω^- is +1, and their corresponding antiparticles have baryon number as –1. For all other matter particles, like mesons, leptons, and quanta of forces, like photon, gluons, *W* and *Z* bosons, the baryon number is assigned to be zero.

8.6.6 Isospin (*I*)

Besides having same spin, proton and neutron have many similarities. Particularly, it is well known that nuclear forces (strong forces) are charge-independent. That means strong binding force between *n–p*, *p–n*, and *n–n* is essentially the same. The small difference between them is attributed to the electromagnetic interactions, which arise due to difference in charges of proton and neutron. These facts suggested the existence of new kind of nuclear spin called *isospin*. Stating differently, in the world of strong interactions only, we can think of proton and neutron as the two orthogonal states of the same particle called *nucleon*[9].

The isospin is a quantity introduced to describe groups of particles (hadrons) which have nearly same mass. A general isospin state is described by two quantum numbers, I and I_3, where I_3 can have values ranging from $-I$ to I separated by unity. Projections of isospin along the third axis, I_3, appear as different charge states of the corresponding particle. For a group of two particles or doublet (like proton and neutron), $I = \frac{1}{2}$, and for triplet (like Σ^+, Σ^- and Σ^0) $I = 1$. Similar to spin, there are $2I + 1$ components of the isospin along third axis, which are distinguished by I_3. Therefore, for a doublet (*p–n*) $2I + 1 = 2$, so the possible values of $I_3 = \frac{1}{2}$ (for proton) and $I_3 = -\frac{1}{2}$ (for neutron).

Among mesons, K^+ and K^0 form isospin doublet. Similarly, for a triplet (Σ^+, Σ^0, Σ^-), isospin is described by isospin vector, i.e. $I = 1$. Therefore, $I_3 = +1$ corresponds to Σ^+, $I_3 = 0$ corresponds to Σ^0 and $I_3 = -1$ corresponds to Σ^-. Among mesons, pions (π^+, π^0, π^-) form isospin vector.

One may observe an interesting relation among the electric charge, third component of isospin and baryon number

$$Q = I_3 + \frac{B}{2}$$

which gives the charge (Q) present on each member of the isospin multiplet. For example, baryon number of *p–n* multiplet is +1 and I_3 for proton is $+\frac{1}{2}$. Therefore, $Q = \frac{1}{2} + \frac{1}{2} = +1$, indicating that +1 is the charge present on proton. For neutron I_3 is $-\frac{1}{2}$. Therefore, $Q = -\frac{1}{2} + \frac{1}{2} = 0$, indicating that no charge is present on neutron.

9. This is very similar to the ordinary spin-up and spin-down states of an electron, which degenerate till some magnetic field is applied.

8.6.7 Strangeness (*S*)

Soon after the discovery of the hyperons, it was realized that these particles were behaving in a strange manner. They are produced through the strong interactions, and have large production cross-sections. However, they have lifetimes characteristics of the weak interactions. Their decays are always by way of weak interaction (or in few cases, weak electromagnetic interactions). Another strange fact about these particles is that they are always produced in pairs, e.g. *K*-meson in association with Λ or Σ-hyperon. This effect was called associated production by A. Pais. For instance, *K*-meson and Λ-baryon were discovered in a pion-nucleon reaction,

$$\pi^- + p \to \Lambda(\to p + \pi^-) + K^0(\to \pi^+ + \pi^-)$$

These are called *strange* particles as they are produced in pairs through strong interaction (lifetimes ~ 10^{-23} s), but decay through weak interactions (lifetimes ~ 10^{-8} s). This situation led to use of the expression "Strange particles" to describe *K*-mesons and hyperons, as all this was certainly puzzling. This finally led to suggest that a new quantum number might be associated with these new particles, which was called *strangeness* (*S*) by Gell–Mann and Nishijima in 1953 to account for the strange behaviour of *K*-mesons and hyperons. This quantum number was conserved in the strong interactions, but not in weak decays. A nonzero strangeness, is assigned to them in contrast to the nucleons, pions and photon, which have zero strangeness, $S = 0$. For the remaining particles, strangeness is assigned using the Gell–Mann–Nishijima relation given below.

8.6.8 Gell–Mann–Nishijima Scheme

Gell–Mann–Nishijima observed that the electric charge (in units of *e*) of a hadron can be related to its isospin, baryon number and strangeness as follows:

$$Q = I_3 + \frac{B + S}{2} \tag{8.1}$$

For instance, for proton, we have $Q = 1$, $I_3 = ½$, $B = 1$ and $S = 0$, which obviously satisfy this relation. Similarly, quantum numbers of positively charged pion, i.e. $Q = 1$, $I_3 = 1$, $B = 0$ and $S = 0$, satisfy it. Using the relation, strangeness of various particles can be assigned. For example, for Λ-hyperon, using $Q = 0$, $I_3 = 0$ and $B = +1$ Gell–Mann–Nishijima relation gives strangeness $S = -1$. Similarly, for Ω^--hyperon, carrying $Q = -1$, $I_3 = 0$ and $B = +1$, the relation yields

$$S = 2(Q - I_3) - B = 2(-1 - 0) - 1 = -3$$

In Table 8.8, we have given masses and various quantum numbers like charge, baryon number, isospin and strangeness of the mesons. Similarly, properties of the baryons are given in Tables, 8.9(a) and 8.9(b).

TABLE 8.8 Quantum numbers of mesons

Particle	*Mass* (MeV)	*Spin* ($\hbar$)	*Charge* (Q)	*Isospin* (I)	*Isospin* (I_3)	*Baryon number* (B)	*Strangeness* (S)	*Hypercharge* (Y)
K^+	493.7	0	+1	1/2	+1/2	0	+1	1
K^0	497.6	0	0	1/2	−1/2	0	+1	1
π^+	139.6	0	+1	1	+1	0	0	0
π^0	135.0	0	0	1	0	0	0	0
π^-	139.6	0	−1	1	−1	0	0	0
$\overline{K}^0$	497.6	0	0	1/2	+1/2	0	−1	−1
K^-	493.7	0	−1	1/2	−1/2	0	−1	−1
η	547.5	0	0	0	0	0	0	0

TABLE 8.9(a) Quantum numbers of baryons $J^P = \frac{1^+}{2}$

Particle	*Mass* (MeV)	*Spin* ($\hbar$)	*Charge* (Q)	*Isospin* (I)	*Isospin* (I_3)	*Baryon number* (B)	*Strangeness* (S)	*Hypercharge* (Y)
p	938.6	1/2	1	1/2	+1/2	1	0	1
n	939.6	1/2	0	1/2	−1/2	1	0	1
Λ	1115.6	1/2	0	0	0	1	−1	0
Σ^+	1189.4	1/2	1	1	+1	1	−1	0
Σ^0	1192.6	1/2	0	1	0	1	−1	0
Σ^-	1197.4	1/2	−1	1	−1	1	−1	0
Ξ^0	1314.7	1/2	0	1/2	+1/2	1	−2	−1
Ξ^-	1321.2	1/2	−1	1/2	−1/2	1	−2	−1

TABLE 8.9(b) Quantum numbers of $J^P = \frac{3^+}{2}$ baryons

Particle	*Mass* (MeV)	*Spin* ($\hbar$)	*Charge* (Q)	*Isospin* (I)	*Isospin* (I_3)	*Baryon number* (B)	*Strangeness* (S)	*Hypercharge* (Y)
Δ^{++}	1232	3/2	2	3/2	3/2	1	0	1
Δ^+	1232	3/2	1	3/2	+1/2	1	0	1
Δ^0	1232	3/2	0	3/2	−1/2	1	0	1
Δ^-	1232	3/2	−1	3/2	−3/2	1	0	1
Σ^{*+}	1382.8	3/2	1	1	+1	1	−1	0
Σ^{*0}	1383.7	3/2	0	1	0	1	−1	0
Σ^{*-}	1387.2	3/2	−1	1	−1	1	−1	0
Ξ^{*0}	1531.8	3/2	0	1/2	+1/2	1	−2	−1
Ξ^{*-}	1535.0	3/2	−1	1/2	−1/2	1	−2	−1
Ω^-	1672.5	3/2	−1	0	0	1	−3	−2

Masses of Δ-baryons lie between 1231 MeV and 1233 MeV. Due to their large decay width of 118 MeV (equivalently very small lifetimes $\approx 10^{-23}$ s) of Δ baryons, their masses are not known with greater precision than given here.

8.6.9 Hypercharge (*Y*)

Gell–Mann–Nishijima relation can also be written as

$$Q = I_3 + \frac{Y}{2} \tag{8.2}$$

where

$$Y = B + S$$

is called *hypercharge*. All the $2I + 1$ members of a given isospin multiplet (also called *isomultiplet*) have the same hypercharge *Y*. For example, for the triplet π^+, π^0 and π^-, the hypercharge is zero. For *K*-mesons, $I = 1/2$, there are two multiplets, one consisting of the particles K^+ and K^0, and the other of the corresponding antiparticles $\overline{K}^-$ and $\overline{K}^0$, the hypercharge of the former pair is +1 and that of the latter pair is –1. Leptons, photon, gluons do not carry hypercharge, i.e. their hypercharge is zero. It is interesting to note that all the mesons and baryons shown in the tables given above carry similar patterns for electric charge and hypercharge. For these hadrons, one may also note that *Y* is also equal to twice the average charge of the members of the isospin multiplet.

Charm, Bottom and Top Quantum Numbers

Subsequent discovery of new particles, like *D*, D_s, *B* and Λ_c hadrons, introduced new quantum numbers, called *charm* (*C*), bottom (*b*), and top (*t*)[10]. Including these quantum numbers, definition of the hypercharge has been extended to

$$Y = B + S + C + b + t,$$

thus extending the Gell–Mann–Nishijima formula to

$$Q = I_3 + \frac{B + S + C + b + t}{2} \tag{8.3}$$

keeping the form (8.2) unaffected. This permits us to give an alternative definition of the hypercharge, as twice the difference between the electric (*Q*) charge and isospin component (I_3) of the particle, i.e.

$$Y = 2(Q - I_3) \tag{8.4}$$

A particle and its antiparticle have opposite value of both *Q* and I_3, hence the hypercharge will be numerically equal in magnitude but opposite in sign, or zero.

8.7 CONSERVATION LAWS

All the interactions are controlled by certain conservation laws. In Section 4.4, conservation laws for nuclear reactions have been discussed. However, there exist a few conservation laws, which are valid for some interactions and not for other interactions. For example, parity, isospin, and other flavour quantum numbers are not conserved in the weak interactions. There happens to exist a rule: the stronger an interaction the more symmetrical it is.

10. Some people like to call beauty and truth in place of bottom and top quantum numbers.

The familiar conservation laws of classical physics, i.e.

- Conservation of energy
- Conservation of charge
- Conservation of linear momentum
- Conservation of angular momentum

also hold good for particle physics, provided we have made relativistic formula for to mass, momentum, etc.

8.7.1 Conservation of Charge

The electric charge is exactly conserved in particle decays and their reactions. It is due to this conservation law that electron, the lightest charged particle cannot decay, for instance, via

$$e^- \rightarrow \nu_e + \gamma$$

Note that final state has null charge in contrast to the initial state particle. That is why, electron is treated as stable particle, which is supported by the experimental lower limit on lifetime of electron for this decay

$$\tau_e > 4.6 \times 10^{26} \text{ years}$$

Some other conservation laws, which are relevant to quantum physics in general and particle physics in particular are discussed below.

8.7.2 Conservation of Lepton Number

This law states that in any particle reaction/decay, the total lepton number is always conserved. Consider the decay of μ^+-meson,

$$\begin{array}{cccc} \mu^+ \rightarrow & e^+ + & \nu_e + & \overline{\nu}_\mu \\ -1 & -1 & 1 & -1 \end{array} \qquad \text{(lepton number)}$$

In this decay, lepton number before and after the decay is –1 and hence lepton number is conserved. One may verify the electron-lepton number and muon-lepton number are also separately conserved in this decay process.

Let us now consider the decay of neutron

$$n \rightarrow p + e^- + \overline{\nu}_e$$

is allowed, since lepton number before the decay is zero and it is also zero (0 + 1 – 1 = 0) after the decay. However, the following decay

$$n \rightarrow p + e^- + \nu_e$$

is not allowed, since total lepton number of products is 0 + 1 + 1 = 2, whereas the lepton number before the neutron decay is zero. So conservation of lepton number does not permit this decay to occur.

8.7.3 Conservation of Baryon Number

This law states that in any particle reaction/decay, the total number of baryons is always conserved. Consider the decay

$$\Lambda \rightarrow p + \pi^-$$

baryon number of Λ is +1 and that of p is also +1, while that for π^-, it is zero. Therefore, the baryon number before and after decay is +1 and hence this decay is allowed. However, in the following decay

$$\Lambda \rightarrow \overline{p} + \pi^+$$

the baryon number before the decay is +1 while after the decay it is –1, hence this decay is not allowed. Similarly, for the reaction

$$\pi^+ + n \rightarrow K^0 + K^+$$

baryon number of the reactants is +1 while that of the products it is zero. Hence, this reaction is not allowed. No decay or process has yet been observed in which baryon number is not conserved. The stability of proton, which is the lightest baryon, provides a critical test of the exact conservation of baryon number.

Conservation laws mentioned above are valid in all the four interactions, i.e. strong, weak, electromagnetic and gravitational interactions. No exception to these conservation laws has been found in any particle reactions. However, there are certain conservation laws, which are valid for strong interactions only. Electromagnetic interactions can violate isospin conservation. Several quantum numbers, like isospin, strangeness, charm, parity, C-parity, may not be conserved for weak interactions.

8.7.4 Conservation of Isospin (I)

According to law of conservation of isospin, the total isospin for particles subjected to strong interactions is always same before and after the particle reaction/decay. Remember that like angular momentum addition, isospin adds vectorially

$$\vec{I} = \vec{I}_1 + \vec{I}_2$$

that means I can have discrete values starting from $| I_1 - I_2 |$ to $(I_1 + I_2)$ differing by unity. For example, isospin 1/2 when added to isospin 1, would give 1/2 and 3/2 as total isospin. Consider the following decay:

$$\Delta^+ \rightarrow p + \pi^0$$

$$3/2 \quad 1/2 \quad 1 \qquad \text{(total isospin, } I\text{)}$$

in which total isospin of the final state (1/2 + 1 = 3/2) match with that of the parent particles, so this decay is allowed to occur through the strong interactions. Now consider the decay of π^0

$$\pi^0 \rightarrow \gamma + \gamma$$

Before the decay isospin is 1 and after the decay, it is zero. Hence, this decay is not allowed by strong interaction. However, this decay is possible via the electromagnetic interactions, as total isospin may not be conserved in these interactions.

8.7.5 Conservation of Component of Isospin (I_3)

According to this conservation law, for strong and electromagnetic interactions I_3, the third component of isospin before and after particle reaction/decay always remains the same. Consider the following reaction:

$$p + p \rightarrow \Lambda^0 + K^0 + p + \pi^+$$

1/2 1/2 0 – 1/2 1/2 1 (third component of isospin, I_3)

Total I_3 component before the reaction is 1/2 + 1/2 = 1, After the reaction total I_3 component is 0 – ½ + ½ + 1 = 1. The component I_3 is same before and after the reaction, and hence this reaction is possible. Similarly, for the following decay:

$$\Delta^+ \rightarrow p + \pi^0$$

1/2 1/2 0 (third component of isospin, I_3)

However, for the decay

$$\Lambda \rightarrow p + \pi^-$$

0 1/2 –1 (third component of isospin, I_3)

Component I_3 before the decay is zero and after the decay 1/2 – 1 = –1/2, therefore, I_3 is not conserved in this reaction. Hence, this reaction cannot proceed via strong or electromagnetic interactions but can take place through weak interactions.

8.7.6 Conservation of Hypercharge

According to this conservation law, for strong and electromagnetic interactions, the total hypercharge remains the same before and after the particle reaction/decay. Consider the reaction

$$p + p \rightarrow \Lambda^0 + K^0 + p + \pi^+$$

Hypercharge before the reaction is 1 + 1 = 2 and after the reaction, it is 0 + 1 + 1 + 0 = 2. Hypercharge before and after the reaction is same (= 2), hence this reaction is possible via strong or electromagnetic interactions. However, weak interactions can violate hyperchange conservation like $\Lambda \rightarrow p + \pi^-$.

8.7.7 Conservation of Parity (*P*)

According to conservation of parity, if a process is subjected to a reflection in mirror (also called inversion), so that the sign of all its Cartesian coordinates are changed

$$\vec{x} \rightarrow \vec{x}' = -\vec{x}$$

the resulting process will be indistinguishable from the original one. In other words, the conservation of parity requires that if an event is possible, its reflection in a mirror represents an equally probable event. It was found that in strong and electromagnetic interactions parity is always conserved. Until 1956, it was thought that parity must always be conserved. However, Lee and Yang in 1956 suggested that parity is not conserved in weak interactions. Examples of such cases are β-decay, decay of muons, kaons, etc.

8.7.8 Conservation of Charge Conjugation Parity (*C*)

According to this symmetry principle, a particle is converted into its corresponding antiparticle,

$$\text{Particle} \rightarrow \text{Antiparticle}$$

or

$$\text{Antiparticle} \rightarrow \text{Particle}$$

In other words, a given process or event should be indistinguishable from one involving the corresponding antiparticles. The charge conjugation operator is normally denoted by *C*. Like parity, charge conjugation symmetry is applicable to strong and electromagnetic interactions only. For weak interactions it fails. A combination of *C*-parity with isospin is called *G*-parity which is conserved only in the strong interactions. Like *P*-parity, *C*-parity is also multiplicative in nature. Combining the two parities, *CP* can also be defined for particles and their processes.

8.7.9 Time Reversal (*T*) Symmetry

According to this symmetry principle, an event on a particle scale should be exactly reversible in time, i.e.

$$t \rightarrow t' = -t$$

If a motion picture film was made of an event of say a particle formation or decay, the event would appear to be equally possible regardless of whether the film is run forward or backward.

8.7.10 *CPT* Theorem

According to postulates of advanced quantum mechanics (or quantum field theory), all interactions should be invariant under the combination of *C*, *P* and *T*, i.e. *CPT* operation,

$$\psi_{\text{particle}}(\vec{x}, t) \rightarrow \psi_{\text{antiparticle}}(-\vec{x}, -t)$$

This is also known as *CPT* theorem. Consequences of this theorem have been discussed in Section 8.4.3.

8.7.11 Permutation Symmetry and Spin Statistics Theorem

The elementary particles are also called identical particles, i.e. no observations can distinguish particles of the same kind. This is typical quantum feature which has no classical analogue. The symmetry under permutation (or exchange of identical particles) operation in the composite state allows particular combinations only. This is intimately connected with structure of their wave functions. As per the spin-statistics theorem, quantum mechanical wave function of a system of identical bosons must be symmetric under the exchange of any pair of the particles, i.e.

$$\psi_B(x_1, x_2, x_3,\ldots) = \psi_{\text{B}}(x_2, x_1, x_3,\ldots)$$

where x_i denotes collectively space–time co-ordinates and internal quantum numbers. Thus, bosons follow Bose–Einstein statistics. Similarly, the quantum mechanical wave function of a system of identical fermions must be antisymmetric under the exchange of any pair of the particles, i.e.

$$\psi_F(x_1, x_2, x_3,\ldots) = -\psi_F(x_2, x_1, x_3,\ldots)$$

and follow Fermi–Dirac statistics.

For instance, for two particle states, we have the following permitted states:

- Symmetric state for the bosons

$$\psi_S = \frac{1}{\sqrt{2}}[\psi(1, 2) + \psi(2, 1)]$$

- Antisymmetric state for the fermions

$$\psi_A = \frac{1}{\sqrt{2}}[\psi(1, 2) - \psi(2, 1)]$$

8.8 WEAK DECAYS OF STRANGE PARTICLES

In general, leptons and hadrons given in Section 8.5 decay through weak interaction. These are classified in the following three varieties.

Leptonic Decays

In such decay modes, only leptons are involved. For example,

$$\text{Muon decays: } \mu^- \rightarrow e^- + \overline{\nu}_e + \nu_\mu$$

$$\text{Tau-lepton decays: } \tau^- \rightarrow \mu^- + \overline{\nu}_\mu + \nu_\tau$$

Semileptonic Decays

In these decay modes, leptons as well as hadrons are involved. For example,

$$\text{Pion decays: } \pi^- \rightarrow \mu^- + \overline{\nu}_\mu$$

$$\text{Neutron-beta decays: } n \rightarrow p + e^- + \overline{\nu}_e$$

Nonleptonic Decays

These decays are also called hadronic decays, as only hadrons are involved in such decays. For example,

$$K\text{-decays: } K^0 \rightarrow \pi^+ + \pi^-$$

$$\text{Lambda decays: } \Lambda \rightarrow p + \pi^-$$

Sometimes a photon is also emitted in hadronic weak decays. Such decays take place by combined effects of weak and electromagnetic interactions. For example,

$$\Lambda \rightarrow n + \gamma$$

8.9 PARTICLE SYMMETRIES

The concept of symmetry has acquired great and fundamental significance in classical as well as in modern physics, more particularly in particle physics. The term symmetry means that a system, state or quantity remains unchanged (or invariant) as a result of a particular transformation, such as a change in space coordinates, or change in time, etc. Symmetries also predict degeneracy between different physical states, which can be connected through the corresponding transformations. Every conservation law is related to a particular invariance (or symmetry) principle. For example, conservation of total energy is due to the invariance of a system under shift in time, called *time translation*. Conservation of linear or angular momentum is due to invariance under displacement in space (space translation) or rotation in space respectively. These transformations can be abstract also, i.e. they may have no relation to actual space, time, etc. Examples of these are like conservation of charge, lepton number, baryon number, isospin etc. Besides these continuous symmetries, there also exist discrete symmetries like parity and *C*-parity.

In the following, we will describe briefly some aspects of special symmetries in particle physics, which are SU(2) symmetry, SU(3) symmetry, and higher symmetries.

8.9.1 SU(2) Symmetry

Besides having same spin, proton and neutron have many similarities. Particularly, it is well known that nuclear forces (strong forces) are charge-independent. That means strong binding force between *n–p, p–n,* and *n–n* is essentially the same. The small difference between them is attributed to the electromagnetic interactions, which arise due to difference in charges of proton and neutron. These facts suggested the existence of new kind of nuclear spin, called *isospin*, originally proposed by Heisenberg in 1934. Stating differently, in the world of strong interactions, we can think of proton and neutron as the two orthogonal states of the same particle, called *nucleon*. As has been mentioned before, a general isospin state is described by two quantum numbers, I and I_3, where I_3 can have values ranging from $-I$ to I separated by unity. Thus, for isospin I, there are $2I + 1$ degenerate states available. Isospin conservation is a consequence of the invariance of the strong interaction Hamiltonian under the rotations in isospin space. Isospin symmetry also reflects in the physical states belonging to an isospin multiplet, which have the same mass. Consequently, the nucleon doublet, i.e. proton and neutron, have the same mass due to the strong isospin SU(2) symmetry[11]. SU(2) denotes special (S) unitary (U) group in two complex dimensions, that means it is a group of 2 × 2 unitary matrices U with unity determinant. A unitary matrix is that whose inverse is given by complex conjugate of its transpose. Such a matrix, can be written as

$$U = \exp\left(i\sum_{j=1}^{3}\sigma_j\theta_j\right)$$

where σ_j are the well-known Pauli matrices,

$$\sigma_1 = \begin{pmatrix} 0 & 1 \\ 1 & 0 \end{pmatrix}, \sigma_2 = \begin{pmatrix} 0 & -i \\ i & 0 \end{pmatrix}, \sigma_3 = \begin{pmatrix} 1 & 0 \\ 0 & -1 \end{pmatrix}$$

and θ_j denotes the angles of rotation in isospin space. SU(2) allows all integral $(2I + 1)$ multiplets;

1, 2, 3, 4, 5,…

11. It may be remarked that there also exists weak isospin symmetry needed to develop quantum field theory of weak interactions, but it is beyond the scope of the book.

respectively for $I = 0, 1/2, 1, 3/2, 2, \ldots$ states. The fundamental particle-multiplet of the SU(2) is two-dimensional which is given by two-component column matrices. For example, we can imagine proton and neutron to be described by the following isospin states of the isospin symmetry:

$$p = \begin{pmatrix} 1 \\ 0 \end{pmatrix}, n = \begin{pmatrix} 0 \\ 1 \end{pmatrix}$$

on which the Pauli matrices can operate.

Strong interactions respect the isospin symmetry leading to conservation of isospin in strong decays or reactions. This is quite similar to the conservation of angular momentum due to the invariance of the Hamiltonian under the rotations in ordinary space (more precisely in spin space). Isospin symmetry is broken by the electromagnetic and weak interactions. However, the electromagnetic interactions conserve the third component of the isospin. Breaking of isospin symmetry permits us to differentiate between a proton and a neutron.

Similarly, strong force treats three pions the same way. It is reflected in the cross-sections being the same for scattering of different pions on proton and neutron. In the isospin space, we can describe three pions as

$$\pi^+ = \begin{pmatrix} 1 \\ 0 \\ 0 \end{pmatrix}, \pi^0 = \begin{pmatrix} 0 \\ 1 \\ 0 \end{pmatrix}, \pi^- = \begin{pmatrix} 0 \\ 0 \\ 1 \end{pmatrix}$$

Looking at Tables 8.8 and 8.9(a, b) giving properties of hadrons, one may notice the states belonging to one isomultiplet are nearly mass degenerate. The small mass difference in the so-called isomultiplets is assigned to the difference in the electric charges of these states.

8.9.2 SU(3) Symmetry

The number of observed hadrons is more than hundred and more of them are yet to be seen. All the observed hadrons are extended objects having size at the Fermi scale. Obviously, all of the hadrons could not be treated as elementary. A certain pattern, similar to the Mendeleev periodic table of atoms, among them was found independently by Gell–Mann and Y. Ne'eman. Consider a system of coordinates in which x-axis and y-axis are used to represent third component of isospin (I_3) and Hypercharge ($Y = B + S$) respectively. If we draw mesons of same spin and parity, say $J^P = 0^-$, they create geometric symmetric octet patterns, called weight diagrams (Figure 8.2). Similar pattern is also created by $J^P = 1/2^+$ baryons, see Figure 8.3. Gell–Mann called this scheme[12] 'Eightfold Way' as term octet was occurring again and again. In fact, all the observed baryons, known till then, seemed to appear in the grouping of 1 (singlet), 8 (octet) and 10 (decuplet), whereas the mesons fall into groups of 1 and 8. The observed patterns, shown in Figures 8.2 and 8.3, were naturally explained through the SU(3) symmetry, a mathematical generalization of the SU(2) isospin symmetry. SU(3) is the group of 3×3 unitary matrices with unit determinant, and its fundamental multiplet is obviously three-dimensional. In contrast to SU(2), SU(3) permits selective multiplets, say

$$1, 3, 3^*, 6, 6^*, 8, 10, 10^*, 15, \ldots$$

12. M. Gell–Mann called this symmetry 'Eightfold Way', borrowing the name from the Buddhist way to Nirvana.

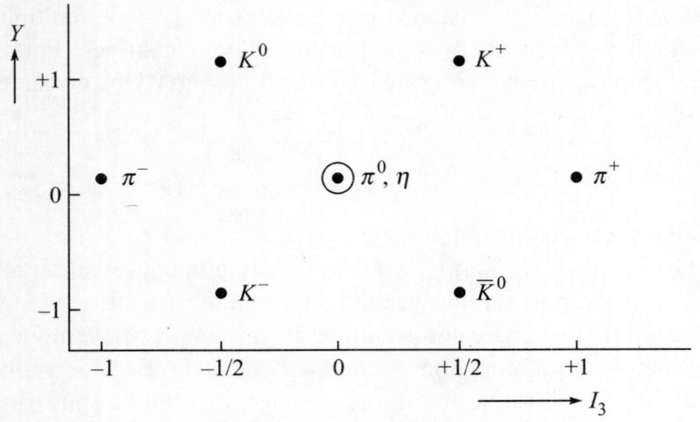

Figure 8.2 Weight diagram of $J^P = 0^-$ mesons.

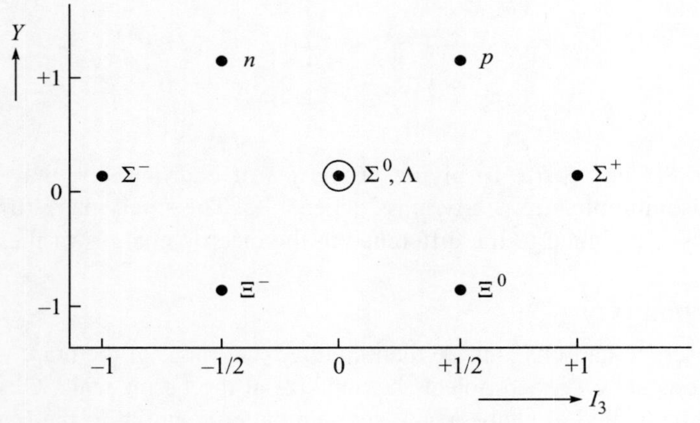

Figure 8.3 Weight diagram of $J^P = \frac{1}{2}^+$ baryons.

The SU(3) could explain several regularities observed in the hadron world. In fact, the SU(3) symmetry scheme achieved wonderful confirmation with the discovery of Ω^--hyperon, which was predicted by Gell-Mann, as a missing partner of the already observed nine spin 3/2 baryons filling the decuplet. However, this symmetry is not exact and is broken even by the strong interactions, though in a particular way. Assuming that the SU(3) breaking also follows eightfold way, several properties of the hadrons could be explained. For instance, Gell-Mann–Okubo obtained a well-satisfied mass formula:

$$2N + 2\Xi = 3\Lambda + \Sigma$$

$$1129 \text{ MeV} \quad 1135 \text{ MeV}$$

where particle symbol denotes the average mass of the particles in that isospin multiplet.

8.9.3 Higher Symmetries

Later higher symmetries, described by SU(4), SU(5), SU(6) and higher groups, have also been employed to study the hadronic behaviour. For instance, SU(6), which combines the SU(3) symmetry with the SU(2) spin symmetry, could explain reasonably the ratio of nucleon magnetic moments

$$\mu_p/\mu_n = -3/2$$

$$\text{Experimental value} = -1.46$$

SU(4) and SU(5) are used to describe spectra of hadrons carrying charm and bottom quantum numbers.

8.10 QUARKS

The unitary symmetries were only to act as intermediate step to further explore the structure of matter. With the successes of the SU(3) symmetry, new questions were also raised. For instance, why nature allows only 1, 8, and 10 multiplets for the hadrons, while SU(3) group does possess lower multiplets like 3, 3*, 6 and 6*. This led to the proposal that the hadrons may have constituents which fill these multiplets. Historically, M. Gell–Mann[13] in 1964 suggested the idea of quarks to explain the existence of a large number of baryons (of half-integral spin) and meson (of integral spin) states. He further suggested that the hadrons are composed of three varieties of quarks, up (u), down (d) and strange (s). The variety of quarks is now formally called *flavour*. Like leptons, all the quarks carry spin ½ $\hbar$. Quarks fill the fundamental multiplet 3 of the SU(3), called *flavour symmetry*. Accordingly, antiquarks were assigned to 3* multiplet:

$$3 = \begin{pmatrix} u \\ d \\ s \end{pmatrix}, \quad 3^* = (\bar{u}, \bar{d}, \bar{s})$$

With this scheme, he explained the existence of all the multiplets, he then observed baryons (as three-quark states) and mesons (as quark–antiquark states). Thus, regularities in the observed hadron spectrum could be understood through the following group-theoretical decompositions (generalization of concept of addition of angular momenta) of the direct-products into the direct-sum of the multiplets under the SU(3) group:

$$\text{Mesons } (q\bar{q}) \Rightarrow 3 \otimes 3^* = 1 \oplus 8, \tag{8.5}$$

$$\text{Baryons } (qqq) \Rightarrow 3 \otimes 3 \otimes 3 = 1 \oplus 8 \oplus 8 \oplus 10. \tag{8.6}$$

Thus, mesons form octet and singlet, whereas the baryons belong to singlet, octet and decuplet.

Charm, Bottom and Top Quarks

Discovery of ψ(3096 MeV) meson in 1974 required a new quark, named *charm,* as it could not be accommodated within the three-quark scheme. Similarly, observation of Y(9460 MeV) meson in 1978 demanded the existence of a fifth quark, named *bottom.* Some of the hadrons, like D-and

13. The idea of quarks was introduced in 1962 independently by M. Gell–Mann and G. Zweig. However, Zweig called them *Aces*.

B-mesons and Λ_c^+, Σ_c^{++}, Λ_b^0 baryons containing one of these quarks have been produced in the high-energy experiments. These developments indicated that nature seems to take the basic pattern of the first generation of two quarks and two leptons, and recreate it at higher energies. So the quark-lepton analogy, namely the number of quarks and leptons be the same, suggested the existence of a sixth quark, called *top*, which has been created by colliding high energy (500 GeV) p and $\overline{p}$ beams at Fermilab in 1994. In all, there are six types of quarks, three with two thirds the charge of a proton and the other three with one thirds the charge of an electron. All six quarks have different masses and other quantum numbers (Table 8.10). Note that mass of the top-quark, which is an elementary particle, is even larger than that of the gold atom.

Like leptons, the quarks are divided into three generations: (u, d), (c, s), and (t, b). A generation consists of a pair of leptons (one charged and one neutral) and a pair of quarks (carrying charges of 2/3 and –1/3). Stable matter is made of the first generation which contains the electron, the electron neutrino, the *u*-quark, and the *d*-quark. The reason for the existence of other generations are not known except that minimum number of generations must be three to allow *CP*-violation, which seems to play a significant role to explain matter generation in the universe, and to explain small amount of antimatter in it. There are six antiquarks also. They are $\overline{d}$, $\overline{u}$, $\overline{s}$, $\overline{c}$, $\overline{b}$ and $\overline{t}$. Their basic properties are shown in Table 8.11.

8.11 QUALITATIVE DESCRIPTION OF QUARK MODEL

By looking at the quantum numbers of the hadrons and the quarks, one can understand easily their quark composition. For example, positive π meson has one up (u) quark and one antidown ($\overline{d}$) quark. Quark compositions of the octet mesons are given in Table 8.12.

Similarly, a neutron contains two *d*-quarks and one *u*-quark, and proton carrying positive charge is made up of two *u*-quark and *d*-quark[14]. It is obvious from the quark combinations to build these nucleons because the *u*-quark carries electric charge[15] two thirds of the charge on a proton, and *d*-quark carries one third of the charge on an electron. Obviously, baryon number of quarks is 1/3. For example, charge, baryon number, and strangeness of the proton are given by sum of the respective properties of the constituent quarks

$$\text{Charge on proton} = \frac{2}{3} + \frac{2}{3} - \frac{1}{3} = +1$$

$$\text{Baryon number} = \frac{1}{3} + \frac{1}{3} + \frac{1}{3} = 1$$

$$\text{Strangeness} = 0 + 0 + 0 = 0$$

14. Three antiquarks can also combine to form antibaryon states. For instance, two anti-*u*-quarks and an anti-*d*-quark would give rise to antiproton.

15. Before the introduction of the quark model, charge of an electron and that of a proton were taken to be the basic units of negative and positive charges, respectively. Now it has been replaced with one third of the proton (or electron) charge.

TABLE 8.10 Properties of quarks

Quark	*Name*	*Mass* (GeV)	*Charge* (*e*)	*Spin* ($\hbar$)	*Baryon number* (*B*)	*Isospin* *I*	I_3	*Strangeness* (*S*)	*Charm* (*C*)	*Bottom* (*beauty*) (*B*)	*Top* (*truth*) (*t*)
d	Down	0.33	$-\frac{1}{3}$	$\frac{1}{2}$	$\frac{1}{3}$	$\frac{1}{2}$	$-\frac{1}{2}$	0	0	0	0
u	Up	0.33	$\frac{2}{3}$	$\frac{1}{2}$	$\frac{1}{3}$	$\frac{1}{2}$	$\frac{1}{2}$	0	0	0	0
s	Strange	0.51	$-\frac{1}{3}$	$\frac{1}{2}$	$\frac{1}{3}$	0	0	–1	0	0	0
c	Charm	1.60	$\frac{2}{3}$	$\frac{1}{2}$	$\frac{1}{3}$	0	0	0	+1	0	0
b	Bottom (beauty)	4.7	$-\frac{1}{3}$	$\frac{1}{2}$	$\frac{1}{3}$	0	0	0	0	–1	0
t	Top (truth)	174.2	$\frac{2}{3}$	$\frac{1}{2}$	$\frac{1}{3}$	0	0	0	0	0	+1

TABLE 8.11 Properties of antiquarks

Quark	*Name*	*Mass* (GeV)	*Charge* (*e*)	*Spin* ($\hbar$)	*Baryon number* (*B*)	*Isospin* *I*	I_3	*Strangeness* (*S*)	*Charm* (*C*)	*Beauty* (*B*)	(*Truth*) (*T*)
$\bar{d}$	Down	0.33	$\frac{1}{3}$	$\frac{1}{2}$	$-\frac{1}{3}$	$\frac{1}{2}$	$\frac{1}{2}$	0	0	0	0
$\bar{u}$	Up	0.33	$-\frac{2}{3}$	$\frac{1}{2}$	$-\frac{1}{3}$	$\frac{1}{2}$	$-\frac{1}{2}$	0	0	0	0
$\bar{s}$	Strange	0.51	$\frac{1}{3}$	$\frac{1}{2}$	$-\frac{1}{3}$	0	0	1	0	0	0
$\bar{c}$	Charm	1.6	$-\frac{2}{3}$	$\frac{1}{2}$	$-\frac{1}{3}$	0	0	0	–1	0	0
$\bar{b}$	Beauty	4.7	$\frac{1}{3}$	$\frac{1}{2}$	$-\frac{1}{3}$	0	0	0	0	1	0
$\bar{t}$	Truth	174.2	$-\frac{2}{3}$	$\frac{1}{2}$	–	0	0	0	0	0	–1

TABLE 8.12 Quark composition of octet mesons

Particle	*Quark content*	*Charge (Q)*	*Baryon number (B)*	*Isospin (I_3)*	*Strangeness (S)*
K^+	$u\bar{s}$	$+\frac{2}{3}+\frac{1}{3}=+1$	$+\frac{1}{3}-\frac{1}{3}=0$	$\frac{1}{2}+0=\frac{1}{2}$	$0+1=1$
K^0	$d\bar{s}$	$-\frac{1}{3}+\frac{1}{3}=0$	$+\frac{1}{3}-\frac{1}{3}=0$	$\frac{1}{2}+0=\frac{1}{2}$	$0+1=1$
π^+	$u\bar{d}$	$+\frac{2}{3}+\frac{1}{3}=+1$	$+\frac{1}{3}-\frac{1}{3}=0$	$\frac{1}{2}+\frac{1}{2}=+1$	$0+0=0$
π^0	$d\bar{d}$, $u\bar{u}$	$-\frac{1}{3}+\frac{1}{3}=0, +\frac{2}{3}-\frac{2}{3}=0$	$+\frac{1}{3}-\frac{1}{3}=0$	$\frac{1}{2}-\frac{1}{2}=0$	$0+0=0$
π^-	$d\bar{u}$	$-\frac{1}{3}-\frac{2}{3}=-1$	$+\frac{1}{3}-\frac{1}{3}=0$	$-\frac{1}{2}-\frac{1}{2}=-1$	$0+0=0$
$\overline{K}^0$	$s\bar{d}$	$-\frac{1}{3}+\frac{1}{3}=0$	$+\frac{1}{3}-\frac{1}{3}=0$	$0+\frac{1}{2}=\frac{1}{2}$	$-1+0=-1$
K^-	$s\bar{u}$	$-\frac{1}{3}-\frac{2}{3}=-1$	$+\frac{1}{3}-\frac{1}{3}=0$	$0-\frac{1}{2}=-\frac{1}{2}$	$-1+0=-1$
η	$s\bar{s}$, $d\bar{d}$, $u\bar{u}$	$-\frac{1}{3}+\frac{1}{3}=0, +\frac{2}{3}-\frac{2}{3}=0$	$+\frac{1}{3}-\frac{1}{3}=0$	$0+0=0, \frac{1}{2}-\frac{1}{2}=0$	$0+0=0$

Similarly, quark contents of various baryons and antinucleons are given in Tables 8.13, 8.14 and 8.15. Higher spin states were found to be orbitally excited states of these quark configurations, somewhat similar to the excited states of hydrogen atom.

TABLE 8.13 Quark composition of octet baryons

Particle	*Quark content*	*Charge* (Q)	*Baryon number* (B)	*Isospin* (I_3)	*Strangeness* (S)
p	*uud*	$+\frac{2}{3}+\frac{2}{3}-\frac{1}{3}=+1$	$+\frac{1}{3}+\frac{1}{3}+\frac{1}{3}=+1$	$\frac{1}{2}+\frac{1}{2}-\frac{1}{2}=\frac{1}{2}$	$0+0+0=0$
n	*udd*	$+\frac{2}{3}-\frac{1}{3}-\frac{1}{3}=0$	$+\frac{1}{3}+\frac{1}{3}+\frac{1}{3}=+1$	$\frac{1}{2}-\frac{1}{2}-\frac{1}{2}=-\frac{1}{2}$	$0+0+0=0$
Λ	*uds*	$-\frac{2}{3}-\frac{1}{3}-\frac{1}{3}=0$	$+\frac{1}{3}+\frac{1}{3}+\frac{1}{3}=+1$	$\frac{1}{2}-\frac{1}{2}+0=0$	$0+0-1=-1$
Σ^+	*uus*	$+\frac{2}{3}+\frac{2}{3}-\frac{1}{3}=+1$	$+\frac{1}{3}+\frac{1}{3}+\frac{1}{3}=+1$	$\frac{1}{2}+\frac{1}{2}+0=+1$	$0+0-1=-1$
Σ^0	*uds*	$+\frac{2}{3}-\frac{1}{3}-\frac{1}{3}=0$	$+\frac{1}{3}+\frac{1}{3}+\frac{1}{3}=+1$	$\frac{1}{2}-\frac{1}{2}+0=0$	$0+0-1=-1$
Σ^-	*dds*	$-\frac{1}{3}-\frac{1}{3}-\frac{1}{3}=-1$	$+\frac{1}{3}+\frac{1}{3}+\frac{1}{3}=+1$	$-\frac{1}{2}-\frac{1}{2}+0=-1$	$0+0-1=-1$
Ξ^0	*uss*	$+\frac{2}{3}-\frac{1}{3}-\frac{1}{3}=0$	$+\frac{1}{3}+\frac{1}{3}+\frac{1}{3}=+1$	$\frac{1}{2}+0+0=\frac{1}{2}$	$0-1-1=-2$
Ξ^-	*dss*	$-\frac{1}{3}-\frac{1}{3}-\frac{1}{3}=-1$	$+\frac{1}{3}+\frac{1}{3}+\frac{1}{3}=+1$	$-\frac{1}{2}+0+0=-\frac{1}{2}$	$0-1-1=-2$

TABLE 8.14 Quark composition of antinucleon

Particle	*Quark content*	*Charge* (Q)	*Baryon number* (B)	*Isospin* (I_3)	*Strangeness* (S)
$\overline{p}$	$\overline{u}\,\overline{u}\,\overline{d}$	$-\frac{2}{3}-\frac{2}{3}+\frac{1}{3}=-1$	$-\frac{1}{3}-\frac{1}{3}-\frac{1}{3}=-1$	$-\frac{1}{2}-\frac{1}{2}+\frac{1}{2}=-\frac{1}{2}$	$0+0+0=0$
$\overline{n}$	$\overline{u}\,\overline{d}\,\overline{d}$	$-\frac{2}{3}+\frac{1}{3}+\frac{1}{3}=0$	$-\frac{1}{3}-\frac{1}{3}-\frac{1}{3}=-1$	$-\frac{1}{2}+\frac{1}{2}+\frac{1}{2}=+\frac{1}{2}$	$0+0+0=0$

The quark model shattered two basic concepts about fundamental particles. Firstly, a proton and a neutron were excluded from the family of fundamental particles and secondly, the charge of an electron did not remain the minimum possible value for the electric charge of any particle. Quarks were considered to be the key components in the quark model. Phenomenological analysis of hadron properties, like masses, magnetic moments and decays, using the quark model has been very useful. For instance, $n \to p + e^- + \overline{\nu}_e$, at quark level, involves the conversion of d-type of quarks through $d \to u + e^- + \overline{\nu}_e$.

TABLE 8.15 Quark composition of decuplet baryons

Particle	*Quark content*	*Charge (Q)*	*Baryon number (B)*	*Strangeness (S)*
Δ^{++}	*uuu*	$+\frac{2}{3}+\frac{2}{3}+\frac{2}{3}=+2$	$+\frac{1}{3}+\frac{1}{3}+\frac{1}{3}=+1$	$0+0+0=0$
Δ^{+}	*uud*	$+\frac{2}{3}+\frac{2}{3}-\frac{1}{3}=+1$	$+\frac{1}{3}+\frac{1}{3}+\frac{1}{3}=+1$	$0+0+0=0$
Δ^{0}	*udd*	$+\frac{2}{3}-\frac{1}{3}-\frac{1}{3}=0$	$+\frac{1}{3}+\frac{1}{3}+\frac{1}{3}=+1$	$0+0+0=0$
Δ^{-}	*ddd*	$-\frac{1}{3}-\frac{1}{3}-\frac{1}{3}=-1$	$+\frac{1}{3}+\frac{1}{3}+\frac{1}{3}=+1$	$0+0+0=0$
Σ^{*+}	*uus*	$+\frac{2}{3}+\frac{2}{3}-\frac{1}{3}=+1$	$+\frac{1}{3}+\frac{1}{3}+\frac{1}{3}=+1$	$0+0-1=-1$
Σ^{*0}	*uds*	$+\frac{2}{3}-\frac{1}{3}-\frac{1}{3}=0$	$+\frac{1}{3}+\frac{1}{3}+\frac{1}{3}=+1$	$0+0-1=-1$
Σ^{*-}	*dds*	$-\frac{1}{3}-\frac{1}{3}-\frac{1}{3}=-1$	$+\frac{1}{3}+\frac{1}{3}+\frac{1}{3}=+1$	$0+0-1=-1$
Ξ^{*0}	*uss*	$+\frac{2}{3}-\frac{1}{3}-\frac{1}{3}=0$	$+\frac{1}{3}+\frac{1}{3}+\frac{1}{3}=+1$	$0-1-1=-2$
Ξ^{*-}	*dss*	$-\frac{1}{3}-\frac{1}{3}-\frac{1}{3}=-1$	$+\frac{1}{3}+\frac{1}{3}+\frac{1}{3}=+1$	$0-1-1=-2$
Ω^{-}	*sss*	$+\frac{2}{3}+\frac{2}{3}-\frac{1}{3}=+1$	$+\frac{1}{3}+\frac{1}{3}+\frac{1}{3}=+1$	$-1-1-1=-3$

8.11.1 Observation of Quarks

The model faced several difficulties also alongside the successes achieved. For instance, the first problem was, "Do isolated quarks exist or these are always bound?". Other question was the apparent violation of Pauli exclusion principle. The first main problem was that though three quarks form a proton, it is impossible to separate them. Many careful and ingenious experiments have been performed with high-energy accelerators to liberate quarks, but without any success. Other searches for free quarks, for example in cosmic rays, stellar spectra, terrestrial and lunar rock samples, and deep ocean sediments, have also failed. Thus, *Hadron-spectroscopy* is different from the *atomic* and *nuclear spectroscopies*. Whereas the atomic and nuclear constituents could be isolated, quarks remain permanently confined in hadrons. In hadronic collisions it is energetically favourable to produce more hadrons rather than ejecting the quarks.

Indirect confirmation of the quark structure of proton has been obtained in deep inelastic lepton–nucleons collisions. Actually, what we see inside the proton depends on the scale on which one observes it. If a proton is scattered with radiation of a wavelength of a few times

10^{-15} m, say through 50 MeV lepton, it appears just a point-like object. It might contain all sorts of its constituents, but no structure is revealed at this scale. Boosting the energy of the probe to a few GeV, resolution gets increased and one finds the quark layer. High energy (E ~ a few GeV) beams of leptons, having de Broglie wavelength ($\lambda = \hbar c/E \sim 10^{-18}$ m) << nucleon size ($\sim 10^{-15}$ m), reveal constituents (also called *partons*) of the proton. The results obtained from neutrino-nucleon scattering experiments, called *deep inelastic scattering*, reveal that the proton (and neutron) contains three point-like quasi-free quark constituents. These experiments indeed confirmed that these quarks do have fractional charges, besides having spin ½. Further investigations also showed the existence of gluons, with which quarks are glued (bound) together to form hadrons.

8.11.2 Colour: A New Degree of Freedom

In order to see the second problem faced by the quark model, consider Δ resonance spin-up state made up of three spin-up *u* quarks,

$$\Delta^{++}{}_{(I=3/2,\, I_3=3/2;\, S=3/2,\, S_3=3/2)} = u\uparrow u\uparrow u\uparrow$$

which is totally symmetric state under the interchange of any two quarks. This is in strict violation to the spin-statistics theorem and hence violating the Pauli exclusion principle. To overcome this violation, it was suggested that the antisymmetry of the baryon, being fermion, could be restored only if each quark is allowed an extra degree of freedom, called *colour.* These degrees have been named to make things a bit colourful, red, green and blue[16]. However, we do not need colour quantum number for the composite hadrons. That means hadrons are colourless (colour singlet).

Using the $SU(3)_{colour}$ group, where quark colours, red (r), blue (b) and green (g), belong to the fundamental multiplet of 3, and antiquark conjugate-colours ($\bar{r}$, $\bar{b}$, $\bar{g}$) belong to 3^*, the colour neutral states of the hadrons could be constructed as the decompositions given in Eq. (8.6) contain a singlet. Respective colour wave functions are:

$$|\,\text{Mesons}>_{\text{colour}} = \frac{1}{\sqrt{3}}|\,\bar{r}r + \bar{b}b + \bar{g}g>$$

$$|\,\text{Baryons}>_{\text{colour}} = \frac{1}{\sqrt{6}}|\,rbg + bgr + grb - rgb - brg - gbr>$$

To have agreement between theory and experiment for all these phenomena, we clearly need three colours. So each of these six quarks occurs in three varieties. Note that the colour assignment has nothing to do with the standard perception of colour.

16. This has nothing to do with ordinary visual colours.

8.12 RECENT DEVELOPMENTS

Particle physics has been an exciting and challenging frontier of physics. In the last few decades, lots of new developments have taken place in this subject, brief description of which is given below.

8.12.1 Quantum Field Theory of Interactions

Strong, electromagnetic and weak interactions are now described by quantum field theories. One may notice that all matter particles are fermions, whereas all the quanta of forces are bosons. In these theories, all elementary particles are described by their corresponding fields, which interact with each other guided by the so-called *principle of gauge invariance*. For instance, in the theory of electromagnetic interactions, known as *Quantum Electrodynamics* (QED), photon is a manifestation of electromagnetic field and electron is a quantum of the electron field, and photon, though itself electrically neutral, acts as the messengers of the electromagnetic interactions. QED was developed in 1948 independently by Schwinger, Feynman and Tomonaga. In this theory, electromagnetic interaction between two charges takes place through exchange of photon. This photon is virtual in the sense that energy ΔE can be borrowed to create the quantum, provided it is done only for a short time Δt, where $\Delta E \cdot \Delta t \sim \hbar$ (uncertainty principle). So an electron may be imagined as continually emitting and absorbing the photons. Sometimes these photons get absorbed by a nearby electron, thus giving rise to the electromagnetic force. This theory has achieved several excellent experimental confirmations: anomalous magnetic moment of leptons, Lamb shift say between $2s\frac{1}{2}$ and $2p\frac{1}{2}$ levels of hydrogen atom etc. It also predicts that the Coulomb law gets modified at very short distances due to effects like vacuum polarization (screening of electric charge).

In mid-sixties, electroweak field theory, called *quantum flavourdynamics* (QFD), was developed by Glashow, Wienberg and Salam in which W- and Z-bosons act as the mediator of the weak interactions, besides unifying the weak and electromagnetic interactions. Many predictions of this theory have been confirmed experimentally. For instance, it predicted the neutral current weak phenomena like $\nu + p \rightarrow \nu + p$, and possibility of parity violation in atomic forces, which have been observed. Similarly, for the strong interaction also, the observation of quasi-free nature of quarks inside the proton suggested a field theory of strong interactions should be possible. Guided by the QED, a quantum field theory of strong interactions, known as the *Quantum Chromodynamics* (QCD), has been developed, in which the colour acts as the strong charge. Basically, QCD postulates the specific quanta, called *gluons*—mediating the strong interaction between the quarks, just as the photons mediate electromagnetic interaction. There are, however, some fundamental differences also. In contrast to the single electric charge, the strong colour charges are three in number, which belong to the fundamental multiplet of the SU(3) colour symmetry group. QED has only one photon, whereas the QCD requires 8 gluons.

Combining these three theories of strong, weak and electromagnetic interactions gives us the so-called 'Standard Model' of the fundamental interactions. Up to now all its predictions have been found to be in excellent agreement with experimental results. However, there are reasons to believe that the standard model needs to be extended further. Supersymmetry (SUSY) and Grand Unified Theories provide some of the possible generalizations of this model.

8.12.2 Unification of Fundamental Forces

Unification of fundamental forces has been a major challenge. However, every step at the unification programme has yielded some bonus. For instance, unification of electricity and magnetism, developed by Maxwell around 1860, provided optics as bonus. The fact that strong, electromagnetic and weak interactions are described by similar quantum field theories suggest their common origin. Theoretical physicists have already developed so-called **Grand Unified Theories** (GUTs) which unify all the three interactions and predict the violation of baryon number[17] (B). Such theories also predict the existence of heavy quanta (around 10^{15} GeV), called *lepto-quark*, which could make proton decay violating the baryon number conservation, for example, in the mode $p \to \pi^0 + e^+$. On the experimental side, careful investigations are being made to look for probable proton decay. At present, no such event has been observed so far and the only limit on its lifetime (> 10^{33} years). Theoretically, proton instability acquires a deeper significance for explaining the origin and development of the universe. Particularly, baryon number violation (along with possible C and CP violation) can play an important role in understanding the problem of matter–antimatter symmetry observed in the universe.

At present much theoretical work is going in developing the *supersymmetry* (SUSY) and *supergravity* (SUGRA) and *superstring* field theories. As a last frontier, final unification of strong, weak and electromagnetic interactions with gravity is also being developed which is called *Theory of Everything* (TOE). However, it is far from completeness as it is suffering from serious difficulties, since gravity has not been understood at the quantum level around Planck mass scale $\sqrt{\hbar c/G_N} = 1.22 \times 10^{19}$ GeV. We may now hope to understand the creation of universe back up to Planck time ($\sqrt{\hbar G_N/c^5} = 0.54 \times 10^{-43}$ s) or Planck length ($\sqrt{\hbar G_N/c^3} = 1.61 \times 10^{-35}$ m), where quantum gravity takes over the domain. Thus, during the last few decades, close links between particle physics and cosmology have been established. Existence of dark matter and dark energy in the universe has provided new challenges to develop these links further. Status of unification theme is given in Table 8.16.

TABLE 8.16 Status of unification of fundamental interactions

Theory	*Interactions unified*	*Bonus*	*Status*
QED	Electricity and magnetism	Optics	Very successful
QFD	Electromagnetism and weak	Neutral currents	Successful
GUT	Strong, electromagnetism and weak	Proton decay	Promising
TOE	Strong, electromagnetism, weak and gravity	Cosmology	Probable

As a result of the successes of particle physics, we now believe that we have a good understanding of the natural phenomena ranging from microcosm (very small) to macrocosm (very large objects). For the first time, we are able to form a unified view of the cosmos. It has been learnt finally that nature's imagination is much richer than our vision. It may have many surprises in its store. For instance, quarks and leptons, today's fundamental particles, may have another layer of structure. Only future experiments would decide such issues.

17. It also predicts violation of lepton number (L); however, it may conserve B–L quantum number.

NUMERICAL PROBLEMS

Section 8.3

Unsolved Problem

1. Prove that fine structure constant $\alpha = \frac{1}{137}$.

Section 8.4

Unsolved Problems

1. Mean lifetime of Λ-baryon is 2.63×10^{-10} s. Determine its decay width.

$\left[\textbf{Hint.}\ \Gamma = \frac{\hbar}{\tau}\right]$

2. Using time–energy uncertainty relation $\Delta E\,\Delta t \approx \hbar$, show that range ($R$) of a force mediated by quanta carrying rest mass (m) is given by $R = \frac{\hbar}{mc}$.

3. Calculate the range of weak forces using mass of W-boson in the range formula.

Section 8.5

Solved Problems

1. An antiproton comes to rest and annihilates with a proton. They produce π^+, π^- and π^0 of equal energy. What is the average kinetic energy of each pion in MeV?

Solution: The decay is

$$p + \overline{p} \rightarrow \pi^0 + \pi^- + \pi^+$$

$$\text{Total energy of } p \text{ and } \overline{p} \text{ system} = 938 + 938 = 1876 \text{ MeV}$$

$$\text{Out of this, the energy consumed in creating 3 pions} = 134.9 + 139.5 + 139.5 = 413.9 \text{ MeV}$$

$$\text{Energy left to be shared by 3 pions} = 1876 - 413.9 = 1462.1 \text{ MeV}$$

$$\text{Therefore, average kinetic energy of each pion} = \frac{1462.1}{3} = 487.4 \text{ MeV}$$

2. A particle X has two decay modes with partial decay rates r_1 s^{-1} and r_2 s^{-1}. What is the inherent uncertainty in mass of particle X?

Solution: The total decay rate of particle X is

$$\lambda = r_1 + r_2$$

So, the mean lifetime of the particle is

$$\tau = \frac{1}{\lambda} = \frac{1}{r_1 + r_2}$$

Therefore, inherent uncertainty in the mass of the particle is given by the relation $\Gamma \times \tau \cong \hbar$. Hence,

$$\Gamma \cong \frac{h}{\tau} = \hbar(r_1 + r_2)$$

Unsolved Problems

1. The Σ^0-hyperon decays to $\Lambda^0 + \gamma$ with a mean lifetime of 7.4×10^{-20} s. Estimate its width in keV.

   [Ans. 8.9 keV]

2. Find the distance covered by a muon having kinetic energy of 5 MeV in free space before it decays. Half-life of muon = 2.197×10^{-6} s and its mass = 105.66 MeV.

   [Ans. 202.8 m]

3. Show that in natural units 1 kg = 5.63×10^{26} GeV.

Section 8.6

Unsolved Problems

1. π^--mesons exist in three charged states +1, 0 and –1 (in units of proton's charge). Find out its strangeness. **[Ans.** 0]
2. Σ-baryons exist in three charged states –1, 0 and +1. Show that the Σ has strangeness $S = -1$.

Section 8.7

Solved Problems

1. Explain why the following reactions cannot occur through strong interactions even if the kinetic energy of the incident proton is several GeV.

   (i) $p + n \rightarrow K^+ + \Xi^- + \Sigma^+$ (ii) $p + n \rightarrow \Lambda^0 + \Sigma^+$

Solution: Since these reactions involve only strongly interacting particles, so all the conservation laws must hold good. If some law is violated, the reaction cannot take place in spite of the energy of the incident proton. In these reactions conservation of strangeness, isospin and I_3 are violated. Hence, these reactions cannot occur.

2. Which of the following reactions are possible?

   (i) $\pi^+ + n \rightarrow \Lambda^0 + K^+$ (ii) $\pi^+ + n \rightarrow K^0 + K^+$

   (iii) $\bar{\nu}_e + p \rightarrow n + \mu^+$ (iv) $\bar{\nu}_e + p \rightarrow n + e^+$

Solution: The first reaction is

$$\pi^+ + n \rightarrow \Lambda^0 + K^+$$

	π^+	n	$\rightarrow$	Λ^0	K^+
Baryon number:	0	+1	$\rightarrow$	+1	0
Q:	1	0	$\rightarrow$	0	1
I_3:	1	$-\frac{1}{2}$	$\rightarrow$	0	$\frac{1}{2}$
S:	0	0	$\rightarrow$	–1	+1

In this reaction baryon number, charge, strangeness, isospin, etc. are all conserved, so the reaction is possible through strong interaction.

The second reaction is

$$\pi^+ + n \rightarrow K^0 + K^+$$

Baryon number: $0 \quad +1 \rightarrow 0 \quad 0$

In this reaction baryon number is not conserved, so the reaction is not possible through known interactions.

The third reaction is

$$\overline{\nu}_e + p \rightarrow n + \mu^+$$

Lepton number L_e: $+1 \quad 0 \rightarrow 0 \quad 0$

Lepton number L_μ: $0 \quad 0 \rightarrow 0 \quad -1$

For this reaction L_e and L_μ both are not conserved, so this reaction is not possible through any known interaction.

The last reaction is

$$\overline{\nu}_e + p \rightarrow n + e^+$$

Lepton number L_e: $+1 \quad 0 \rightarrow 0 \quad +1$

Baryon number: $0 \quad +1 \rightarrow 1 \quad 0$

Q: $0 \quad 1 \rightarrow 0 \quad 1$

In this case, all conservation laws hold, so the reaction is possible through weak interactions.

3. Determine whether the following reactions are allowed or forbidden.

(i) $p + p \rightarrow K^+ + \Sigma^+$ (ii) $p + \pi^- \rightarrow \Sigma^0 + \eta^0$

(iii) $p + p \rightarrow p + p + p + \overline{p}$

Solution: The first reaction is

$$p + p \rightarrow K^+ + \Sigma^+$$

Charge: $+1 \ +1 \rightarrow +1 \ +1$ is conserved

Baryon number: $+1 \ +1 \rightarrow 0 \ +1$ is not conserved

Strangeness: $0 \ 0 \rightarrow +1 \ -1$ is conserved

Isotopic spin I_3: $\frac{1}{2} \ \frac{1}{2} \rightarrow \frac{1}{2} + 1$ is not conserved.

Hence, this reaction is not possible.

The second reaction is

$$p + \pi^- \rightarrow \Sigma^0 + \eta^0$$

Charge: $+1 \ -1 \rightarrow 0 \ 0$ is conserved

Baryon number: $+1 \ 0 \rightarrow +1 \ 0$ is conserved

Strangeness: $0 \ 0 \rightarrow -1 \ 0$ is not conserved

Isotopic spin I_3: $\frac{1}{2} \ -1 \rightarrow 0 \ 0$ is not conserved.

Hence, this reaction is not possible through strong interactions.

The third reaction is

$$p + p \rightarrow p + p + p + \overline{p}$$

Charge: $+1 + 1 \rightarrow +1 + 1 +1 -1$ is conserved

Baryon number: $+1\ +1 \rightarrow +1\ +1\ +1\ -1$ is conserved

Strangeness: $0\ 0 \rightarrow 0\ 0\ 0\ 0$ is conserved

Isotopic spin I_3: $\frac{1}{2}\ \frac{1}{2} \rightarrow \frac{1}{2}\ \frac{1}{2}\ \frac{1}{2}\ -\frac{1}{2}$ is not conserved

Hence, this reaction is possible through strong interactions.

Unsolved Problems

1. Show that $K^+ \rightarrow \pi^+ + \pi^0$ violate parity conservation.

2. A particle of mass (M) at rest decays to two particles carrying masses m_1 and m_2. Determine the kinetic energy of decay products, using energy–momentum conservation laws.

3. State the conservation law, which forbids the following reactions:

(i) $p + p \rightarrow \gamma$ (ii) $p \rightarrow p + e^-$
(iii) $e^- \rightarrow \gamma + \nu_e$ (iv) $n \rightarrow p + \pi^-$
(v) $p \rightarrow p + \gamma$ (vi) $K^- \rightarrow \pi^0 + e^-$
(vii) $\mu^- \rightarrow e^- + \gamma$ (viii) $\Xi^0 \rightarrow \Sigma^- + p$

4. State which of the following processes are allowed and which are forbidden. Give reasons.

(i) $\Xi^0 \rightarrow \Sigma^0 + \nu$ (ii) $\mu^- + p \rightarrow \Lambda^0 + \nu_\mu$
(iii) $\Sigma^0 \rightarrow \Lambda^0 + \mu^+ + \nu_\mu$ (iv) $K^- + d \rightarrow \pi^+ + \Sigma^-$
(v) $K^+ \rightarrow \pi^0 + e^+ + \nu_e$

[**Ans.** (i) allowed, (ii) allowed, (iii) forbidden by energy conservation $m_\mu > m_\Sigma - m_\Lambda$, (iv) forbidden by baryon conservation, (v) allowed].

5. Why the following reactions are allowed?

(i) $p + p \rightarrow p + \Lambda + \Sigma^+ + \overline{\Xi}^0$ (ii) $p + \overline{p} \rightarrow \Sigma^0 + \overline{\Lambda}$
(iii) $\gamma + n \rightarrow n + \pi^0$ (iv) $\pi^- + p \rightarrow \Sigma^0 + K^0$
(v) $\pi^+ + p \rightarrow \Xi^0 + K^+ + K^+$ (vi) $\pi^- + p \rightarrow \Lambda + K^+ + K^-$

6. Why the following reactions are forbidden through strong interactions?

(i) $p + p \rightarrow \Sigma^+ + p$ (ii) $\overline{p} + p \rightarrow \Lambda + \Lambda$
(iii) $\pi^- + p \rightarrow \Sigma^+ + K^-$ (iv) $K^+ + p \rightarrow \Sigma^+ + \pi^+$

7. State the conservation law, which forbids the following decays:

(i) $p + \overline{p} \rightarrow \gamma$ (ii) $n \rightarrow p + e^-$
(iii) $e^- \rightarrow \gamma + \nu_e$ (iv) $n \rightarrow p + \pi^-$
(v) $p \rightarrow p + \gamma$ (vi) $K^- \rightarrow \pi^0 + e^-$
(vii) $\mu^- \rightarrow e^- + \gamma$ (viii) $\Xi^0 \rightarrow \Sigma^- + p$

8. The Ξ^0-hyperon is seen to decay as $\Xi^0 \rightarrow \Lambda + \pi^0$ but not as $\Xi^0 \rightarrow n + \pi^0$ even though neutron (n) is lighter than Λ-hyperon. Explain why.

Section 8.9

Unsolved Problems

1. Determine the charged state of a resonant baryon having isospin $I = \frac{3}{2}$ and zero strangeness. **[Ans.** ++, +, 0, –]

2. Assume mass operator for isomultiplet to be given by

$$M = a + bI_3$$

where a and b are unknown parameters and I_3 denotes third component of isospin, show that decuplet baryons satisfy the following equal spacing relation:

$$m_{\Delta^{++}} - m_{\Delta^+} = m_{\Delta^+} - m_{\Delta^0} = m_{\Delta^0} - m_{\Delta^-}$$

3. Show that a set of Pauli matrices σ_i satisfy the following:

$$\sigma_i \sigma_j - \sigma_j \sigma_i = 2i\sigma_k$$

where $i, j, k = 1, 2, 3$ cyclically

4. Σ^{*+} is an hyperon observed to have mass 1.385 GeV and decay width 35 MeV. Its dominant decay mode is $\Sigma^{*+} \to \Lambda + \pi^+$. Is this decay strong, electromagnetic or weak? Give reasons for your answer.

Section 8.10

Unsolved Problems

1. Λ_c^+-baryon has $I = 0$, $S = 0$ and $C = 1$. What is its quark content?

[Ans. *udc*]

2. What values of electric charge are possible for a baryon in quark model?

[Ans. ++, +, 0, –]

3. What values of electric charge are possible for a meson in quark model?

[Ans. +, 0, –]

4. Given that meson masses are expressed in terms of its constituent quark masses (m_i) and spin (s_i)

$$M = m_1 + m_2 + \alpha \frac{\vec{s}_1 \cdot \vec{s}_2}{m_1\, m_2}$$

Determine the masses of pseudoscalar and vector mesons, taking $m_d = m_u = 320$ MeV, $m_\tau = 480$ MeV and $\frac{\alpha}{m^2 u} = 640$ MeV.

REVIEW QUESTIONS

Short Answer Type

1. What are baryons?
2. What is the difference between π-mesons and K-mesons?
3. What are baryon and lepton conservation laws?
4. Assign the isospin quantum numbers to nucleonic doublet and pionic triplet.
5. Give the quark structure of a proton, neutron and meson.
6. Explain the concept of charge conjugation.
7. Name different types of quarks.
8. Explain the concept of isospin.
9. What are the quarks make-ups of neutron and proton?
10. What is CP violation?
11. What are quarks? Outline the basic assumptions and properties of quarks.
12. Classify the various types of interactions in relation to elementary particles.
13. Give the charges of any four types of quarks.
14. Explain why the decay

$$\Sigma^0 \rightarrow \Lambda^0 + \gamma$$

is observed, but not

$$\Sigma^0 \rightarrow p + \pi \text{ or } \Sigma^0 \rightarrow n + \pi^0$$

15. What are resonances?
16. State CPT theorem.
17. What do you understand by isospin?
18. How do you differentiate between leptons and baryons?
19. What are leptons?
20. What is strangeness?
21. What are quarks? Are they coloured?
22. μ-meson belongs to:
    (i) Leptons (ii) Baryons
    (iii) Photons (iv) Pions
23. Meson is a bound state of
    (i) 3 quarks (ii) 2 quarks
    (iii) 5 quarks (iv) None of these
24. Write three properties of quarks.
25. Mesons are:
    (i) Leptons (ii) Fermions
    (iii) Bosons (iv) None of the above

Long Answer Type

1. What are quarks? Explain quark flavours.
2. Explain which of the following reactions are allowed and forbidden under the conservation of strangeness, conservation of baryon number and conservation of charge.
   (i) $\pi^+ + n \rightarrow \Lambda^0 + K^+$ (ii) $\pi^+ + p \rightarrow K^0 + K^+$
   (iii) $\pi^- + p \rightarrow \Lambda^0 + K^0$ (iv) $\pi^- + p \rightarrow \pi^0 + K^0 + \Lambda^+$
3. Discuss the classification of particles on the basis of isospin.
4. What are fundamental particles? Discuss the Gell–Mann–Nishijima scheme of classification of particles.
5. Discuss the quark structure of nucleons.
6. Write down the four basic interactions in nature giving their relative strengths and also name the exchange particles responsible for them.
7. Explain the concept of associated production.
8. Discuss the concept of associated production and give the decay modes of some of the strange particles.
9. Construct quark structure of a nucleon and a pion.
10. Compare the properties of leptons and baryons.
11. Explain the concept of strangeness and associated production. Discuss the Gell–Mann–Nishijima relation.
12. Explain in detail the quark model and explain the various types of quarks (flavours) along with their properties.
13. Explain the terms isotopic spin and strangeness. In what respect are they important in the classification of elementary particles?
14. Write a brief note on different types of interactions which the elementary particles can undergo. Give examples.
15. Give reasons to explain the associated production of strange particles and charmed particles.
16. What are quarks? Give the elementary theory of structure of a few hadrons on the basis of quark model.
17. Write notes on:
   (i) Isospin quantum number
   (ii) Strange particles and their decay modes
18. What are quarks? Give the qualitative description of quark model.
19. Give the quark model of
   (i) mesons (ii) proton and antiproton
   (iii) neutron and antineutron
20. Describe briefly the following intrinsic quantum numbers in connection with elementary particles:
   (i) Lepton number (ii) Baryon number
   (iii) Isospin (iv) Hypercharge
   (v) Strangeness (vi) Decay process of mesons.

   What conservation laws govern them? Give at least one example in support of each conservation law.

21. Discuss the quantum numbers associated with elementary particles. Give the corresponding conservation laws. Give at least one example in support of each conservation law.
22. Discuss the various types of interactions between elementary particles giving their characteristic coupling constant and lifetimes. What are the interaction carriers in electromagnetic and strong interactions between nuclei?
23. Discuss the following intrinsic quantum numbers associated with elementary particles:
    (i) Charge number (ii) Baryon number
    (iii) Lepton number (iv) Isospin quantum number
    (v) Hypercharge (vi) Strangeness
    (vii) Decay process of mesons

    Explain with example.
24. What are quarks? Explain the quark model. Also give two factors which do not support the existence of quarks.
25. Classify the various types of elementary particles in reference to their lepton number, baryon number and isospin.
26. What is charge conjugation? Explain with example.
27. What is meant by colour of a quark? Give the colours associated with quarks.
28. Discuss the quark model and explain how mesons and baryons are formed using quarks.
29. Explain:
    (i) Isospin (ii) Gell–Mann–Nishijima scheme
30. Define parity, charge conjugation and time reversal. State CPT theorem.
31. A positron and an electron having negligible kinetic energies annihilate each other to produce two photons. Calculate the frequencies of the two photons.
32. Which of the following can occur? State conservation laws violated by others.
    (i) $p + p \rightarrow n + p + \pi^+$ (ii) $p + p \rightarrow p + \Lambda^0 + \Sigma^+$
    (iii) $e^+ + e^- \rightarrow \mu^+ + p^-$
33. What are leptons? Name any three leptons and their antiparticles. Briefly discuss the properties of leptons. Discuss the statement, μ-meson is also a lepton.
34. What is charge conjugation?
35. Discuss in detail the strange particles.
36. What is quark model? Is there any evidence in its support?
37. Discuss the decay modes of mesons.
38. What are hyperons and leptons? Give their decay modes.
39. Write notes on:
    (i) Isospin (ii) Hypercharge
    (iii) *K*-mesons
40. What are elementary particles? Give a brief account of the discovery and properties of elementary particles.

Chapter 9

Cosmic Rays

9.1 INTRODUCTION

Cosmic rays are high-energy charged particles, originating from outer space, that travel at nearly the speed of light and strike the earth from all directions. Mostly cosmic rays are bare nuclei of atoms, ranging from the lightest to the heaviest elements in the periodic table. Cosmic rays also include high-energy electrons, positrons, and other subatomic particles.

The energy of cosmic rays is usually measured in units of MeV (million electron volts) or GeV (billion electron volts). Most of the galactic cosmic rays have energies between 100 MeV (corresponding to a velocity for protons of 43% of the speed of light) and 10 GeV (corresponding to 99.6% of the speed of light). The highest energy of cosmic rays measured to date is more than 10^{20} eV, equivalent to the kinetic energy of a baseball travelling at approximately 100 mph.

9.2 DISCOVERY OF COSMIC RAYS

At the start of the twentieth century, French physicist Henry Becquerel discovered that certain elements are unstable, and transmute into other elements, and in the process, emit what appeared to be particles. These "particles" were given the name "radiation", and the process itself referred to as "radioactive decay". It was noticed that when a gold leaf electroscope was exposed to these radiations, it spontaneously discharges. The rate of discharge of an electroscope is then used as a measure of the level of radiation. The electroscope thus became a standard instrument for studying radiation and radioactive materials in the first decade of the twentieth century.

However, physicists noticed that electroscopes were found to discharge slowly even in the absence of radioactive matter. This residual discharge could not be attributed to leakage. Background radiation seemed to contribute to residual discharge.

To study the source of this background, Victor Hess in 1912 found that an electroscope discharged more rapidly as he ascended in a balloon. He attributed this to a source of radiation entering the atmosphere from above, and in 1936 he was awarded the Nobel Prize for his discovery. For some time it was believed that the radiation was electromagnetic in nature (hence the name cosmic "rays"), and some textbooks still incorrectly include cosmic rays as part of the electromagnetic spectrum. However, during the 1930s it was found that cosmic rays must be electrically charged because they were affected by the earth's magnetic field.

From the 1930s to the 1950s, before man-made particle accelerators reached very high energies, cosmic rays served as a source of particles for high-energy physics investigations, and led to the discovery of subatomic particles that include the positron and muons. Although these discoveries continued since the dawn of the space age, the main focus of cosmic ray research has been directed towards astrophysical investigations from where cosmic rays originate, how they get accelerated to such high velocities. What role do the rays play in the dynamics of the Galaxy. What is inferred about the composition of matter in region outside the solar system. To measure cosmic rays directly, before they are slowed down and broken up by the atmosphere, research is carried out by instruments carried on spacecraft and high altitude balloons, using particle detectors similar to those used in nuclear and high-energy physics experiments.

9.3 THE LATITUDE EFFECT

In the earlier cosmic ray studies, it was widely believed that the primary cosmic rays were photons of high energy coming from outside the earth's atmosphere. It was thought that interactions of these photons with orbital electrons of O_2, N_2, etc. present in the atmosphere were through Compton effect, producing high-energy electrons. However in 1929, Bothe and Kolhorster, as a result of absorption of cosmic rays in thick layers of gold, suggested that the primary cosmic rays were electrically charged particles and not photons. If this is the case then the cosmic rays should be affected by earth's magnetic field. Charged particles coming from extraterrestrial sources would suffer maximum deflection if they approach in the direction of earth's geomagnetic equator. The extent of this deflection should decrease as the particles approach more towards the polar region. Thus, if the cosmic rays do indeed consist of charged particles, a variation of intensity with the geomagnetic latitude should be observed.

Comprehensive investigations by A.H. Compton and his coworkers in 1930 showed the definite existence of the latitude effect. They showed that as we move from the earth's magnetic north to the earth's magnetic south, the intensity of cosmic rays remains nearly constant until magnetic latitude of about 50° is reached. The intensity then begins to drop and reaches a minimum value around the earth's magnetic equator. Beyond this, the intensity again rises as shown in Figure 9.1.

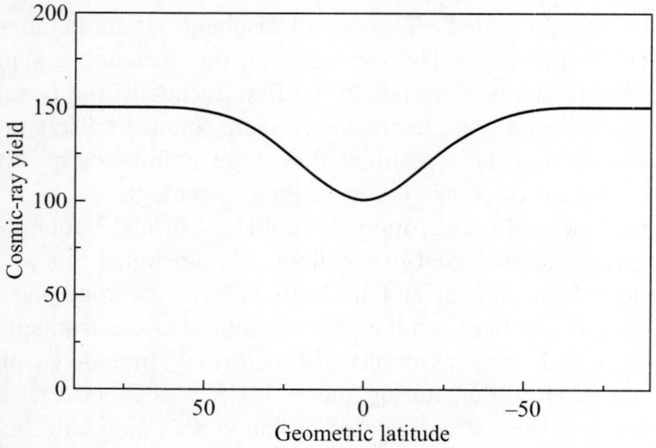

Figure 9.1 Variation of cosmic-ray yield (arbitrary units) versus geometric latitude.

9.4 EAST–WEST ASYMMETRY EFFECT

In 1930, B. Rossi proposed that if cosmic rays contain a large number of electrically charged particles, there should be a difference in the intensity of the cosmic rays coming from easterly and westerly directions. It is due to the fact that lines of force of the earth's magnetic field have a direction from the earth's geomagnetic south to the north outside the earth. If the vertically incident particles are positively charged, then according to Fleming's left hand rule, the particles are deflected towards the east. Thus, an observer on earth finds a greater intensity approaching approximately from westerly direction than from an easterly direction. However, if the cosmic rays are mostly negatively charged particles, earth's magnetic field will deflect them towards west, and will appear to come from east. This effect is known as *east–west asymmetry effect*.

By making use of two or more GM counters arranged in coincidence to act as a cosmic ray telescope, it is possible to count cosmic rays coming from a specific direction. With the arrangement of GM counters, the existence of east–west asymmetry in cosmic rays has been proved. From the measurements reported in 1935, near the magnetic equator in Peru, scientists found that about 14% more cosmic rays are approaching from the west, indicating that cosmic rays are mostly positively charged particles.

9.5 THE ALTITUDE EFFECT

In the beginning of this chapter, it was pointed out that one of the essential arguments in favour of the extraterrestrial origin of cosmic rays was the marked increase of the intensity observed with increasing altitude for a given latitude. With the improved apparatus, a more careful and detailed examination of the effect of altitude has been made at several different geomagnetic latitudes. The intensities obtained for cosmic rays are shown in Figure 9.2. These curves reveal an interesting point. The total cosmic ray intensity increases with increasing altitudes. And is maximum at about 25 km for a given latitude. The intensity of cosmic rays starts falling off with decreasing height.

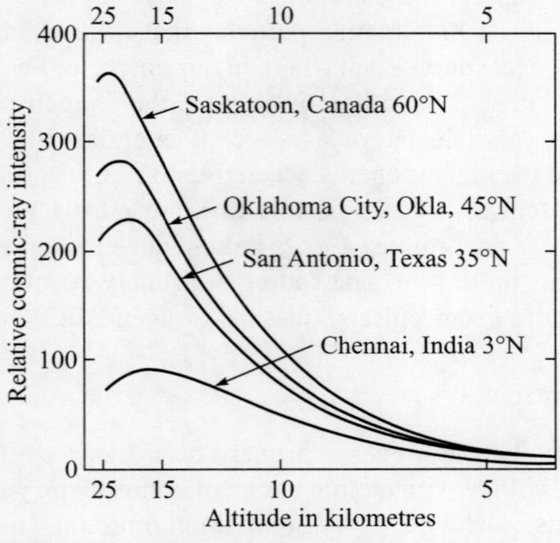

Figure 9.2 Variation of relative cosmic-ray intensity (arbitrary units) versus altitude in kilometres.

The general shapes of these curves can be explained in the following manner. When the cosmic rays reached the upper layers of atmosphere at a height of about 25 km, the primary cosmic rays, consisting of mostly protons, interact with the nuclei of the atmospheric gases to produce large number of secondary particles, like pions, muons, positive and negative electrons, and γ-rays photons. These secondaries produced at a height of about 25 km, move towards the earth's surface. As they move down, some of them lose energy and get stopped in the atmosphere. So, the number of secondaries goes on decreasing as we move from a height of 25 km towards the earth's surface. An instrument which measures the total number of ionization, thus register an increase of intensity of cosmic rays as it is taken to higher altitudes. This increase in the intensity is registered up to about 25 km. At this height, the number of secondaries is maximum. If the instrument is taken to still higher altitudes, the number of secondaries decreases, it will register lesser number of ionizing particles.

Another point that is revealed from the four curves at 3° N, 35° N, 45° N and 60° N show that as we approach the magnetic equator, the cosmic-ray intensity decreases. This is due to the fact that a proportion of the cosmic-ray particles, which can overcome the earth's magnetic field do not have sufficient energy to reach the earth's surface near the magnetic equator.

9.6 PRIMARY AND SECONDARY COSMIC RAYS

From a large number of studies of the properties of cosmic-ray particles, cosmic rays have been divided into following two groups:

1. Primary cosmic rays.
2. Secondary cosmic rays.

9.6.1 Primary Cosmic Rays

Primary cosmic-ray particles are the particles which are initially incident on the outer boundaries of the earth's atmosphere from all the directions.

Primary cosmic rays are stable charged particles that have been accelerated to enormous energies by some astrophysical source somewhere in our universe. They must be stable (lifetimes greater than few million years), in order to survive the long trips through interstellar or intergalactic space. These particles have a range of energies 10^9 eV (1 GeV) to 10^{20} eV (10^8 TeV). In comparision, the highest energy accelerator on earth accelerates protons up to about 20 TeV. The particles present in primary cosmic rays are mostly protons or hydrogen nuclei. About 95% of all cosmic rays are protons, 4% are helium nuclei and rest 1% are nuclei from other stellar synthesized elements up to iron. The source of primary cosmic rays is still not known. It can be a black hole, a neutron star, pulsars, quasars, or even Big Bang itself.

9.6.2 Secondary Cosmic Rays

At an altitude of about 25 km, high energy primary cosmic-ray particles, usually protons and sometimes heavier nuclei, collide with atomic nuclei of atmospheric gases like oxygen, nitrogen, etc. During these collisions, nuclei generally break up into protons, neutrons, pions and muons. Other particles such as antiprotons, antineutrons, kaons, hyperons, etc. are also produced, but their numbers are relatively small. Thus, the primary cosmic rays create a jet of large number of elementary particles, which continue travelling in almost the same direction as the primary cosmic rays. The particles in the jet can themselves create more particles as they hit other nuclei of oxygen or nitrogen in the air. These secondary particles produced by interaction of primary cosmic rays are known as *secondary cosmic radiations*.

In some cases, secondary cosmic radiations are further divided into two categories:

(i) Hard component of cosmic rays.
(ii) Soft component of cosmic rays.

Hard Component of Cosmic Rays

This component is more predominant at higher altitudes (15–25 km) and is about 70% of all the secondary cosmic rays. This component mostly consists of very high energy positive and negative pions, muons, neutrinos, some protons and neutrons, etc. Because of very high energy, they can penetrate a few metres thick lead sheets.

Soft Component of Cosmic Rays

This component is more predominant at lower altitudes or at sea level. It is about 25% at an altitude of about 15 km. It mostly consists of positive and negative electrons and photons. This is a low energy component of secondary cosmic rays. They can penetrate 10–20 cm thick lead sheets.

9.7 COSMIC-RAY SHOWERS

In 1929, D. Skobelzyn observed in a cloud chamber that when a cosmic-ray particle passes through a chamber, set of associated tracks was often obtained indicating the passage of particles in group. Subsequently, a variety of similar effects were observed with both cloud chamber and GM counters. This set or group of particles is referred to as a cosmic-ray showers as shown in Figure 9.3. Several types of showers, some of them of great complexities, have been observed. The formation of cosmic-ray showers was explained by Bhabha and Heitler and also by Carlson and Oppenheimer in the following way.

Suppose the particle entering a cloud chamber is a high energy photon, it is converted into an electron–positron pair in the first metal plate of the cloud chamber. Both electron and positron carry about half of the energy carried by incident photon. The electron–positron pair, while slowing down in the metal plate emits Bremsstrahlung photons. These Bremsstrahlung photons are further converted into electron–positron pair. Thus, one photon entering the plate produces two secondary electrons. These secondary electrons in turn emit Bremsstrahlung photons, which form more electron–positron pairs, and so on. The same phenomenon also occurs if instead of a photon, an electron or a proton, or an alpha particle, etc. enters the cloud chamber. Thus, there is a very rapid increase in the number of negatively and positively charged electrons as a result of the entry of a single electron or a photon in the cloud chamber. When these particles emerge from the first plate of the cloud chamber, their numbers may be in hundreds or even in thousands. Corresponding to a single charged particle moving in the cloud chamber, there is one visible track. Therefore, a large number of tracks are thus observed between the first and second plate.

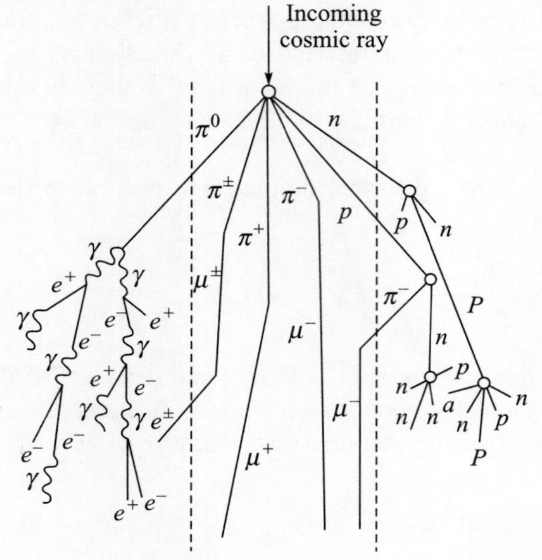

Figure 9.3 Schematic diagram of cosmic-ray showers.

Same process continues when these particles enter the second plate, though the energy of the particles entering the second plate is less compared to the case when they were in the first plate. They cause again large number of tracks between the second and third plate. This process continues.

9.8 ORIGIN OF COSMIC RADIATIONS

In the beginning of the investigations of cosmic radiations, it was thought that cosmic radiations are originating from the sun. But later on, it was found that cosmic radiation intensity remained almost constant during day and night. Also cosmic radiation intensity was found to be same, if we point our cosmic radiation detectors towards the sun or opposite to it. Therefore, it was concluded that if sun is contributing to cosmic radiations, this contribution is very small.

Recent experiments have shown that cosmic radiations originate from following three sources:

1. Sun—solar cosmic radiations.
2. Our galaxy—galactic cosmic radiations.
3. Extragalactic space—extragalactic cosmic radiations.

9.8.1 Solar Cosmic Radiations

Solar cosmic rays are those that originate from the sun. They have relatively low energy (10–100 keV). The average composition of the solar cosmic radiations is similar to that of the sun itself. The proof that some component of cosmic radiations originates from the sun comes from the fact that intensity of solar cosmic radiations increases after some solar event such as solar flares, etc.

9.8.2 Galactic Cosmic Radiations

These cosmic rays might be originating from extrasolar sources within our own galaxy such as rotating neutron stars, supernovae, or radio galaxies, quasars etc. in other galaxies. They are mostly protons, electrons, and fully ionized nuclei of light elements and have energy between 10^9 eV and 10^{14} eV.

The magnetic fields of the earth, the sun, and the galaxy itself tend to change the paths of the galactic cosmic rays, so that when one is detected, nothing can be inferred about the direction of its origin. Because mostly galactic cosmic rays are of extremely high energy, they must have originated through very energetic processes. Some are believed to have been accelerated by the shock waves of supernovae.

Some of the isotopes observed in galactic cosmic rays have half-lives that are comparable to the time interval of their formation, and isotopic ratios can, therefore, carry some information about the amount of time that has passed since they were formed.

9.8.3 Extragalactic Cosmic Radiations

Extragalactic cosmic rays are very high-energy particles that flow into our solar system from beyond our galaxy. The energies these particles possess are in excess of 10^{15} eV. Unlike solar or galactic cosmic rays, little is known about the origins of extragalactic cosmic rays. Partly, this can be attributed to a lack of statistics as the amount of cosmic rays reaching the earth's surface originating from extragalactic sources is about one particle per square metre per year.

REVIEW QUESTIONS

Short Answer Type

1. Write down the composition of cosmic rays.
2. What is the composition of cosmic-ray primaries?
3. What is a cosmic-ray shower?
4. Why is the intensity of cosmic rays minimum at magnetic equator?

Long Answer Type

1. What is the origin of cosmic rays? Give their composition.
2. What are cosmic rays? Discuss the nature and composition of cosmic rays.
3. What are cosmic rays? How are they detected and how is their nature established?
4. What are cosmic rays? Explain the latitude, longitude and altitude effects on cosmic rays.
5. What are primary and secondary cosmic rays? Given an account of production and properties of π- and μ-mesons.
6. Explain how cosmic rays were discovered. What is the effect of latitude on their intensity? Discuss east–west asymmetry of cosmic rays.

7. Discuss the nature of hard and soft cosmic rays. Why does the nature of cosmic rays vary with altitude?
8. What are cosmic rays? Discuss the origin of cosmic rays. What are soft and hard cosmic rays?
9. What are the constituents of primary and secondary cosmic rays? Give their origin and production in atmosphere. Also discuss the phenomenon of cosmic-ray showers.
10. Write notes on:
    (i) Latitude effect. (ii) East–west effect.
11. What are cosmic rays? Discuss their origin and describe cosmic-ray showers.

Appendix A

Useful Constants

TABLE A.1 Important physical constants

Mass	*Symbol*	*Value*
Electron rest mass	m_e	9.1094×10^{-31} kg
		$= 5.4858 \times 10^{-4}$ amu
		$= 0.5110$ MeV/c^2
Proton rest mass	m_p	1.6726×10^{-27} kg
		$= 1.007276$ amu
		$= 938.272$ MeV/c^2
	$\frac{m_p}{m_e}$	1836.153
Neutron rest mass	m_n	1.6749×10^{-27} kg
		$= 1.008665$ amu
		$= 939.565$ MeV/c^2
Mass of deuteron	m_d	2.0136 amu
		$= 1875.61$ MeV/c^2
Mass of α-particle	m_α	4.0015 amu
		$= 3727.3892$ MeV/c^2
Mass of muon	m_μ	105.6584 MeV/c^2
Mass of π-meson	$m_{\pi^\pm}$	139.6 MeV/c^2
	m_{π^0}	135.0 MeV/c^2

TABLE A.2 Conversion factors

Factor	*Unit*	*Value*
1 atomic mass unit	amu	1.6605×10^{-27} kg
		$= 1.4923 \times 10^{-10}$ J/c^2
		$= 931.494$ MeV/c^2
1 kilogram	kg	8.9876×10^{16} J
		$= 5.61 \times 10^{29}$ MeV/c^2
1 electron volt	eV	1.602×10^{-19} J

(*Contd.*)

TABLE A.2 Conversion factors (*Contd.*)

Factor	*Unit*	*Value*
1 curie	Ci	3.7×10^{10} dps
1 barn	b	10^{-28} m^2
1 fermi	fm	10^{-15} m
1 joule	J	6.242×10^{18} eV $= 6.242 \times 10^{12}$ MeV
1 torr		1 mm of Hg
300 K		$\frac{1}{40}$ eV
1 gauss	G	10^{-4} T
1 tesla	T	10^4 G

TABLE A.3 Values of prefixes

Prefix	*Value*
yotta	10^{24}
zetta	10^{21}
exa	10^{18}
peta	10^{15}
tera	10^{12}
giga	10^9
mega	10^6
kilo	10^3
hecto	10^2
deca	10^1
deci	10^{-1}
centi	10^{-2}
milli	10^{-3}
micro	10^{-6}
nano	10^{-9}
pico	10^{-12}
femto	10^{-15}
atto	10^{-18}
zepto	10^{-21}
yacto	10^{-24}

TABLE A.4 Combination of constants

Constant	*Symbol*	*Value*
Avogadro's number	N_0	6.023×10^{26}/kilomole
Planck's constant	h	6.6261×10^{-34} J s $= 4.1357 \times 10^{-21}$ MeV s

(*Contd.*)

TABLE A.4 Combination of constants (*Contd.*)

Constant	*Symbol*	*Value*
	$\hbar = \frac{h}{2\pi}$	1.0546×10^{-34} J s
		$= 6.5821 \times 10^{-22}$ MeV s
Electron charge	e	1.6021×10^{-19} C
Bohr radius	a_0	5.2918×10^{-11} m
Compton wavelength of electron	λ_c	2.426×10^{-12} m
Compton wavelength of proton	$\lambda_{c,p}$	1.3214×10^{-15} m
Boltzmann's constant	k	1.3807×10^{-23} J/K
		$= 8.6173 \times 10^{-5}$ eV/K
Gravitational constant	G	6.6744×10^{-11} N m^2/kg^2
Permeability of free space	μ_0	1.257×10^{-6} T m/A
	$\frac{\mu_0}{4\pi}$	10^{-7} T m/A
Permittivity of free space	ε_0	8.8542×10^{-12} C^2/N m^2
	$\frac{1}{4\pi\varepsilon_0}$	8.988×10^9 N m^2/C^2
Speed of light in free space	c	2.9979×10^8 m/s
Fine structure constant	α	$\frac{1}{137.04}$
Bohr magneton	μ_B	9.2741×10^{-24} J/T
Nuclear magneton	μ_N	5.05082×10^{-27} J/T
Classical electron radius	r_e	2.8179×10^{-15} m

Appendix B

The Periodic Table

1 H																	2 He
3 Li	4 Be											5 B	6 C	7 N	8 O	9 F	10 Ne
11 Na	12 Mg											13 Al	14 Si	15 P	16 S	17 Cl	18 Ar
19 K	20 Ca	21 Sc	22 Ti	23 V	24 Cr	25 Mn	26 Fe	27 Co	28 Ni	29 Cu	30 Zn	31 Ga	32 Ge	33 As	34 Se	35 Br	36 Kr
37 Rb	38 Sr	39 Y	40 Zr	41 Nb	42 Mo	43 Tc	44 Ru	45 Rh	46 Pd	47 Ag	48 Cd	49 In	50 Sn	51 Sb	52 Te	53 I	54 Xe
55 Cs	56 Ba	57 La	72 Hf	73 Ta	74 W	75 Re	76 Os	77 Ir	78 Pt	79 Au	80 Hg	81 Tl	82 Pb	83 Bi	84 Po	85 At	86 Rn
87 Fr	88 Ra	89 Ac	104 Rf	105 Db	106 Sg	107 Bh	108 Hs	109 Mt	110 Ds	111 Rg							

58 Ce	59 Pr	60 Nd	61 Pm	62 Sm	63 Eu	64 Gd	65 Tb	66 Dy	67 Ho	68 Er	69 Tm	70 Yb	71 Lu
90 Th	91 Pa	92 U	93 Np	94 Pu	95 Am	96 Cm	97 Bk	98 Cf	99 Es	100 Fm	101 Md	102 No	103 Lr

Appendix **C**

Table of Elements

Element	*Symbol*	*Atomic number*
Hydrogen	H	1
Helium	He	2
Lithium	Li	3
Beryllium	Be	4
Boron	B	5
Carbon	C	6
Nitrogen	N	7
Oxygen	O	8
Fluorine	F	9
Neon	Ne	10
Natrium, sodium	Na	11
Magnesium	Mg	12
Aluminium	Al	13
Silicon	Si	14
Phosphorus	P	15
Sulphur	S	16
Chlorine	Cl	17
Argon	Ar	18
Kalium, potassium	K	19
Calcium	Ca	20
Scandium	Sc	21
Titanium	Ti	22
Vanadium	V	23
Chromium	Cr	24
Manganese	Mn	25
Ferrous, iron	Fe	26
Cobalt	Co	27
Nickel	Ni	28
Cuprum, copper	Cu	29
Zinc	Zn	30

Element	*Symbol*	*Atomic number*
Gallium	Ga	31
Germanium	Ge	32
Arsenic	As	33
Selenium	Se	34
Bromine	Br	35
Krypton	Kr	36
Rubidium	Rb	37
Strontium	Sr	38
Yttrium	Y	39
Zirconium	Zr	40
Niobium	Nb	41
Molybdenum	Mo	42
Technetium	Tc	43
Ruthenium	Ru	44
Rhodium	Rh	45
Palladium	Pd	46
Argentum, silver	Ag	47
Cadmium	Cd	48
Indium	In	49
Stannum, tin	Sn	50
Stibium, antimony	Sb	51
Tellurium	Te	52
Iodine	I	53
Xenon	Xe	54
Caesium	Cs	55
Barium	Ba	56
Lanthanum	La	57
Cerium	Ce	58
Praseodymium	Pr	59
Neodymium	Nd	60
Promethium	Pm	61
Samarium	Sm	62
Europium	Eu	63
Gadolinium	Gd	64
Terbium	Tb	65
Dysprosium	Dy	66
Holmium	Ho	67
Erbium	Er	68
Thulium	Tm	69
Ytterbium	Yb	70
Lutetium	Lu	71
Hafnium	Hf	72
Tantalum	Ta	73

Element	*Symbol*	*Atomic number*
Wolfram, tungsten	W	74
Rhenium	Re	75
Osmium	Os	76
Iridium	Ir	77
Platinum	Pt	78
Aurum, gold	Au	79
Hydrargyrum, mercury	Hg	80
Thallium	Tl	81
Plumbum, lead	Pb	82
Bismuth	Bi	83
Polonium	Po	84
Astatine	At	85
Radon	Rn	86
Francium	Fr	87
Radium	Ra	88
Actinium	Ac	89
Thorium	Th	90
Protactinium	Pa	91
Uranium	U	92
Neptunium	Np	93
Plutonium	Pu	94
Americium	Am	95
Curium	Cm	96
Berkelium	Bk	97
Californium	Cf	98
Einsteinium	Es	99
Fermium	Fm	100
Mendelevium	Md	101
Nobelium	No	102
Lawrencium	Lr	103
Rutherfordium	Rf	104
Dubnium	Db	105
Seaborgium	Sg	106
Bohrium	Bh	107
Hassium	Hs	108
Meitnerium	Mt	109
Darmstadtium	Ds	110
Roentgenium	Rg	111

Appendix **D**

Myths and Realities

Myth: Einstein played a major role in the development of nuclear weapons.
Reality: He never participated in the Manhattan project which built the first atomic bombs.

Myth: Protons and neutrons cannot be further subdivided.
Reality: They are believed to be made of smaller particles called *quarks*. It is speculated that quarks are made of even smaller particles.

Myth: Hundreds of thousands of people were killed as a result of the Chernobyl nuclear disaster in Russia.
Reality: Approximately 50 emergency workers died as a result of radiation poisoning (known as ARS—Acute Radiation Syndrome). Later on nine children died as a result of thyroid cancer.

Myth: We can increase or decrease the half-life of radioisotope.
Reality: No, each radionuclide has its own characteristic half-life. No operation or process can alter it.

Myth: The Bohr atom can explain all kinds of spectra.
Reality: The Bohr atom deals only with atoms that have one electron, principally the hydrogen atom. Ionized helium is explained to a lesser degree.

Myth: The Sun has a radioactivity zone.
Reality: The Sun has, inside its convection zone, a radiation zone, that is, a zone in which energy is transported by radiation instead of by convection. Radioactivity, the decay of nuclear particles, is not a phenomenon that takes place in or on the Sun.

Myth: Man-made nuclear radiation differs from natural radiation.
Reality: Radiation emitted from man-made radionuclides is exactly of the same form as radiation emitted from naturally occurring radioactive materials (namely alpha, beta or gamma radiation). As such, the radiation emitted by naturally occurring materials cannot be distinguished from radiation produced by materials in the nuclear fuel cycle.

Myth: Natural nuclear radiation will not hurt you, but human-made radiation will.
Reality: Radiation is the same whether it is natural or man-made. There is no difference in what occurs as a result of exposure to natural versus man-made radiation.

Myth: Nuclear power plants releases dangerous amounts of radiation into the atmosphere.

Reality: Nuclear power plants do emit some radiation, but the amounts are environmentally insignificant and pose no threat to the biological life.

Myth: Your food becomes radioactive if it is irradiated with gamma rays.

Reality: Actually, this statement and "irradiation of food creates harmful chemicals in the food" are the most common arguments against food irradiation. The truth is, the food does not become radioactive and there is no scientific evidence that any harmful chemicals are produced.

Myth: The radiation from the dental X-ray causes severe headache.

Reality: Large doses of radiation can cause harmful effects like nausea and vomiting. Small doses of radiation from more common everyday activities do not cause any harm. Radiation doses from diagnostic medical exams like dental X-ray do not cause headaches or otherwise make you sick.

Index